Praise

CHASING S

"At last, the story of tracking the ocean's most charismatic and controversial predator, compellingly told by the man who has learned more about the Atlantic great white shark than any other person alive. You *must not miss* this fantastic book! I could not put it down."

—Sy Montgomery, *New York Times* bestselling author of *The Soul of an Octopus*

"When my late husband, Peter Benchley, started writing *Jaws*, he thoroughly researched available data on white sharks, but in the early 1970s, scant scientific data existed. *Chasing Shadows* chronicles the research we've learned in the decades since and provides fascinating details about the great white shark in the northwest Atlantic, why it is an evolutionary marvel, and how it differs from other species of sharks. As a diver and shark advocate for the past fifty years, I reveled in reading the arc of transformation, from an era of ignorance to today, where shark conservation is applauded and admired. It offers a stirring but candid account of shark conservation success that addresses the complexities, risks, and ramifications for both people and prey."

—Wendy Benchley, ocean conservationist and marine policy advocate

"An exhilarating and illuminating ode to the ocean's most misunderstood gangsters. As seals have gradually returned to the shores of Cape Cod since the 1970s, so too have white sharks, spurring high drama and anxiety in waters thick with swimmers and surfers. Told through the eyes of a leading white shark biologist, this nuanced account of the restoration of an apex predator to its historic range demystifies an elusive creature that's less monster than fish. A fascinating story filled with wonder and awe, *Chasing Shadows* seeks to reconcile our primal fear with our ecological conscience, exploring what it truly means to coexist with wild things."

—Emily Voigt, author of *The Dragon Behind the Glass: A True Story of Power, Obsession, and the World's Most Coveted Fish*, a PEN America Literary Award finalist

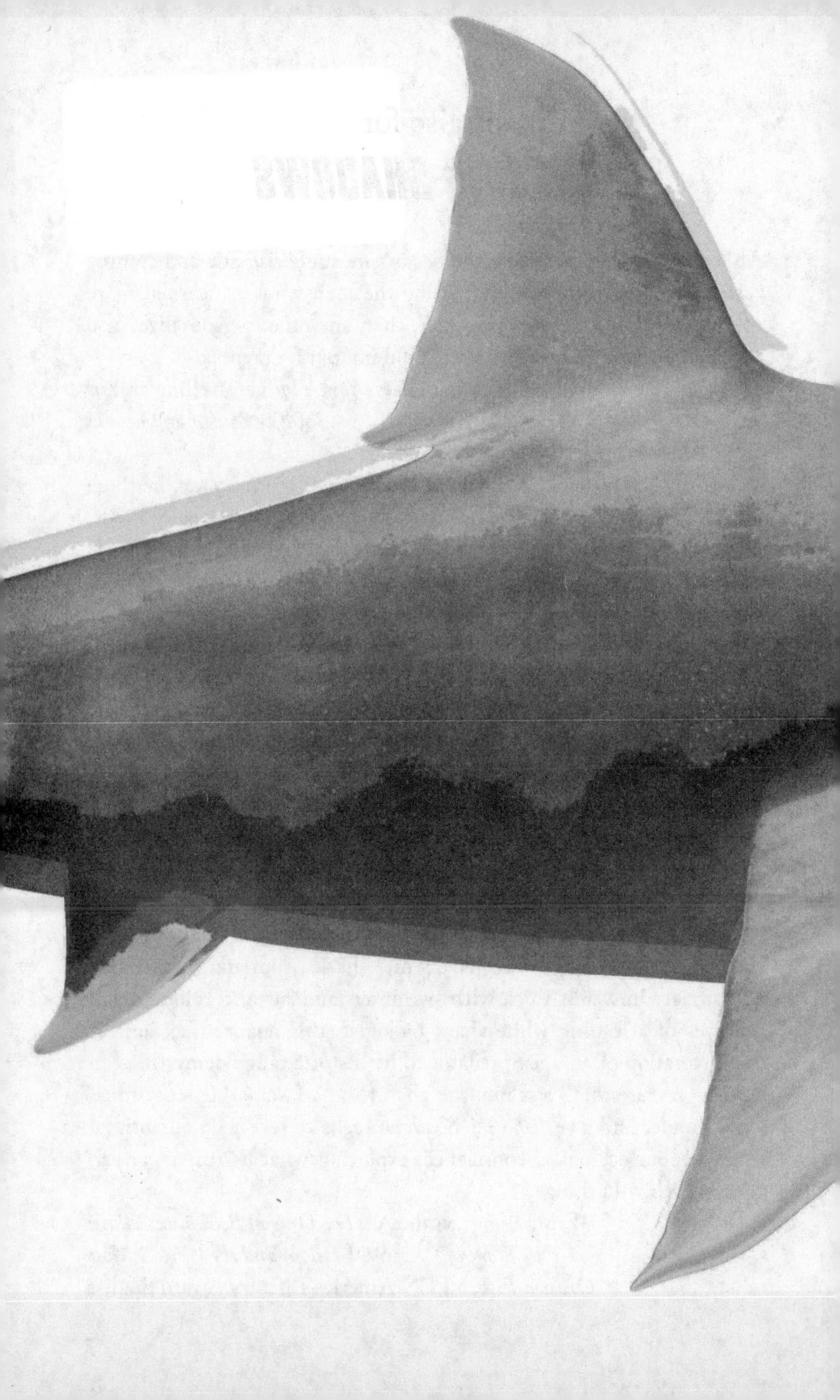

UNRAVELING THE MYSTERIES OF THE GREAT WHITE SHARK

CHASING SHADOWS

GREG SKOMAL

WITH RET TALBOT

wm

WILLIAM MORROW

An Imprint of HarperCollins*Publishers*

Quotation from an article from the *Dayton Daily News* on page 64 reproduced with permission.

CHASING SHADOWS. Copyright © 2023 by Greg Skomal. Addendum © 2024 by Greg Skomal. All rights reserved. Printed in the United States of America. No part of this book may be used or reproduced in any manner whatsoever without written permission except in the case of brief quotations embodied in critical articles and reviews. For information, address HarperCollins Publishers, 195 Broadway, New York, NY 10007.

HarperCollins books may be purchased for educational, business, or sales promotional use. For information, please email the Special Markets Department at SPsales @harpercollins.com.

A hardcover edition of this book was published in 2023 by William Morrow, an imprint of HarperCollins Publishers.

FIRST WILLIAM MORROW PAPERBACK EDITION PUBLISHED 2024.

Designed by Elina Cohen
Title page illustration courtesy of Shutterstock/ Nerthuz
Maps and spot illustrations by Karen Talbot

Images in the photo insert are courtesy of the author, except:
p. 2, (top) courtesy of the NOAA, (bottom) courtesy of *Hartford Courant*/TCA; p. 3, (top) courtesy of Chuck Stillwell/NOAA, (middle) courtesy of Wes Pratt/NOAA, (bottom) courtesy of Wes Pratt/NOAA; p. 4, (top) courtesy of Greg Skomal/NOAA, (bottom) courtesy of the NOAA; p.5, (middle) courtesy of Wes Pratt/NOAA, (bottom) courtesy of Greg Skomal/NOAA; p. 6, (bottom) courtesy of Nick Caloyianis; p. 7, (bottom left) courtesy of Wayne Davis, (bottom right) courtesy of John King; p. 9, (top and bottom) courtesy of Wayne Davis, (middle) courtesy of Megan Winton; p. 10, (top) courtesy of Erin Burke/MDMF; p. 11, (top) courtesy of Brian Skerry, (bottom) courtesy of John Chisholm; p. 12, (middle) courtesy of MDMF, (bottom) courtesy of Ed Lyman/ MDMF; p. 13, (top) courtesy of Shelly Negrotti, (bottom left and bottom right) courtesy of John Chisholm; p. 14, (top) courtesy of Derek Perry/MDMF, (middle left) courtesy of Megan Winton; p. 15, courtesy of Wayne Davis

Library of Congress Cataloging-in-Publication Data has been applied for.

ISBN 978-0-06-309084-2

24 25 26 27 28 LBC 5 4 3 2 1

TO KIMBERLY: FOR TURNING THE WORLD SO IT'S FACING THE WAY THAT I'M GOING.

—Greg

TO ANYONE WHO CANNOT LIVE IN A WORLD DEVOID OF WILDNESS, AND TO WHITE SHARKS, WHO EMBODY IT.

—Ret

This is a work of narrative nonfiction. We have done our best to accurately re-create events, locales, and conversations based on memory, interviews, and research. Many conversations in the book come from Greg's recollections or are reconstructed from public records. The dialogue, in most cases, is not meant to be a word-for-word transcript. A good-faith effort was made to interview most of the living individuals mentioned in the text, but in some cases, details are based on perception, memory, and research. In a few cases, we relied on conjecture based on the best available information and science at the time of publication to help bring to life the remarkable story of the resurgence of an apex predator in New England's waters. While much remains unknown about the white shark in the northwest Atlantic, our goal in writing this book is to contribute meaningfully to our collective understanding of this magnificent animal, its natural history, and the history of its relationship to both humans and its environment.

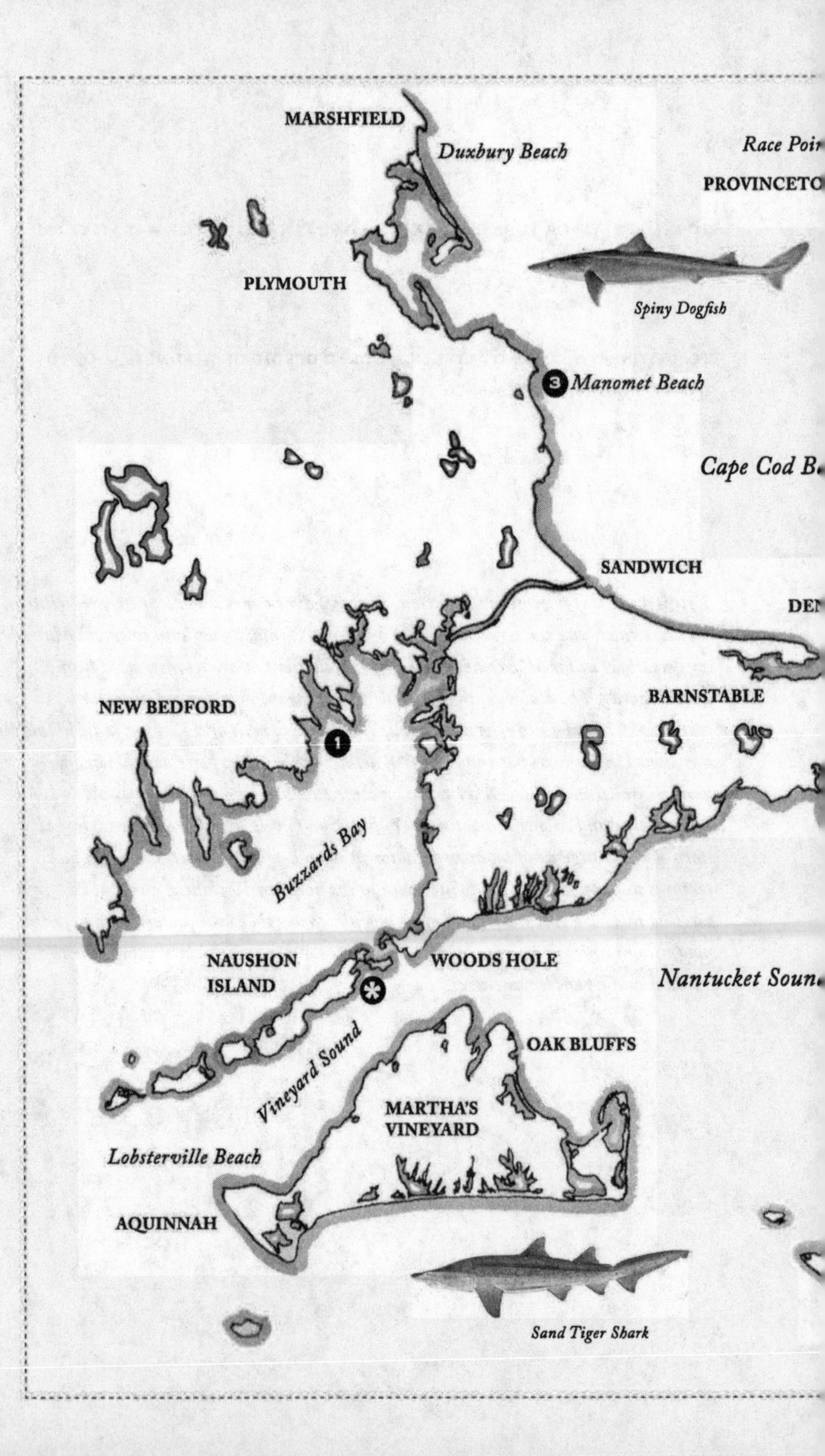
MARSHFIELD
Duxbury Beach
Race Poi
PROVINCETO
PLYMOUTH
Spiny Dogfish
Manomet Beach
Cape Cod B
SANDWICH
DE
NEW BEDFORD
BARNSTABLE
Buzzards Bay
NAUSHON
ISLAND
WOODS HOLE
Nantucket Soun
OAK BLUFFS
Vineyard Sound
MARTHA'S
VINEYARD
Lobsterville Beach
AQUINNAH
Sand Tiger Shark

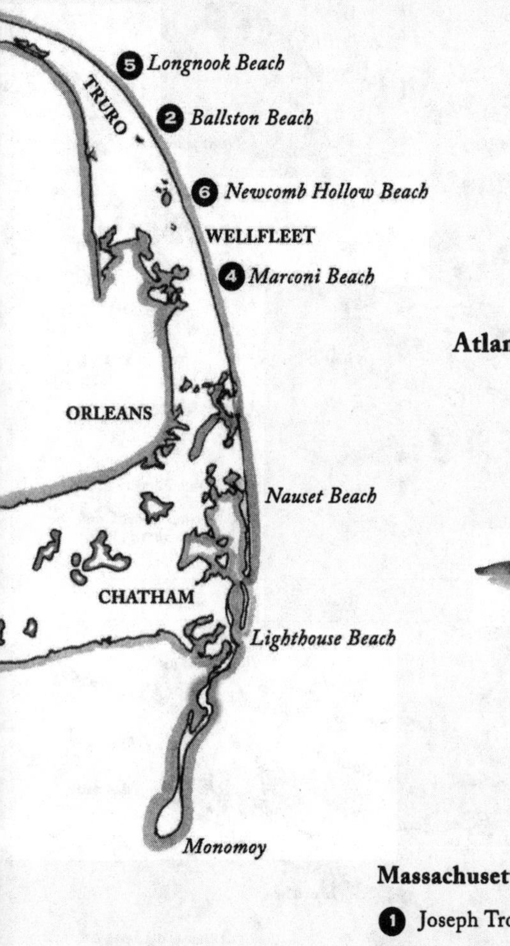

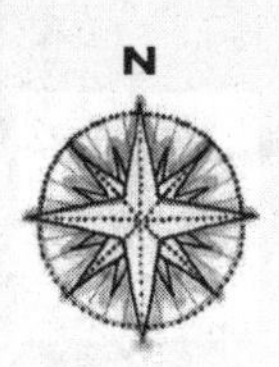

Atlantic Ocean

Basking Shark

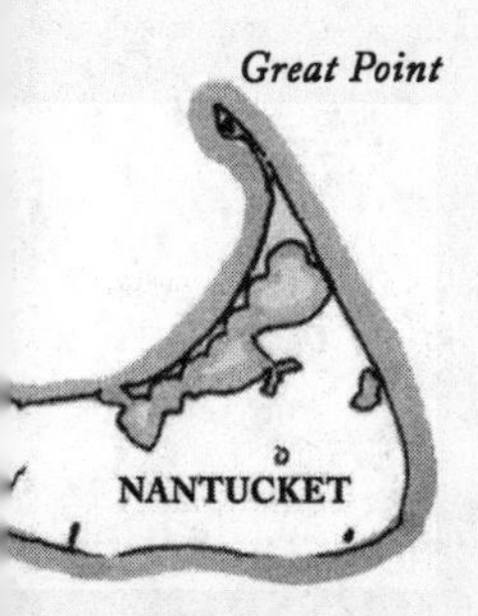

Massachusetts White Shark Attacks

1. Joseph Troy Jr. (1936)
2. Chris Myers (2012)
3. Ida Parker & Kristin Orr (2014)
4. Cleveland Bigelow (2017)
5. William Lytton (2018)
6. Arthur Medici (2018)

* Gretel (2004)

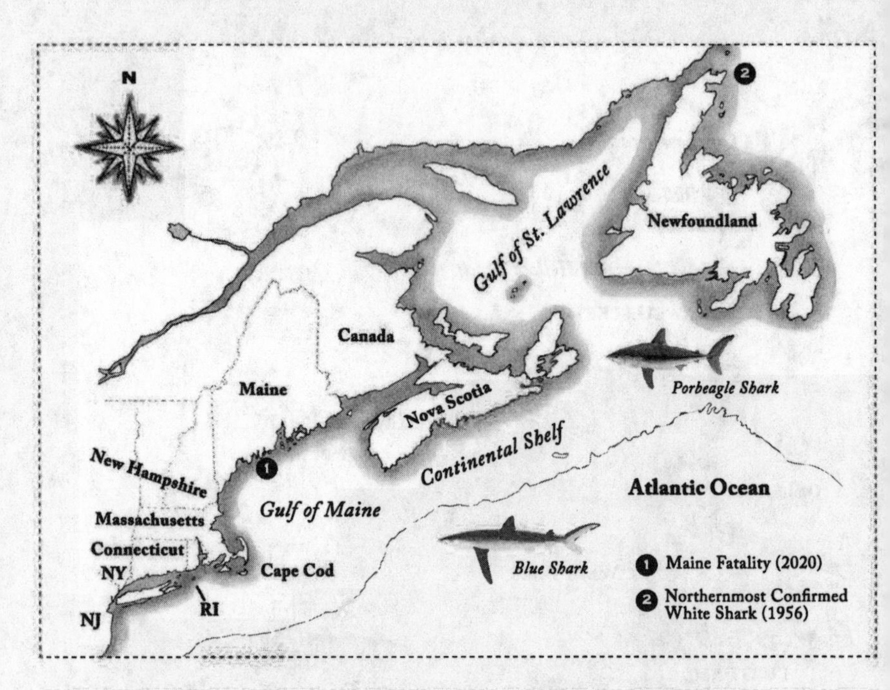
N
Newfoundland
Gulf of St. Lawrence
Canada
Maine
Nova Scotia
Porbeagle Shark
New Hampshire
Continental Shelf
Atlantic Ocean
Gulf of Maine
Massachusetts
Connecticut
NY
Cape Cod
Blue Shark
RI
NJ
1 Maine Fatality (2020)
2 Northernmost Confirmed White Shark (1956)

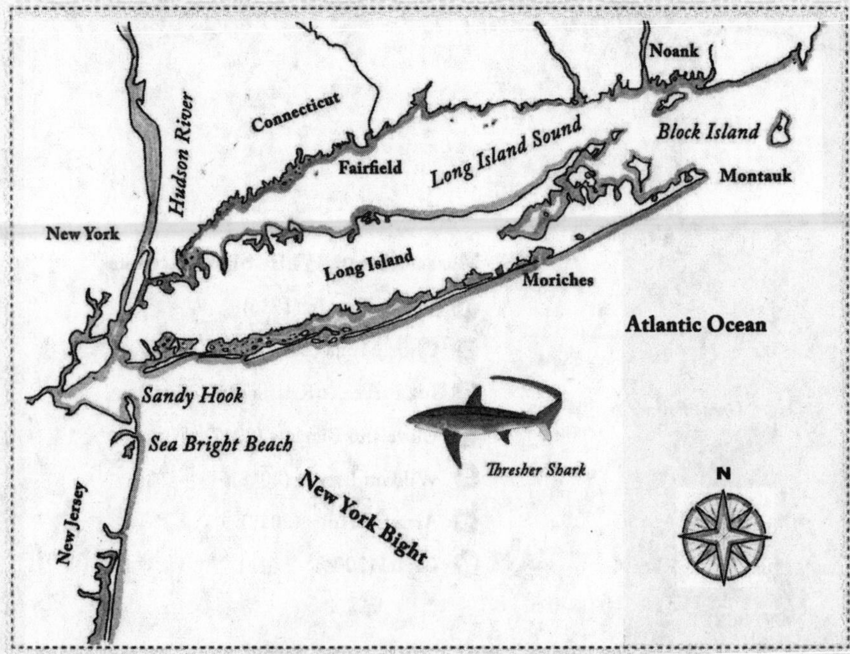
Noank
Connecticut
Hudson River
Block Island
Long Island Sound
Fairfield
Montauk
New York
Long Island
Moriches
Atlantic Ocean
Sandy Hook
Sea Bright Beach
Thresher Shark
New Jersey
New York Bight
N

CONTENTS

MAPS vi

FOREWORD BY JOHN "JACK" CASEY xi

PART 1: RARE EVENT SPECIES

CHAPTER 1: WHITE SHARKS AND WHALES 3

CHAPTER 2: THE REAL MATT HOOPER 25

CHAPTER 3: UNLIKELY PARTNERSHIPS 55

CHAPTER 4: LIVING IN THE KEY OF G 78

CHAPTER 5: GOING TO SEA 108

PART 2: PREDATOR OR CELEBRITY

CHAPTER 6: THE VINEYARD 133

CHAPTER 7: LIGHTS, CAMERA . . . WHITE SHARK! 160

CHAPTER 8: IT BEGINS 193

CHAPTER 9: THE NEW HOTSPOT 215

PART 3: INTERACTION

CHAPTER 10: SEVERE BUT NOT FATAL 243

CHAPTER 11: PREDATOR AND PREY 276

CHAPTER 12: A SLAP IN THE FACE 307

ACKNOWLEDGMENTS 340

ADDENDUM 341

INDEX 342

FOREWORD

JOHN "JACK" CASEY

I began working on sharks more than sixty years ago. The research began in 1961, at the Sandy Hook, New Jersey, laboratory. A rare unprovoked shark attack off New Jersey the previous year resulted in public concern that the menhaden purse seine fishery might attract sharks close to shore. In response, the menhaden fishing industry and the Smith Research Corporation made their research vessel, R/V *Cape May*, available to the Sandy Hook Lab to conduct research on sharks off the northeast coast. Lucky to be in the right place at the right time, I was assigned as the leading scientist to conduct a survey of sharks using longlines along the Atlantic coast from Long Island, New York, to Cape Henlopen, Delaware. The resulting catch of more than three hundred large sharks received widespread publicity. Concerned about the public's response, we did not immediately tell people that we caught a number of young white sharks within a quarter mile of popular beaches, leading us to suspect that the area may well be a nursery for this infamous species.

An unexpected result of our shark survey was that hundreds of recreational fishermen requested information on how to fish for sharks and, in addition, they wanted to know how they might help in our research. In the decades that followed, I established what is now considered the largest shark tagging program in the world. This ongoing program has resulted in more than 300,000 sharks tagged and released off the American and European coasts, something I never anticipated. I just wanted to get a better understanding of sharks at a time when we knew very little about shark populations in the area. It was an era when the only good shark was a dead shark and when management and conservation were not on anybody's radar. I realized early in my career that people who spend their lives on the ocean have a wealth of experience, so I tapped into the skills and knowledge of fishermen. They bolstered our efforts and truly enhanced our research. I also worked with notable scientists like Frank Mather, Frank Carey, Eugenie Clark, and others from all over the world. From these collaborations came new tools and new techniques, many of which are primitive by today's standards. We were starting almost from scratch in those days, so everything we learned about shark migrations in the Atlantic was new and exciting.

In 1966, the tagging program was transferred to a new National Marine Fisheries Service laboratory on the University of Rhode Island Bay Campus. There, with staff biologists Chuck Stillwell, Wes Pratt, and Nancy Kohler, we began working closely with students and volunteers. As I look back, I realize now that working with students was perhaps the most rewarding aspect of my career. They came from all walks of life, and they all loved sharks. They came to volunteer, to work, and to study, and we learned as much from them as they did from us. Over the years, students came and went. Some kept working on sharks, while others pursued careers as college professors, medical doctors, teachers, and marine biologists. Every student became part of the team—and Greg Skomal was, I am proud to say, one who went on to a successful career studying sharks.

Collectively, we worked at sea and at shark tournaments; we processed samples, analyzed data, and wrote research papers. We went from being up to our elbows in the bodies of sharks to getting pounded by the ocean, crammed in a van, or smashed into a small hotel room. We didn't always get along, nobody does, but we got work done and I know we produced a lot of shark science. In so many ways, the "shark project" was my extended family. We worked together, and often enough, we socialized together. Early on, Greg lived with my son Michael and an assortment of other characters just a block from my home. Like every graduate student, Greg wasn't paid much, so it was pretty typical for the boys to come over for dinner. My wife, Dee Dee, was the consummate mother, and she enjoyed cooking for the motley crew as we reveled in stories about sharks, fishing, and life. Those were happy times.

Greg moved on to the state of Massachusetts in 1987. I thought it was a good move for him. He established the Massachusetts Shark Research Program, which became a mini version of my Apex Predators Program. He continued to focus on life history, eventually expanding his research to physiology and ecology, all the while still working with us as a collaborator. Over the years, we kept in touch, even after I retired in 1995. To me, it never really felt like Greg left our program, and his studies have taken him well beyond what we could have imagined sixty years ago.

In the US and other parts of the Western world, people rarely need to worry about being injured or killed by a wild animal. To many, spiders and snakes are about the most dangerous and, in all but a few cases, unjustly so. In reality, we are far more likely to be hurt by ourselves or something we created, like cars or pollution, than anything living in our backyard. Regardless, the fear of sharks remains pervasive—particularly the so-called "man-eating" white sharks. When I studied white sharks, we focused on life history and distribution because shark attack was not a major concern. On a statistical scale, these attacks are still rare events, but they have increased over the last few decades, in particular along the coast of New England. What has changed? Water temperatures? Prey distribution and abundance? More sharks? More people? Better reporting? Likely a bit of all these things. Can science explain these changes relative to any danger to humans but also in the context of better understanding the biology of this species? And, with that understanding, will we appreciate the role white sharks play in the marine ecosystem? These are the questions tackled by Greg Skomal.

Greg has been obsessed with the white shark since he was a child. He will tell you that he was in the right place at the right time in his career when white sharks returned to the coastal waters of Cape Cod and New England. I've always lived by the mantra that sometimes it's better to be lucky than good, but Greg is both. Over the past thirteen years, he has personally tagged more than three hundred white sharks with tag technologies that didn't exist when I studied these animals. In doing so, he has enhanced our knowledge of this species in this part of the world and beyond. His extensive experience at sea and in the laboratory has given us insights that only he can provide. In this book, he shares some of those findings and provides a synthesis of information, both current and historical, that will serve as the basis for future research. While doing so, he also shows us that chasing a childhood dream is well worth the effort.

The white shark will always be a source of fascination for humanity. Greg's research takes our understanding of this species to another level so we can all better appreciate one of the most feared creatures in the sea.

PART 1

RARE EVENT SPECIES

The shark is perhaps the world's least popular creature. Detested and feared by all, the animal has been hunted down relentlessly, and historically has been the victim of the most fiendish torture the human mind can devise.

—Ann Weeks, NOAA, the quarterly publication of the National Oceanic and Atmospheric Administration, 1974

CHAPTER 1

WHITE SHARKS AND WHALES

It is my personal belief that the responsibility for this horrible shark attack rests squarely upon the shoulders of the aforementioned officials for their utter lack of attention and inaction regarding the growing shark problem on Cape Cod of the last few years.

—Barnstable County, Mass., commissioner Ron Beaty

This has turned into Amity Island real quick out here.

—Joe Booth, local surfer

SEPTEMBER 15, 2018—41°55'24.6"N, 69°58'20.5"W (7:55 A.M.)

The morning sun cut through the green-hued water, casting bright downshafts of light that undulated along the slate-gray back of the eleven-foot shark. It moved effortlessly—or so it appeared—but also with inimitable purpose. If a fish could swagger, this is surely how it would look. From the side, its pointed snout, large dorsal fin, and crescent-shaped tail disclosed its identity, but it's the color—the abrupt demarcation between dark upper and glaring white underside—that is unmistakable to even the most casual observer.

The white shark is commonly called *the* "great white," because the mere mention of its name conjures up larger-than-life imagery. In the face of those jaws or that dorsal fin—those are, of course, the most common tropes—anything short of hyperbole falls flat. The shark's confidence becomes arrogance. Its efficiency, aggression. It's frequently called "mythical," and while it does occupy a central role in the mythologies of many cultures, there really is no need to idealize or embellish. The actuality is enough.

The hard reality is that this animal is about as close to perfection as any animal in existence today. Everything about the white shark is purpose built to excel in one of earth's most challenging environments, as it has done for millions of years. The white shark swam here when Glacial Lake Cape Cod was still draining—when humans emerged and when the dinosaurs went extinct. The white shark's ancestors managed to survive every mass extinction since they arrived on the scene more than 450 million years ago. To call the white shark an evolutionary success story would be like saying Leonardo da Vinci was a reasonably competent artist.

The white shark is no myth. It is no accident or aberration. It is the result of a sequence of precise adaptations achieved over millions of years, and it has culminated in this—a near-perfect, albeit little-understood, animal swimming steadily north along the Outer Cape on a mid-September morning.

SEPTEMBER 15, 2018—PLYMOUTH TOWN WHARF, MASSACHUSETTS (7:35 A.M.)

It was just past seven thirty in the morning and Michelle Collins was in the galley of the whale-watching vessel *Tails of the Sea,* stowing food for the day's trip. She glanced at the clock on the bulkhead and then at her watch. Michelle was not a morning person, and the fact that I still wasn't there was not making her morning any easier. Michelle had guided many whale-watching trips aboard the *Tails*, as those familiar with Captain John Boats' largest whale watching vessel commonly call her, but this trip was different.

The *Tails* captain, Russ Burgess, poked his head in the galley door.

"Is Greg here yet?" Michelle asked him.

"I haven't seen him," Russ replied.

Michelle made her way topside and then onto the wharf. Tendrils of fog hung low over the harbor. It was the coldest morning of the month to date, but despite the damp chill, the forecast called for temperatures climbing into the seventies and a bright sun burning off the fog. It was the last weekend of the summer, and it should have been a near-perfect day for viewing white sharks. "Where the hell is Greg?" she asked herself.

Michelle looked down the wharf toward a line of eager passengers who had paid three times the going rate for a whale-watching tour. Some had driven a couple of hours or more to get there, while others had spent the night in Plymouth, planning to make a weekend of it. A few had even flown in for the trip. The White Sharks and Whales trip happens only twice a year, in September, and it always sells out. It's a joint fundraiser for Whale and Dolphin Conservation (WDC), the organization for which Michelle works, and the Atlantic White Shark Conservancy. She glanced again at the passengers, knowing that, while they hoped to see a whale, and they were certainly glad to support the nonprofits, most are really there to tick something big off their bucket list—an opportunity to see a white shark in the wild. The Conservancy thought that having me narrate the encounter would be the icing on the cake.

What Michelle couldn't see from her vantage point on the dock beside the *Tails* was me, pulling my fifteen-year-old black Toyota pickup into the parking lot on the far side of the wharf. She didn't know that I had left home an hour ago with my daily mug of Earl Grey and a toasted bagel with cream cheese. I had assumed an hour would be plenty of time to get to Plymouth on a Saturday morning, but my wife will tell you I'm always late. This wasn't my first time guiding the White Sharks and Whales trip, and while many shark biologists would not choose to spend a day of their weekend on a whale-watching vessel with a bunch of self-identifying "sharkophiles," I really enjoyed it—but I'm as much a sharkophile myself as I am a scientist.

SEPTEMBER 15, 2018—PLYMOUTH TOWN WHARF, MASSACHUSETTS (7:50 A.M.)

Emerging from my truck, I exchanged the empty mug for my camera, binoculars, and a few extra layers. I locked the truck and started off at a quick pace toward the wharf, walking by the now-silent gauntlet of seafood restaurants, takeout stands, and gift shops that have replaced commercial fishing boats, totes of seafood, and the pungent smell of bait on so many "historic" New England wharves. I passed the empty Lobster Hut, which called itself a "piece of seaside Americana," and the Cabby Shack, with its palm trees and outdoor seating, where summer tourists were instructed to act like "those in the know" and "ask for the super-sized, three-and-a-half-pound Big Man's Lobster." Just past Cabby's there was what Google identifies as "a watering hole with seafood" called the Shanty Rose. It occupied an old war surplus building that was moved to that spot in the 1940s as part of the operations of the wholesale fish company Reliable Fish. A cornerstone of Plymouth's working waterfront, Reliable Fish operated in one form or another on that spot for nearly a century, until, just a few years ago, its owner finally gave up the ghost to "America's Hometown" tourist economy, where a bar is more valuable on a city-owned wharf than a fishmonger.

Over the past several years, I had become accustomed to being a central figure in shark research in New England. This was because New England is a relatively small place, I had been studying sharks in this region for many years, and now white shark news had become ubiquitous in the local media—in essence, I'd emerged as the go-to expert for most every story. *The big fish in a small pond*, as my sister would say. Local reporters had me on speed dial and, not infrequently, would reach out with shark news that I hadn't even heard yet. It was probably also because I'd become somewhat of a regular on the immensely popular *Shark Week,* on the Discovery Channel. As I approached the vessel, I could see the passengers who were boarding the *Tails* in hopes of seeing some of Cape Cod's white sharks up close, and I was inspired by their broad smiles and excitement. For a few seconds, it brought me back many years to the anticipation I felt as I boarded a boat in Rhode Island in my quest to see a live shark for the first time.

Michelle's reaction to my somewhat tardy arrival told me that she was relieved to see me. Although this was her ninth White Sharks and Whales trip, not to mention the countless hours she had spent on the *Tails* during regular whale-watching trips, I sensed that Michelle didn't want to have to deal with both the whales and the sharks during this trip. Michelle started as an intern at WDC after graduating from college in 2012. She spent five days a week on the boat during her first season. The next summer, she traded a bunk in WDC's intern housing for an opportunity to work on the boat again. When she took an office job with the organization in 2014, she reduced her days at sea to just two per week, but her passion for being on the water and helping people connect with marine life, especially whales, only continued to grow. In short, she knew her stuff, but that didn't include white sharks. My job that day was to introduce these passengers to white sharks, and to tell them what we knew and what we didn't know. I didn't mind taking the time on a Saturday to do this because I enjoyed being with fellow shark enthusiasts and I knew they had all spent a lot of money to be there. In these situations, I make every effort to be easygoing and approachable. I take photos with passengers and share my stories. I might be the guy they see on TV, but I'm still just a biologist with the Massachusetts Division of Marine Fisheries. Simply put, I might be viewed as some kind of celebrity shark expert, but I don't view myself as such, and I'd rather be the kind of guy you just want to grab a beer with.

These trips take a team to be successful. Michelle would be responsible for narrating the marine mammal encounters, I would handle the sharks, and we generally had someone else on board to talk seabirds. Russ would get the boat to the animals, and the crew would be responsible for the passengers' comfort and safety. One person who was essential when it came to finding white sharks was not on board: Wayne Davis. To find the sharks, Russ Burgess, the captain of the *Tails*, relied on Wayne, who is the spotter pilot I'd been working with for many years. The only year we didn't see a shark on the White Sharks and Whales trips was the first year, when Wayne was grounded by fog. Once Wayne spots a shark from the air, he directs Russ to it. Often that means taking the *Tails* into the shallows just off the beaches of the Outer Cape, making it very different than the usual whale-watching itinerary, which heads offshore to Stellwagen Bank, a National Marine Sanctuary and one of the primary feeding grounds for humpback, fin, pilot, minke, and endangered right whales. Navigating a 110-foot boat into less than ten feet of water to look at white sharks so close to the beach could be viewed by some as a dicey operation, but Russ had more than twenty years of experience in these waters.

Russ shrugs: "The boat only draws six and a half feet."

By 8 A.M., the passengers had all boarded the *Tails*, and everyone was excited to get under way. Despite the chill, Cynthia Caezza was wearing just a white Washington Redskins T-shirt, a bejeweled-glitter white ball cap, and lipstick as bright as a life preserver. She grew up in Milford, Massachusetts, about twenty-five miles southwest of Boston. She's been

"obsessed with sharks" since she was five and her mother took her to see *Jaws* in the theater. "It traumatized me," she said. "I'm still traumatized, but it's become one of my favorite movies." While growing up, her family went to the seashore at Nantasket Beach in Hull, just up the coast from Plymouth. She remembers being terrified even though her mother and aunts would tell her there were no sharks in Massachusetts Bay. "They're only in Florida," they would say. "Don't worry about it." At the time, the last fatal shark attack in Massachusetts may have been in 1936, but that didn't stop Cynthia from worrying—and, to this day, she remains terrified of sharks.

"I'll go in up to my ankles in the ocean," she says. "I know that the chances of a shark attacking me are low. I know that they feed at twilight and early in the morning. I know they're attracted to sparkly things. I know all that stuff. I'm very aware of all of the things you need to be aware of when you do go into the ocean. But you know, I'm still terrified, and that stems from *Jaws* in 1975. That never went away."

Cynthia's obsession with white sharks goes well beyond fear, however. "I love them, respect them, and I love to learn anything that I can about them," she says. She's a member of a lot of shark groups on social media, everything from several "fans of *Jaws*" groups to shark conservation groups, so it was no surprise that the Facebook algorithm served her up a sponsored post about the White Sharks and Whales trip. As soon as she saw it, she immediately tagged her high school friend Tracee. "Do you want to go?" she wrote. Tracee lives in Washington State and was just as obsessed with sharks as Cynthia—or maybe more, having booked a trip to South Africa to photograph white sharks the previous year. "When she saw that I'd tagged her, Tracee bought a plane ticket to Boston ten minutes later," Cynthia said. "I bought two tickets for the boat trip later that day."

Cynthia was standing beside me on the deck of the *Tails*. I signed a copy of a book I'd written called *The Shark Handbook* for her, and she beamed as Tracee snapped our picture with Cynthia's phone. I blushed as she told her friend that seeing me in person was "like seeing the Beatles." *To me, he's the Paul McCartney of marine biology!* I sometimes hear these kinds of comments from folks who love sharks, and while it feels great, I also feel awkward and don't really know how to respond. At this point I have a canned response for moments like these: "I wish my kids felt that way!"

The fog was already burning off as the boat's four diesel engines grumbled into gear. I'd been working with sharks for more than thirty-five years, and during that time I've logged a lot of hours on boats, so leaving the dock is always a comforting sound. Cynthia, on the other hand, was anxiously double-checking her stash of saltines, ginger ale, Pepto Bismol, scopolamine for nausea (she already had the patch behind her ear), baby wipes, and, perhaps for a worst-case scenario, a box of white plastic trash bags.

The channel leaving Plymouth Harbor is long and circuitous as it wends its way out beyond the breakwater. Russ had done this trip so many times, he could have done it in his sleep if it weren't for the weekend warriors out on their pleasure boats. He steered the *Tails* toward the long, sandy neck of Plymouth Beach, where the channel doglegs northwest and then parallels the beach to sandy Beach Point. Rounding the point, Russ shot south of Bug Light and made for the expanse of Cape Cod Bay and Stellwagen Bank National Marine Sanctuary.

SEPTEMBER 15, 2018—WELLFLEET, MASSACHUSETTS (8:30 A.M.)

The sixty-five-room Wellfleet Motel is situated conveniently along the east side of Route 6 in the town of Wellfleet, on the Outer Cape. Owned by the same family for nearly fifty years, its clean rooms are a bit dated, but nothing that can't be overlooked for the budget price tag, friendly vibe, and fresh-squeezed orange juice served with breakfast. Most important to Arthur Medici and Isaac Rocha, however, was the fact that the motel is just a quick drive from the beaches of Cape Cod National Seashore, where the two young men were planning to spend the weekend surfing on their boogie boards.

Arthur and Isaac woke early, grabbed a quick breakfast, and were soon headed for the beach. It was about a ten-minute drive to Newcomb Hollow Beach, where they had an incredible sunset session the evening before. The conditions were near perfect, but unfortunately, they had missed most of it, given that school was back in session and they couldn't leave for the Cape until Isaac, a junior at Everett High School in Everett, Massachusetts, got out of class. Yo, Arthur texted Isaac on Friday afternoon while he was still in class, let's go to Cape Cod. We're gonna grab a hotel and go surfing. Go home and grab your stuff and be ready because I'm coming to your house.

Arthur was six years older than Isaac, and he was also the more experienced boogie boarder. In fact, he was quite talented, and even though he was not a local on the Cape, he'd achieved a level of respect and acceptance from some local surfers that is pretty unusual owing to both his nonresident status and the fact that he's a boogie boarder. Isaac, on the other hand, was really just a beginner—a *grommet,* in surf parlance—but he loved it, and he'd learned a lot from Arthur about riding the waves and so much more. The two had met at Maranatha Christian Church in Revere, where Isaac's father was a pastor, and had become good friends despite their age difference. It was a relationship that was sure to deepen when Arthur proposed to Isaac's sister, Emily.

As a lifelong surfer, Arthur was immediately drawn to Cape Cod when he moved from his native Brazil to Massachusetts to study civil engineering at Boston's Bunker Hill Community College almost four years before. Cape Cod appeals to many people for many reasons, but its world-renowned beaches less than two hours from Boston have been the major draw over the last fifty years. The year-round population of 215,000 booms to more than 500,000 during the busy summer tourist season, and direct spending by travelers exceeds $1.3 billion annually. Tourism generates more than $58 million in state taxes and over $75 million in local taxes.

People have been coming to the Cape for a long time. Even before the Mid-Cape Highway was built in 1953—before the bridges crossing the Cape Cod Canal went up in the 1930s and before the railway was extended from Boston to Provincetown in 1873—people traveled to Cape Cod. According to some historians, the Vikings may have been the first, but we know for a fact that Bartholomew Gosnold arrived in the early seventeenth century. It was he who bestowed the name "Cape Cod" to the tip of the Cape, near present-day Provincetown. Samuel de Champlain charted Cape Cod's harbors a few years later, and Henry Hudson landed on its shores in 1609. The Pilgrims were actually late to the game when they explored the Cape's tip before moving on to Plymouth in 1620. It was also on the Cape where the Pilgrims first interacted with the Wampanoag Native American people, who, of course, had inhabited the region for thousands of years before the first European arrived.

Anyone looking at a map of the east coast of the United States will appreciate that Cape Cod is one of the most distinctive features of the entire seaboard. It's no wonder that early explorers took note and that modern-day travelers seek it out. The Cape juts some thirty miles into the Atlantic Ocean, and while it is actually an extension of the mainland, the Federal Emergency Management Agency classifies everything east of the Cape Cod Canal as an island. The canal itself was proposed as early as 1623, but it did not open until 1914, when it linked a couple of tidal creeks and severed the Cape's terrestrial bond to the rest of the state.

Henry David Thoreau visited Cape Cod for the first time in 1849. He described its shape on a map as "the bared and bended arm of Massachusetts." Like the orthodox stance of a right-handed boxer, a Muhammad Ali or Mike Tyson, Thoreau envisioned Cape Cod as the leading left arm of Massachusetts, "with her back to the Green Mountains, and her feet planted on the floor of the ocean." Her powerful right fist, Thoreau wrote, "keeps guard . . . upon her breast at Cape Ann," ready to thrust forward, "boxing with northeast storms, and, ever and anon, heaving up her Atlantic adversary from the lap of earth."

Like all New England surfers, Arthur had fixated on storms all week. Surfers understand and think about storms in a way that is counterintuitive to the general public. This is especially the case when it comes to hurricanes, and there was a big one brewing in the Atlantic that week. Surfers up and down the east coast, from Florida's Cocoa Beach to Higgins Beach in Maine, had been anticipating hurricane season. When averaged out over decades, September 10 represents the historic peak day and, almost like clockwork, Hurricane Florence had made landfall the day before, near Wrightsville Beach, North Carolina, at 7:35 A.M. It was the first major hurricane of the 2018 hurricane season, and New England surfers had spent the few days before its arrival eagerly texting each other, poring over National Hurricane Center forecasts, studying storm tracks, and checking surf cams up and down the coast waiting for the swell to arrive. On the Cape, they'd been watching wave height and period data produced by the National Data Buoy Center's 44018 buoy, located nine nautical miles north of Provincetown. The day before, the kind of swell and weather they had hoped for all season had finally arrived. It wasn't an all-time epic swell, but it was good enough, and was forecast to remain that way all weekend.

Arthur pulled his black Nissan Altima out of the Wellfleet Motel parking lot and onto Route 6 heading north. Like those aboard the *Tails,* making its way across Cape Cod Bay some thirty miles to the west-northwest, Arthur and Isaac could not have hoped for better weather. The fog had burned off and the temperature was already in the midfifties and climbing as the laid-back grooves of Jack Johnston's "Love Song #16" played on the radio. The sun streamed in through the passenger window, and Isaac tapped his foot to the staccato groove of the surfer-turned-strummer's newest radio release. One music critic writes that it's a song "so laid-back it could be boiled down and used as morphine," which somehow felt appropriate for this perfect morning on the last weekend of summer as best friends shed their weekday woes for the simple joy of becoming one with the surge of swell and the immensity of the Atlantic Ocean. It would be hard to script the last weekend of summer any better.

SEPTEMBER 15, 2018—42°05'43.4"N, 70°13'40.1"W (10:46 A.M.)

Michelle stood beside Russ, who was steering from an auxiliary station on the roof above the *Tails'* wheelhouse. The visibility was best from this elevated vantage point, and Michelle scanned the horizon for whales. The team had decided to whale watch first and give Wayne some time to find sharks. That strategy paid off—that day alone, we saw nine humpbacks, including two calves, as well as several harbor porpoises and a couple of minke whales. We were just over two miles off the forty-five-foot-tall, iron-plated Race Point Lighthouse in Provincetown on the northwest tip of Cape Cod, when Michelle, to her delight, spotted a whale named Scylla. She knew that it was Scylla by the telltale markings on her large fluke, or tail fin.

Michelle had first met Scylla back in 2012 during her first month as an intern with WDC, and over the past seven years, Scylla has become her favorite whale. Scylla was born in 1981 and is, according to Michelle, a "supermom." For starters, Michelle explained over the PA system, most humpbacks don't have their first calf until they are at least eight years old, but Scylla had her first calf at just six and her second calf when she was eight. Then there was the fact that humpbacks are pregnant for roughly a year before giving birth to a newborn that can be fifteen feet long and weigh in excess of two thousand pounds. After birth, the mother will care for her young for ten to twelve months, and during the first six months, the calf will consume up to eighty gallons of its mother's milk per day.

"As you can imagine," Michelle said, "having a calf takes its toll on Mom." This elicited knowing nods from the human moms leaning against the rail watching Scylla. "So, you can understand," Michelle continued, "why it's so extraordinary that Scylla had calves in consecutive years in both 1997 and 1998, and then again in 2000 and 2001." To have calves in back-to-back years means Scylla was both pregnant and taking care of a calf at the same time, an incredible feat.

As Michelle spoke about Scylla, Russ was on the radio with Wayne in his spotter plane above. Wayne said he wasn't finding any sharks along the northern tip of the Cape. If Wayne couldn't find sharks there, Russ was probably going to have to take the *Tails* down the backside of the Outer Cape, which would make for a long run.

Michelle was in touch with her boss, Regina Asmutis-Silva, via text—just to keep her updated on the day. As they left Scylla behind, Michelle texted, Whales went good. Logging, one breach but there are lots here. Headed to the beach now.

The passengers loved the whales, but I sensed their desire to switch gears and see sharks. I went up through the hatch onto the roof of the wheelhouse, grinning broadly in anticipation.

"Ready to go find some sharks?" I asked Russ and Michelle.

SEPTEMBER 15, 2018—NEWCOMB HOLLOW BEACH, OUTER CAPE (10:46 A.M.)

Isaac was frustrated. He and Arthur had arrived at Newcomb Hollow Beach a couple of hours ago and found near-perfect conditions. There was no wind and the blue sky was snapped crisp to the horizon. Like last evening, the surf was better than expected, with clean waves, good regularity, and none of the short chop that can plague the breaks along the Outer Cape. Waves were stomach to chest high with occasional larger sets, and the water was clear—not tropical Caribbean gin-clear blue, but more of a North Atlantic, hazy-IPA crystal. Despite the conditions, Isaac just wasn't finding his stride, missing wave after wave.

Like an angler deciding where to cast, a surfer reads the underwater topography. Sandbars, channels, and deep holes are all ephemeral and dictate how a particular wave will break at a particular time on a particular day. Newcomb was in great shape that year. The sandbars had built up nicely—there had been no big nor'easters to tear them apart. Yesterday, Arthur noted that in front of the parking lot, a sandbar was causing the wave to break right. That's where most of the surfers were. The right break had multiple takeoff points and offered reasonably long rides, which helped to spread out the crowd. It was the obvious place to paddle out. To the south, however, about three hundred yards down the beach, there was another sandbar that was causing the wave to peel left. These lefts were shorter and steeper and drove into a deepwater hole. Arthur and Isaac had surfed there almost alone the night before, and that's where they were surfing all morning.

People interviewed later recalled a good vibe on the beach that day. The beach was more crowded than one expects after Labor Day, but it was mostly a local crowd. Although the beaches on the Outer Cape were not guarded after Labor Day, several off-duty lifeguards were present. Parking, a constant source of irritation in the summer, was a breeze this time of year. It wasn't uncommon for there to be tension between surfers, stand-up paddleboarders, and body boarders like Arthur and Isaac, but a couple of surfers who surfed the left with Arthur and Isaac the day before had nothing negative to say about the "shark biscuits"—Australian slang for body boarders. "These two kids were good," one local who had surfed near them recalled. "And Arthur was insanely good. Double-three-sixties-same-wave good. Their wave count was very, very high. And their smiles . . . Arthur's in particular."

That day, Arthur had some phenomenal waves, but he was also patiently coaching Isaac—helping him find the right takeoff point and anticipate the rhythm of the swell. Arthur was just that kind of person—someone who derives joy from seeing others succeed.

SEPTEMBER 15, 2018—41°57'55.8"N, 69°59'09.6"W (10:47 A.M.)

The shark had been swimming north slowly and deliberately, probing the various bars, channels, and holes in the shallows not far from the beach.

He sometimes headed for deeper water, but he spent about half his time in the shallows—that's where the food was. In a little over twenty-four hours, he had made his way from the southern tip of the Outer Cape at Monomoy to just off the beach at Newcomb Hollow—a distance of nearly thirty miles.

The morning before, he had swum along the shifting barrier beaches off Chatham, where tourists can buy a "Welcome to Chatham" sign adorned with faux serrated shark bites along the perimeter, an illustration of a white shark, and the tagline "Summer Home of the Great White." The shark eventually passed the storm breach caused by the Patriots Day storm of 2007, which turned North Beach into North Beach Island, and by early afternoon he was off Orleans, swimming close to the picturesque beach on the spit of land separating Pleasant Bay from the Atlantic. He continued north, passing Nauset Beach and eventually arriving at Coast Guard Beach, which is often claimed to be one of the most beautiful beaches in America and usually has a crowd justifying the superlative. It was a couple of hours before sunset.

The shark was hunting. There is only one reason for a large shark to expose itself to the same conditions that draw surfers: seals. At eleven feet long, this white shark has traded his juvenile dentition for the teeth of an adult, which indicates a shift in his diet. As a younger shark, he consumed primarily smaller animals like fish and squid, but now he seeks larger meals. His triangular, finely serrated adult teeth are well adapted to rip chunks of flesh from larger prey. This does not, however, mean the shark has become nastier or a "man-eater"; rather, it's about energy efficiency. Everything a white shark does is about tradeoffs that mitigate the risks associated with living in a hostile environment.

When it comes to energy, feeding is one of the costliest activities a shark needs to regularly pursue, and the calculus is simple. Eating a single seal yields a greater energetic payoff than chasing schools of smaller fish. This is why large white sharks seek out places with an ample supply of large-prey items like seals and sea lions, and it explains the location of the famous white shark hotspots around the globe—places like California's Farallon Islands, South Africa's False Bay, Australia's Neptune Islands, and Mexico's Guadalupe Island. A couple of decades ago, it would have been rare to see a white shark this size swimming along the coast of the Outer Cape, but with the return of both gray and harbor seals, the Cape is quickly emerging as a white shark hotspot.

If this shark had come upon a vulnerable seal the day before while swimming north, it may well have diverted from its slow, steady swim. But unlike the undiscriminating, even haphazard, predator his species is made out to be in movies and media reports, this shark would have been cautious. While he is indeed an apex predator, and while he kills with remarkable speed and power using a hunting strategy described as "ram and bite," he is surprisingly guarded—even bordering on circumspect. Becoming one of the ocean's most efficient predators and having climbed to the top of the food web in the Atlantic did not happen as a result of his ancestors rushing headlong into unknown situations. There is, no doubt, a gruesome power in being able to quickly dispatch an 850-pound adult gray seal in a boil and burst of white water frothed with crimson blood, but more important than this physical dominance, the white shark has evolved to use its muscles and many rows of teeth judiciously.

The shark arrived off Newcomb Hollow Beach around 10:15 A.M. and decided to hang around in the late morning ebb of the tide. Armed with an impressive sensory array, he was constantly processing smells, sounds, and visual cues, and he was continually "feeling" his surroundings by detecting changing currents and the movement of other animals in the water.

SEPTEMBER 15, 2018—42°00'55.4"N, 70°00'56.2"W (11:42 A.M.)

Sharks are taking a while, Michelle texted Regina. Since seeing Scylla, Russ had taken the *Tails* south and then southeast from Race Point. They passed Herring Cove and rounded Wood End, where they could see the Pilgrim Monument over Provincetown. No sharks. Russ then took the *Tails* back north and east around the "fist" of Cape Cod and south toward the beaches of the Outer Cape and started the run down the backside. They had gone a little over ten miles south along the Atlantic beaches of the Cape. We are passing the golf ball now, Michelle texted Regina.

The golf ball is the name locals use for the lone remaining radar dome on the site of the former North Truro Air Force Station. The dome stands on a bluff overlooking the Atlantic and is about six miles north of Newcomb Hollow Beach, where Arthur and Isaac were drawing the attention of those on the beach. They were "doing all these cool tricks, doing flips, spinning backwards." Just then the radio on the *Tails* crackled to life. It was Wayne with good news—he'd found some white sharks off Nauset Beach, almost twenty miles to the south. I jumped on the PA system and told the passengers, some of whom were becoming impatient. Russ goosed the engines.

In about a mile, I stared intently as we were passing Longnook Beach, where exactly a month before to the day, sixty-one-year-old William Lytton was attacked by a shark. The juxtaposition wasn't lost on me. As the shark enthusiasts aboard the *Tails* eagerly anticipated seeing their first white shark of the day, possibly of their lives, Lytton was still at the Spaulding Rehabilitation Hospital in Boston, recovering from deep lacerations on his leg and torso. He said he was able to escape the shark by punching it in the gills.

A little before noon, the *Tails* passed where Arthur and Isaac were riding waves at Newcomb Hollow Beach and continued south to where Wayne had spotted several large sharks.

SEPTEMBER 15, 2018—WELLFLEET FIRE DEPARTMENT (12:11 P.M.)

The officer on duty at the Barnstable County's Sheriff's Office answered the phone at 12:11. It was a state 911 dispatcher reporting that a female caller had just tried to report some kind of emergency at Cahoon Hollow Beach. "I think it might have been a shark bite," the dispatcher told the officer, "but the signal dropped."

But after the first cryptic call to the county sheriff's office, additional calls started flooding in.

The sickening reality set in. There had been a shark attack.

It was at Newcomb Hollow Beach, and it was serious. Lack of good cell coverage at the beach, however, made it difficult to get a clear picture of what was happening.

The dispatcher at the Wellfleet Fire Department made a full department recall—asking for every available member of the department to respond to Newcomb Hollow Beach for a "priority two; possible shark bite." Those who heard the call over the radio knew it was a potentially life-threatening emergency.

Almost immediately, two of Wellfleet's three ambulances were responding along with many first responders. "At this time," Wellfleet Fire Department's dispatcher notified those speeding to the beach, "I have a report of an unknown shark bite, male party. They are currently carrying the party up to the parking lot at Newcomb Hollow Beach."

Fragmented reports continued to inundate the 911 system, with each one painting a more complete picture of the chaos occurring on the beach.

"I've just seen a surfer get bit by a shark."

"It was a helluva hit, man. I saw the whole thing happen."

"It looks like they're carrying him off the beach right now."

"Yeah, I think they just tried to do the . . . they're trying to restart his heart again. I think that someone needs to be here now."

Finally, an on-scene first responder was able to reach the Wellfleet fire dispatcher. "I'm on the beach," he reported. "We have an unconscious male, severe leg injury. We'll need MedFlight if available."

The 911 dispatcher who was fielding the calls from bystanders at the beach notified Wellfleet Fire that calls were coming in of a possible code 99—CPR was being administered. A Boston MedFlight helicopter was scheduled, although weather in Boston would not permit a direct flight. The dispatcher notified the paramedics that there would need to be a combination of MedFlight and ground transport to the trauma center in Boston.

Wellfleet Fire's Ambulance 97 reached the Newcomb Hollow Beach parking lot and established a command center. "A-7 has command at the parking lot," they radioed in. "There is a code 99"—a patient in critical condition—"We do not need MedFlight. They're working him off the beach at this time."

"Received," the Wellfleet fire dispatcher responded. "A-7 has command in the parking lot, reporting code 99, working the patient on the beach. No need for MedFlight."

SEPTEMBER 15, 2018—41°48'17.3"N, 69°55'57.7"W (12:36 P.M.)

The sun was shining, and Wayne Davis in his tiny plane had put the *Tails* on its first white shark of the day. I was on top of the wheelhouse with Michelle and Russ. We were about a mile north of Nauset Beach and less than one thousand feet from the beach.

"There's a shark," I announced over the boat's PA system. The passengers, including Cynthia, crowded to the port rail, cell phones in hand and excited murmurs rippling throughout the boat. Most were now wearing the white T-shirt commemorating the trip. This was the moment for which they'd been waiting. "Beautiful shark," I continued.

Cynthia's cell phone video was shot in the vertical orientation. It zoomed in and shows green water with sunlight slanting back off the undulations. You can hear the hum of the diesel engines as the *Tails* cruises slowly in the shallow water. Occasionally the hull rises and comes down on the water with a thudding sound that casts white spray across the bottom of the image.

For the average person, seeing a fish below the surface of the water, no matter how large, is not easy—especially if you don't have polarized sunglasses. I made every effort to tell them where to look. "Swimming right up near the surface coming down the port side. Great visibility. Looks like a nice big one."

Cynthia's cell phone continued to scan the surface of the water for the shark. The crescendo of excitement began to swell as other passengers saw it.

"Yeah, that's a good view of it," I told them. "You folks see it? There we go. That's a nice twelve-foot shark."

"Oh shit!" Cynthia suddenly exclaimed, as the shark came into view in the upper part of the frame. "Oh my god!"

"Get your cameras ready," I continued. The shark was a couple of feet below the surface, swimming in a characteristic fashion with minimal side-to-side head movements and wide, steady sweeps of its powerful tail fin. The angle revealed in crisp detail the striking demarcation between the shark's dark upper and luminous white underside. There was no mistaking this as anything but a white shark, and it was absolutely stunning.

"Oh shit! Fuck!" Cynthia screamed as the cheers and whoops and clapping around her drowned out my voice over the PA. "Get that fin out of the water!" Cynthia screamed. "Let's go!"

"Never gets boring, does it?" I chuckled to myself with a smile broad across my face. That was what it was all about for me. "Perfect," I said, not realizing I was still speaking into the boat's PA system.

While some scientists might eschew, or at the very least only tolerate with disdain, a day spent shark-watching with the general public, I embrace it. My professional goal has always been to be not only a good shark biologist, but also a public ambassador for white sharks. It's a role I've dreamed about since first seeing *Jaws* as a kid. I knew I wanted to be like Matt Hooper, the shark biologist portrayed by Richard Dreyfuss, and I knew I wanted to study white sharks, but New England wasn't a great place to study the species. Most of what was known about white sharks in the Atlantic came from a handful of specimens that fishermen had brought in. I know that there's a lot to be learned from a dead fish, because I spent much of my early scientific career dissecting sharks. But a dead fish can't tell you much about its behavior, its ecology, and its movements. Except for a single 1979 study by marine biologist Frank Carey of the Woods Hole Oceanographic Institution, scientists knew nearly nothing about living white sharks in the Atlantic.

But the 1972 Marine Mammal Protection Act, combined with the 1997 ban on killing white sharks, began to change that—began to change what we know about living white sharks. When I started the Massachusetts Shark Research Program in 1987, I never in my wildest dreams thought I would be doing white shark research, but when seals started recolonizing the Massachusetts coastline, the white sharks followed. I was clearly in the right place at the right time, being the shark biologist for the state where these sharks were now aggregating. As the overlap between white sharks and people in Massachusetts increased over the last decade—as attacks like the one at Longnook Beach in Truro exactly a month before the White Sharks and Whales trip were on the rise—my job took on a new urgency. My primary role as a scientist is to collect, analyze, and present the data, and the data clearly show a conservation success story. But as a state fisheries biologist, I'm also now squarely in the crosshairs of a debate concerning public safety. When it comes to an animal that can harm people, connecting science with public safety gets really tricky. This isn't just fisheries management anymore, where at the end of the day people can go and have a beer. These are human lives, and this is a hypersensitive subject that requires quite a bit of nuanced juggling.

My phone buzzed in my pocket. Along with the passengers on the decks below, I was still reveling in the first white shark sighting of the day, so I was inclined to leave the phone in my pocket. But then it buzzed again. And again. We'd come into an area of reception, and Michelle, who was standing next to me, could see that my demeanor had changed, and my focus had shifted from the water to my phone's screen. Information about the attack at Newcomb Hollow Beach had reached the *Tails*, and in those first moments there were many more questions than answers. I was soon fielding texts from major media outlets like ABC's *Good Morning America* and CNN all the way down to local news such as the *Cape Cod Times*. I made every effort to continue guiding the trip, but I was also desperately trying to learn more details about what had happened. Just like on Newcomb Hollow Beach, however, the cell phone reception made it hard to get a clear picture, and I decided not to notify the passengers yet. I needed more details.

But the passengers already knew. A minute after I spotted the first white shark of the day, at 12:37 P.M., the *Cape Cod Times* tweeted, Rescue crews are at Newcomb Hollow Beach in Wellfleet because a #shark bite was reported. Stay tuned for more on this developing story #CapeCod.

Given the passengers' intense interest in sharks and the hashtags and Twitter feeds they followed, the news of the attack spread quickly. Michelle had gone below to get more snacks for Russ, and everyone's cell phones were buzzing. She returned to the fly bridge and told me. Nonetheless, I continued to work with Russ and Wayne to find more sharks—we saw six more white sharks between 12:41 and 1:24.

At 12:55, Michelle texted Regina again, writing, There may have been a fatal shark attack in Truro. Greg just got a call from the CCtimes. She asked Regina to watch the news for us. By a little after one, I learned that the victim of the attack was a boogie boarder in his midtwenties and that he may have bled out on the beach.

Michelle snapped a picture with her phone of what would turn out to be the last white shark of the day, and she sent the picture to Regina. It's such a conflicting day, she wrote. It had been less than an hour since the *Tails* spotted the first white shark of the trip. It was an hour that many of the passengers will rank as one of the most incredible experiences of their lives—seeing not only one white shark, but many, including a rare encounter between two sharks. It was also an hour in which I would start to come to terms with the fact that something I'd been expecting for years had finally happened. This attack, if fatal, would potentially change everything.

At a little after 2:30, Regina sent Michelle a link to the *Cape Cod Times* article announcing that the "shark attack victim has been pronounced dead." Michelle texted back that they were going to announce the news. We are going to have a moment of silence for him, she wrote.

Although it was late summer and that special New England brand of hot, it was cooler on the water. I was wearing a light jacket with a flannel liner as I leaned on the rail looking aft over the deck of the *Tails* where many of the passengers had gathered. Russ had turned the boat around, and we were headed back to Plymouth. My blue Massachusetts Marine Fisheries ballcap was pulled low over my sunglasses, casting a shadow across my face. I held the microphone in my right hand, not really sure what I was going to say. I knew I had to choose my words carefully.

"I've been in touch with folks onshore who are investigating an incident that involves a young man who was boogie boarding," I told them. "By all accounts so far, and there are not many, he was attacked by a shark off Newcomb Hollow Beach and he suffered injuries that were fatal. I will know a lot more when we get back in and I have a chance to investigate further." I paused. The only sound was that of the *Tails'* diesel engines and the slap of her hull on the cold Atlantic. The sunken faces below me reflected how much the mood had changed. One man who had been boisterously rattling off white shark facts all trip was now silent and appeared shell-shocked. A woman who had smiled broadly when I signed her book appeared to have tears in her eyes. A boy of no more than eight, who had told me earlier that white sharks were not scary, looked bewildered at his mother, who was explaining that something terrible had happened. I continued, "We all revere these animals, but on a rare occasion they remind us that they are powerful predators that make mistakes."

We observed a moment of silence as the boat made its way back to the marina.

SEPTEMBER 15, 2018—41°57'55.8"N, 69°59'09.6"W (6:52 P.M.)

The sun dipped behind the dunes facing onto Newcomb Hollow Beach. The surf was still good, but nobody was surfing like they were the previous night at this time. Officials had cleared the water after Arthur Medici was attacked and killed earlier that day, and nobody was allowed back in.

The shark returned at 6:15 and was swimming just outside the outer bar. He turned quickly so the last rays of refracted light flared off his white belly in the golden-hued greenroom of one of the small barrels that break and surge beachward. The shark turned again making passes along the same bar, perhaps wondering about the changes in noise, vibration, and smell. He certainly was sensing an area devoid of mammals, where earlier that day there had been so many.

If I'd seen the shark in the last moments of daylight, I would have recognized the white spot on the lower part of his tail fin, as well as a telltale white mark on the leading edge of his dorsal fin. I would have recognized the scratch marks on his face, the way Michelle had recognized Scylla by the telltale markings on her fluke. I would have known the shark was one we had named Mueller. Michelle and I may even have agreed that, beyond the unique markings, there were other unique characteristics each animal exhibited that differentiated him or her from others of his or her species. Michelle would be quicker to use the word *personality*, but I admit that, while I resisted the idea at first, my attitude toward these animals has changed as I've gotten to know some of them, seeing them season after season.

Mueller, however, was a relatively new shark to me back in 2018. I'd tagged him just below his dorsal fin on the last day of July with a coded acoustic transmitter that sends a signal to a receiver secured to the base of a yellow buoy that is moored to the bottom. Unlike a satellite tag that can track the shark's movement whenever it's at the surface, this shark's location can only be recorded when he swims near enough to one of the receivers in the array that me and my team had been establishing along the beaches of the Outer Cape. When I eventually downloaded the receiver at Newcomb Hollow Beach in late September 2018, I knew that it was likely not Mueller, nor any of the 150 other white sharks I had tagged between 2009 and 2018, that was responsible for Arthur's death.

SEPTEMBER 15, 2018—WOOD'S SEAFOOD, PLYMOUTH WHARF (7:15 P.M.)

That night after the White Sharks and Whales trip, Cynthia and Tracee had dinner at Wood's Seafood, one of the restaurants on the Plymouth Town Wharf. Cynthia ordered a piece of fried cod, which comes with a side of rice and a slice of lemon. She also got a stuffed clam and a cup of chowder. "It's strange," she said to Tracee, reflecting back over the day. "I don't know how to say this without it sounding bad, but to be on a ship with Greg Skomal when the first fatality in Massachusetts in, I don't know . . . what was it, a hundred years or so?" She paused. "It was just an exciting and thrilling experience to have when you're, you know, as much into sharks as I am. I mean, it's awful and tragic that the person was killed, but to be on the ship with Greg Skomal . . ." She hesitated, as if she was still trying to process it. "Like what are the chances that could ever happen? And to me, um, that's just a moment I'll never forget."

CHAPTER 2

THE REAL MATT HOOPER

The shark. Hammerhead or white tip. Tiger, shovels, or blue whaler. By whatever name, he is a fearsome brute, an almost perfect killing machine with finely balanced instincts of curiosity and caution. He comes out of the remote past virtually intact. A primitive creature that can be traced back in a direct line over 180 million years. He is one of the most puzzling species of the sea. To understand him, he must be met on his own terms, in his own environment. This is the challenge that awaits Cousteau and the men of the *Calypso*.

—"Sharks," first episode of *The Undersea World of Jacques Cousteau*

It was poor man's big-game fishing.

—Frank Mundus

AUGUST 5, 1961—40°20'36.081"N, 73°57'33.3936"W

The young white shark is almost five feet in length, only several inches larger than he was at birth a couple of months earlier. It had been a relatively quiet summer in the western Atlantic to date—just one hurricane had formed, a tropical wave that emerged from Africa's west coast in mid-July, tracking a course west-northwest and then across the Caribbean Sea, hitting Honduras, Utila, and finally Belize. The shark now weighs about fifty pounds. He was born a fully developed, impressive predator, and to get this big so fast, he began—in a phenomenon known as *oophagy*—by feasting on unfertilized eggs produced by his mother while still in the womb. But now he needs to find his own food—no more freebies from Mom.

The shark's mother was probably more than thirty years old when she gave birth. At close to fifteen feet in length, she'd just reached sexual maturity, and it was her first litter. She'd probably carried her ten pups for more than eighteen months before giving birth, although the length of a white shark's gestation period remains one of the species' many mysteries. The mother provided no care for her offspring. Given the chance, she likely would have even viewed them as prey.

After giving birth, the shark returned to her largely solitary life. Although she is more than three decades old and bears scars obtained while hunting, she now also carries deeper, more menacing mating scars. In the coming year, she will likely mate again, meeting a male shark in the mysterious expanse of the Atlantic. It's not known what will ultimately bring them together, nor the mechanism of attraction, the role of the environment or instinct. It will be a relatively brief encounter. He will grasp her with his teeth and insert one of his claspers, modified extensions to his pelvic fins, into her. They will turn, thrashing. Powerful fins. Flashes of white underbellies. Muscles tensed. And then they will go their separate ways. White shark pups in a single litter can have different fathers, but it's not known if she will mate with other males before giving birth—if she will acquire new scars.

Her offspring, the four-month-old shark swimming north along the coast of New Jersey, has no scars. He swims close to the surface, where the sun's rays spider his smooth back with a maze of crisscrossing patterns of light. Occasionally his fin emerges above the surface of the water. This shark, however, is largely unaware that he's one of the ocean's top predators. In fact, he's been traveling somewhat cautiously, remaining in the relative safety of shallow, inshore waters, as many young sharks fall prey to other sharks, or they become bycatch in gill nets or longlines. Ten white shark pups of similar size had been landed recently by surprised scientists on a half-mile, thirty-two-hook longline set a quarter mile off a bathing beach in Sandy Hook, just a few miles north.

The mortality risk to the young shark will decrease rapidly as he grows, but to grow, he must hunt, which is exactly what he's doing in the relatively warm waters off this New Jersey beach in early August. Unlike other predators who may learn hunting strategies from their parents or older animals with whom they associate, the pup learns on its own. He's using instinct and picking up on cues from his environment—learning to use his advanced array of sensors. He must learn to use each sense through experience.

The young shark lacks the girth of an adult white shark, but that makes him particularly well adapted to chasing small prey. He's become an effective hunter and now patrols theses shallows brazenly.

I've spent my entire career studying sharks, and for the last fourteen years, white sharks specifically. I often think of it, especially early on in my career when the white shark remained so elusive, as chasing shadows—that dark smudge on the surface of the water. A fleeting glimpse. I've learned a lot over that time both from shark researchers who came before me and from my own work. I'm constantly reminded, however, of how much we don't know about the species, and that what we do know, we've only learned relatively recently. In part, this is because sharks are difficult to study. We're terrestrial animals. Sharks live in the ocean, which is a dark, dense world that is foreign to us. Sharks don't conform to a human schedule. When a fifteen-foot white shark shows up off New Jersey or Long Island or Cape Cod, how do you know where it came from and where it's going? How do you know if it's been there before? How do you know if it will stick around or, if it leaves, whether it will come back? How do you know how old it is or how long it lives? How do you know if it's sexually mature? Unless you saw it eat, how do you know what it eats or how often it eats? How do you know what, if anything, eats it? How do you know how large its population is and if that population is growing or shrinking?

These questions might seem to be answered simply by observing sharks, but that only begs another question: How do you reliably find a shark you want to study? And if you are lucky enough to find it, how do you observe it without altering its behavior? How do you find it again to see what, if anything, has changed? The challenges of studying animals that roam the ocean freely are many. Biologists who study Yellowstone wolves, another apex predator, can set up a spotting scope miles away from a den and wait for the parents to bring the pups out into the sun on the first warm day of the spring. The biologist can observe the interactions between adult wolves and pups without interfering. But you can't sit on an Atlantic beach with a pair of binoculars and watch a shark in the same way. As hostile an environment as the Greater Yellowstone Ecosystem can be, it's nothing like the ocean. As land beasts, we may get cold or wet or have trouble crossing rough terrain, but we don't need to get our butts kicked by weather on a boat offshore or wear scuba gear to study wolves.

It's even a challenge to know what exactly we knew about white sharks throughout history. It's uncertain when in the history of naming the things around us somebody accurately pointed at a white shark and said, "That's a white shark" and "It's distinctly different from that shark over there and that other one over there." It's a little clearer when someone identified it as *Carcharodon carcharias*, but for many it remained simply "man-eater." And where did this man-eater live? There were reports from all over the world, but were they all the same species? If they were the same species, but were separated by enough ocean or a continent that made interbreeding impossible, were the individual groups distinct genetic populations? Had they adapted to different environments? Evolved independently? Although it seems all white sharks tolerate a similar temperature range, they often face distinctly different challenges based on where they live. For example, they may rely on different prey species or predate in vastly different environments. Successfully hunting along a shallow, sandy beach may require a different stalking strategy than hunting along a rocky shoreline that falls away quickly into deeper water. Would a white shark successfully hunting Cape fur seals off Seal Island, South Africa, be successful patrolling the shallows of Cape Cod for gray seals?

Because sharks are difficult to study, perhaps it's not so surprising that the scientists who pioneered the study of them were themselves a rare breed. They were often unconventional in their methodology. They frequently possessed huge personalities that often led to conflict . . . and camaraderie. I've been interested in sharks for more than fifty years and I've studied them professionally for the past forty. I've been fortunate to know some of those pioneers and to learn directly from a few of them. Part of what I love about studying white sharks—part of what gets me out of bed in the morning—is that they are so hard to study. Every day I must bring my A-game, think outside the box, and, quite frankly, be prepared to fail.

I wouldn't have it any other way.

Meeting Sharks on Their Own Terms

The first shark I saw was on television.

It was 1968, and I was six years old. My family—me, my parents, and my six siblings—had recently moved to the big 1915 brick colonial house on Lookout Drive in Fairfield, Connecticut. From my bedroom on the third floor, I could just make out Long Island Sound. In the other direction was the sweep of the Brooklawn Country Club golf course. Founded in 1895, when this area was both a center of Connecticut industry and one of the most fashionable New England cities, Brooklawn remained a bastion of a bygone era, even as the surrounding neighborhoods lost a touch of their gilt in the waning years of the 1960s. While my house didn't retain all its period details—and was ready for more than a few home improvement projects—it was perfect for our large family. And my father had gotten a good deal on it.

Off the back of the house and overlooking the sixteenth green of the golf course was a large sunroom, where our only television set resided. It was one of those large, boxy affairs that dominates a room, owing to its state-of-the-art cathode ray tube technology. With seven children, ranging in age from fourteen to just under a year, securing the television for a specific show was rarely a foregone conclusion. But on that night in 1968, ABC was airing the first episode of the US version of *The Undersea World of Jacques Cousteau,* and nobody was fighting to turn the channel.

My brother Burt, who shared the large bedroom on the third floor with me, sat beside me. Burt is a year and a half younger, and as the show began, he noticeably recoiled from the television as the otherworldly shape of a hammerhead shark swam out of the darkness toward the camera. The shark was filmed from below and, because of the thick lead glass commonly used in "tube" televisions, the colors were muted, lacking the real-world clarity with which the modern viewer is accustomed. The result was a preternatural feel, heightened by a musical score hovering between dread, anxiety, and the paranormal.

The title "SHARKS" appeared on the screen in a mustardy yellow, as the hammerhead turned and circled back. A clash of cymbals broke the pounding cadence of kettle drums. The intensity swelled and ebbed like a dangerous riptide as the shark circled. Nearly thirty seconds in,

Rod Serling, whose voice was readily recognizable to my parents as the *Twilight Zone*'s narrator, began, as he had done for countless Americans with his science fiction tales, to transport me into another world. It was a world that was as mysterious and beguiling to me as space to an aspiring astronaut.

"To understand him," Serling said of the shark, "he must be met on his own terms, in his own environment."

While I was too young to recognize Serling's assertion as a challenge—a life calling—that's exactly what it was. With more than five decades of hindsight, I see that clearly now. Today I'm close to Cousteau's age when the documentary aired (he was born in 1910 and died in Paris in 1997), and in many ways, I'm following in his pioneering footsteps and now doing some of the same things Cousteau was doing in the late 1960s—observing, studying, and, yes, meeting the shark on his own terms. Today I am also often sharing that experience with millions of people through television. The technology has changed dramatically, and we know a lot more about sharks today than we did in 1968, but the remarkable thing for me is how much we still don't know—how much there is still to learn.

I was a rising high school freshman in 1975 when I saw the movie *Jaws* in the 1920s-era theater in downtown Fairfield. The four letters on the neon-tubed Art Deco marquee above the theater's entrance would come to define a generation's relationship with sharks. In many ways, the movie also defined my relationship with sharks, and especially with one species—*Carcharodon carcharias*, the white shark. Like millions of other Americans that summer, I was told, "Don't go in the water." But instead of being pulled out of the water by fear, I found himself pushed into it. It was like watching the Cousteau documentary about sharks. When others heard Serling say the shark "is a fearsome brute, an almost perfect killing machine, with finely balanced instincts of curiosity and caution," they heard "fearsome," "brute," and "killing machine"—but I heard something else entirely. Likewise, upon seeing *Jaws*, rather than falling prey to the horror movie tropes that director Steven Spielberg so brilliantly employs, I found himself fascinated by the animal itself.

I was particularly drawn to one character in the movie, Matt Hooper, the shark expert played by Richard Dreyfuss, who is brought in to investigate the shark attacks. In particular, there was this scene where Hooper and Chief Brody go out at night on Hooper's boat looking for the shark and come across fisherman Ben Gardner's boat. I thought Hooper's boat was really cool, decked out with all kinds of modern equipment (modern for the time, at least) and set up for diving. To me it was the boat I imagined for a marine biologist. Hooper spends the film chasing around a big "rogue" white shark that is biting people, and while the takeaway for most viewers was that white sharks eat people, I simply thought he was doing a job I wanted to do.

Jaws consolidated my obsession with sharks into an aspiration that was far more concrete than wearing Cousteau's trademark red cap around the house in Fairfield in the same manner that other children wear cowboy hats or play firefighter. I redoubled my efforts to excel in science, and I took advantage of every opportunity I thought would put me on track to becoming a real-life Matt Hooper. As soon as I was old enough, I enrolled in a scuba class at the local YMCA. I also took to swimming and, of course, I read everything I could get my hands on about sharks in general and white sharks specifically. In the mid to late 1970s, however, it wasn't as easy for a high school student to learn much about the current science pertaining to white sharks. Of course the internet wasn't there to deliver PDFs of scientific papers to my home computer—heck, home computers didn't exist—but it was mainly because scientific knowledge about white sharks was, at best, scant. That seemed strange to me. Here was one of the most charismatic marine animals on earth, and scientists knew almost nothing about its movements, biology, or life history.

The Narragansett Lab and Meeting the "Real Matt Hooper"

We first met on the volleyball court at the Northeast Fisheries Science Center's Narragansett Laboratory in the early 1980s. Marine biology degrees were not common in those days, so I was majoring in zoology at the University of Rhode Island. In 1982 I applied for a research assistant job with the Apex Predators Program, which was based out of the Narragansett Lab. The mission of the program was to conduct life history and movement studies on shark species subjected to commercial and recreational fisheries in the western North Atlantic. It seemed like a dream job to me. I advanced as far as getting an interview for the job with a scientist named Harold "Wes" Pratt, but unfortunately, I didn't get the position. Instead, as a sort of consolation prize, I landed a summer research technician job with the Plankton Ecology Program, also out of the Narragansett Lab. I was disappointed not to be working directly with sharks, but that disappointment was tempered by the fact that I'd been accepted into a program to study coral reef ecology at the now-defunct West Indies Lab on St. Croix in the Caribbean for my last semester. St. Croix was a place I had grown to love during family vacations, and it was a world apart from the cold, wet, dreary winter weather on the southern New England coast.

The National Marine Fisheries Service's Narragansett Laboratory is located on Tarzwell Drive at the end of South Ferry Road in Narragansett, Rhode Island, amid a grouping of government and academic buildings on Narragansett Bay, overlooking Conanicut Island and the town of Jamestown. To many marine scientists, the campus was sacred ground, having been home to a marine laboratory since the 1930s and the days of the Narragansett Marine Lab of Rhode Island State College, run by Director Charles Fish and his wife and fellow researcher, Marie Poland Fish. The area was a hub of marine-related research.

Driving to the lab every morning during the summer of 1982 was a big deal for me. Passing through the gate, parking my car in the employee lot, and then walking into the building as a research technician gave me a sense of both purpose and importance, even if I was working in a trailer adjacent to the main building. My job was to assist in the entry, verification, quality control, and retrieval of data essential for providing estimates of plankton population production. Plankton data, including date, species, location, depth, temperature, and more, were collected on board National Marine Fisheries Service (NMFS) survey vessels. The agency used a mainframe computer, and all these data were keypunched or, in some cases, loaded via punch cards. My primary role was to cross-check the resulting data in the computer against raw data sheets using computer printouts. In the trailer working with me were other young folks close to my age, and I started to hang out with them beyond work, especially Glenn Gioseffi and Ronna Lupovitz—it was these folks who ultimately led to my first meeting with the "real Matt Hooper."

Even though I wasn't working with the Apex Predators Program—

what many around the lab simply referred to as "the shark project"—I quickly came to see it as the coolest program on campus. From my office in the trailer, I would see them loading up gear for travel to shark tournaments and research cruises, and I desperately wanted to know more about what they did. I'd learned that the head of the program was a shark researcher named John G. Casey, whom Jerry Kenney, a longtime outdoor columnist for the New York *Daily News*, had identified as "Hooper the Oceanographer" in a 1975 article about *Jaws*. For obvious reasons, Casey was a guy I wanted to get to know.

When I would see John Casey—whom I quickly learned everybody called "Jack"—in the halls of the main building while headed to the bathroom or on a supply run, he struck me as a man of stature, more like a professor than Matt Hooper. In his forties, Jack looked the part of the consummate professional and, at least to my mind, he looked kind of slick. He frequently wore neatly pressed, short-sleeve, button-down shirts and khaki pants, and his dark hair, which was tinged with gray above his ears and thinning on the top, was always combed back. Although he was just of average height and build, he seemed larger than life to me.

It was Glenn and Ronna who suggested we join the daily volleyball game that took place on the dirt patch adjacent to the main building at lunch. At first I was hesitant to join. College had helped draw me out of my shell a bit, but I was still somewhat shy, and I was certainly not the type of guy who was comfortable striking up a conversation with someone like Jack, nor with his colleagues Wes Pratt (even though I had been interviewed by him a couple of months earlier) and Chuck Stillwell. Despite my initial anxiousness, though, the opportunity that the volleyball game presented was just too good to pass up. So, somewhat begrudgingly, I decided to accept Glenn and Ronna's invitation to join them on the court, and I planned to bring a change of clothes to work.

When lunchtime rolled around the next day, Glenn and I went to the dive locker where everyone changed. At first it brought back some of the awkward memories of high school locker rooms, but being there immediately felt right. There was an air compressor, dozens of oxygen tanks lined up, drying racks for gear, and the mixed smell of neoprene and salt water. This was a real dive locker, but I couldn't help but think that I would rather be changing into a wet suit instead of gym shorts. I liked volleyball, but I hadn't exactly played a lot of it, and was worried I would embarrass myself.

"So, how serious are these volleyball players?" I asked Glenn, as I sat on the bench tying my sneakers.

"Oh, some folks are more serious than others," he said. "Not everyone plays well and some guys get pissed-off when that happens, but for the most part, this is a pretty laid-back group." Despite Glenn's reassurances, I still felt out of my comfort zone.

The volleyball court was located out behind the lab where there was no shade, and the noonday sun conspired with the humidity to make even the thought of an athletic endeavor feel exhausting. Jack was standing on the far side of the court chatting it up with the rest of the team. I wasn't sure exactly what the real Matt Hooper should look like, but my first impression of Jack the volleyball player was no more like the shark expert from *Jaws* than the professional scientist I would occasionally run across in the halls of the main building. Jack was wearing a T-shirt with the sleeves cut off. His potbelly bulged slightly over his gym shorts, which were appropriately short for the early 1980s. White socks and sneakers completed the ensemble. The game hadn't started, and while I wiped some sweat from my brow, Jack was already sweating profusely.

Glenn ducked under the net and went to the far side of the court, leaving me standing awkwardly on the sidelines.

"Hey!" Jack called.

I looked up. Jack was looking at me.

"You must be new here," he said. "You're on our team. What's your name? I'm Jack Casey."

"I'm Greg," I squeaked out as I scurried over to his side of the net. "I work with the plankton folks."

Jack served, and the ball returned over the net to me, where I somehow timed it right and delivered it back over the net for a point.

"All right!" Jack said. "That's the way it's done."

It was not a difficult play, and I'm sure nobody else remembers it, but I felt like I'd just hit the winning home run in the World Series.

The daily volleyball games became a highlight of that summer working at the Narragansett Lab. There was a lot of teasing back-and-forth between the players, and it soon became clear to me that Jack was a fun guy with a great sense of humor. He was quick-witted and ready to chide anyone on the court, but he was also not afraid to make fun of himself. After a few games, his self-deprecating humor drew me out of my shell, and I joined in the banter. Getting a laugh from Jack melted my shyness away.

I really didn't get a chance to see the more serious side of Jack until late in the summer. One of my new friends, Shawn Drew, worked in the main building, and while making a supply run, I would often stop and chat with him. One day as Shawn and I were talking, I recognized Jack's voice coming from his office. I peered around the corner and saw Jack, Wes, and Chuck deep in conversation.

Like Jack, I'd come to view both Wes and Chuck as larger-than-life characters within the world of shark research. I didn't know Wes as well, as he would often opt for a walk, swim, or dive at lunch instead of the volleyball court. Wes was a big guy—the archetype of the marine biologist, big and burly with a full beard. In a wet suit and scuba tank, he could easily be cast into any Hollywood film calling for a rakish, strong diver. Jack, who often came up with nicknames for people, called him "the Winch," because if you needed muscle, you needed Wes. Despite this outward appearance, I came to know Wes as a soft-spoken, thoughtful, and gentle man—it was almost like his disposition didn't match his body. In the water, Wes was impossibly fishlike for a man his size.

The three men were chatting about a manuscript they were trying to finish regarding age estimates for the sandbar shark. Like most researchers at the time, they were using rings in the shark's vertebrae, like rings in a tree, to derive age estimates. The persistent problem with this method was that sharks were not trees and researchers didn't know how often rings were formed. Was it one per year, two per year—or not at all related to time? Jack and his team had an advantage, though. They possessed growth information from sharks the program had tagged and recaptured that they could compare to the growth estimates from the vertebrae. Given that they had been tagging sharks for two decades by 1982, they possessed a massive dataset much larger than anything else to which other researchers had access.

"I cannot impress upon you guys more the need to get this paper out," Jack said to Wes and Chuck. His tone made it clear that Jack wanted the paper finished and submitted to a scientific journal sooner rather than later.

"I agree one hundred percent," Chuck said. "With commercial shark fisheries expanding in the south and sandbars being targeted, time is most definitely of the essence." Chuck typically worked on shark food habits, but Jack had brought him into this paper because Chuck had tagged a lot of juvenile sandbar sharks off the coast of Virginia. He was of average build, with chiseled features, but looked almost small standing beside Wes. He sported a cropped beard connected to one of those thin mustaches that pass just over the upper lip. Unlike most other scientists, Chuck looked like he'd worked on a boat all his life and often wore a denim jacket and jeans. Like Jack, Chuck had a great sense of humor, and he had a tendency to nickname people by tacking an "er" to their name. (Later, he came to call me "Gregger.")

"You'll hear no argument from me," Wes said. "We need new age and growth estimates to be vetted by the scientific community so they can be used for management." I knew that Wes's lab was where most of the age and growth work was happening because I'd applied to be his summer technician. In his lab, each shark vertebra was processed for "reading," which meant that it was cleaned up, embedded in paraffin, sliced, and stained so the rings could be counted. Wes was particularly focused on shark conservation, and with the National Marine Fisheries Service about to make some important management decisions regarding fishing for the species, he knew how important it was to get the data peer-reviewed and published.

"Now that we're all in agreement," Jack said with a wry smile, "what is the holdup?"

Over the next few minutes, they debated various issues associated with methodology, figure development, semantics, and manuscript structure, all the while forging forward with potential solutions. They didn't always agree on the path forward, but there was mutual respect and a team-oriented approach. For me, this conversation illustrated Jack's commitment to the work, his leadership skills, and how a real research team works together. It also showed me that these guys were doing really cool, cutting-edge stuff that was directly applicable to sustainable shark management. As I walked back across the parking lot to the trailer, I kept going over the conversation in my head.

By that point, I knew that Jack ran the largest shark tagging program in the world, but before getting to know those guys that summer, I don't think I really grasped the true scope and magnitude of their work. This wasn't just some obscure shark project focused on one esoteric aspect of shark biology. They were focused on all aspects of shark biology, from reproduction to ageing to food habits to migration. As much as I was looking forward to St. Croix, I didn't want my time at the lab to end.

I was still fascinated by white sharks, but it was clear to me that white sharks didn't occupy a lot of Jack's and his team's time. Only two white sharks were tagged by the Apex Predators Program in 1982, compared to over 2,400 blue sharks, 464 sandbar sharks, and 440 dusky sharks. In mid-July I'd seen a headline in the *Boston Globe* that read "Pacific Sharks Experiencing a Population Explosion?" The article quoted a shark expert named John McCosker, who was also the director of San Francisco's Steinhart Aquarium. McCosker said in no uncertain terms that the white shark population off the coast of California was increasing. He pointed specifically to the Farallon Islands, and the fact that the data showed that white shark attacks on sea mammals used to be observed every few months. "Then it got to be an attack every month," he said. "Now it's several times a week." The increase in white sharks, McCosker said, was directly related to the increase in marine mammals.

I'd gotten to know Jack well enough that I mustered up the courage one day after volleyball to ask him if he thought that would happen in the Atlantic.

"There's no doubt that white sharks are here," he told me, "but they're rare throughout most of the western North Atlantic, and I suspect they'll stay that way unless something dramatic happens either with their food source or with the environmental conditions." He wiped the sweat from his face with a towel. "If you want to study white sharks," he said after a moment, "you should consider going to California, but you know, there's a whole lot to learn about sharks right here in New England."

At the end of the summer, my job as a research technician with the Plankton Ecology Program ended and fall semester classes began. It was my senior year, and although my plate was full, especially given that I was headed to the Caribbean in just a couple of months, I decided to volunteer with the Apex Predators Program. Jack agreed to take me on, and so every day after classes at URI, I drove to the lab, ready to assist with any task, no matter how mundane. Most of what I ended up doing involved coding and filing tag cards sent in by cooperative fishermen who were tagging sharks for the program. It was beyond fascinating to me to see the raw data.

In 1982, cooperative fishermen in the program tagged 4,467 sharks and recaptured 134 that had been previously tagged. Ten of those recaptured fishes were blue sharks that had traveled more than two thousand miles from the United States to Europe, Africa, and South America. A recaptured sandbar shark set a record for time between capture and recapture—seventeen years. The tag on a nine-and-a-half-foot tiger shark recaptured on a longline off Martinique in the West Indies showed that it had originally been measured and released by Brad Haskell, a US foreign fisheries observer, off North Carolina in 1980. That shark set a long-distance record for a tagged tiger shark—1,556 miles! Each card I coded and filed piqued my curiosity about these mysterious animals, about which even Jack and Wes and Chuck had so many questions. The more data I pored over, the more irresistible sharks became.

One advantage to working at the lab was that I had better access to published scientific papers. Every paper I'd read in college had to be obtained from the stacks at the URI library, usually after an exhaustive search through hardbound copies of indices like the *Zoological Record* and *Biological Abstracts*. If I wanted a hard copy of the paper, I had to photocopy the journal at a cost of ten cents per page. Given the amount of time, energy, and money that went into reading scientific papers in those days, most of that effort was spent on getting papers associated with coursework and not what I was most interested in, namely sharks. Volunteering for the Apex Predators Program, I was able to immerse myself in the scientific literature surrounding sharks, where the name "John G. Casey" was as common as calamari on a Rhode Island restaurant's menu.

I'd often ask Jack about the details of a certain paper, and while I was fascinated by the science, it was also the stories Jack would tell of the pioneering days of shark research that captured my imagination.

More than once, I wished I'd been born a couple of decades earlier. There were few people who had studied sharks as intensively as Jack, and there was really nobody in the northwest Atlantic who knew more about sharks and their movements. I learned that it was just as important to cultivate relationships with the many local fishermen as it was to do field research and publish papers.

As fall edged into winter on the coast of Rhode Island, damp, cold days outnumbered sunny ones. Heading for the warm waters of St. Croix and the West Indies Lab for my last semester of college sounded like a better and better idea with each dusting of snow or, more commonly, coating of ice, and I was already packing my things when Wes called me into his office. The research assistant position at the Apex Predators Program, the one for which I had previously been turned down, had unexpectedly opened again, and Wes offered it to me. While I suspect that many young marine scientists in my place wouldn't have hesitated to head to the warm, clear waters of the tropics, I was torn. I'd committed to my parents, the West Indies Lab, and myself to head south for the winter. The wheels were in motion and although the train hadn't yet left the station, it was damn close. In St. Croix I would get some valuable field experience and close out my college career in a perfect place I loved as much as anywhere on the face of the earth. But in Rhode Island I would continue working with world-renowned researchers to study sharks . . . and get paid to do it. My parents happened to be spending the following weekend in Newport, so I went to visit and, as my father would say, we "noodled" all the options. It really boiled down to "dream location" versus "dream job."

They ultimately supported my decision to go with the latter.

As a research assistant, I worked directly for Wes. While Wes's research focused on shark reproduction, as well as age and growth, he was also an avid scuba diver and accomplished underwater photographer. For me it was the triumvirate—sharks, diving, and underwater photography. I'd not only chosen sharks, but I'd chosen to surround myself with people I admired and hoped to emulate.

Jack Casey, the Sandy Hook Lab, and the History of Sportfishing

It was hard for me to imagine an accomplished scientist like Jack Casey, someone who had pioneered shark research in the western North Atlantic and was frequently credited as being the inspiration for the character of Matt Hooper, as ever being a student himself—but of course he had been. In 1959, Jack was studying science at the University of Massachusetts, and that summer he took a job in Woods Hole, a small village in Falmouth on Cape Cod, with a relatively new federal program called the Bureau of Sport Fisheries and Wildlife. This program, along with its sister organization the Bureau of Commercial Fisheries, was a precursor to the National Oceanic and Atmospheric Administration's National Marine Fisheries Service, which didn't yet exist. Both bureaus were created by way of the 1956 Fish and Wildlife Act, and both agencies were overseen by the US Fish and Wildlife Service.

Before the summer in Woods Hole, Jack was planning to pursue a career in conservation and wildlife management, but when he was introduced to scuba diving and the marine world in the cold waters off Woods Hole, his plans quickly changed. Upon graduation with a bachelor of science degree in 1960, Jack applied for a full-time position with the Bureau of Sport Fisheries and Wildlife at its new laboratory at Sandy Hook, New Jersey. This was an ideal location for the laboratory because the bureau's marching orders were to "undertake a comprehensive continuing study of the migratory marine fish of interest to recreational fishermen in the United States," and there were few other places in the US where sportfishing had attained such popularity and accessibility to anglers. As the first director explained, the most important aspect of the lab's location was the fact that it was "near the center of one of the greatest concentrations of sportfishing in the world, and one which draws from species of boreal, transitional, and tropical life zones."

Sandy Hook is centrally located within the New York Bight, a bight being the term used to define a natural curve or indentation in the coastline. The New York Bight is about 15,000 square miles of ocean defined on one side by the generally north–south trending shoreline of New Jersey and on the other side by the shore of New York State, extending east out along Long Island to the famed sportfishing capital of Montauk. The two shorelines meet at an almost right angle at the Hudson River, which extends out into the Bight as the Hudson Canyon. Anglers who fished the New York Bight from Montauk to Cape May, New Jersey, welcomed the new laboratory and its mission. They were pleased the federal government was acknowledging the value of recreational fishing, which had grown to a $600 million industry, boasting nearly 6.3 million anglers. More than half of those recreational anglers fished the Atlantic coast, and they liked that some of their tax dollars were being directed to better managing sport fisheries. They liked that the Fish and Wildlife Act acknowledged "the inherent right of every citizen and resident to fish for pleasure, enjoyment, and betterment." Sportfishing anglers were glad the government was finally recognizing and funding it.

Sportfishing in the ocean for big game like tuna and billfish is as American as baseball and apple pie. Although sharks have been a major component of sportfishing for more than the last half century, it wasn't always that way. Throughout the first sixty years of sportfishing history in the United States, sharks of any species were looked down upon, derided, and despised. In almost all cases, shark fishing, when it did occur intentionally, was a means to an end—to rid the waters of a dangerous scavenger that was little studied and even less understood. Even as anglers became concerned with conserving the sportfish they coveted, most continued to view sharks with abject disdain, and in so doing, they helped inform the public's opinion of sharks as evil villains.

In 1898, off California's Catalina Island, Charles Frederick Holder became the first known angler to land a tuna on rod and reel. For many, this was the moment when recreational sportfishing began as a formal pursuit. Holder, who came from a wealthy New England family, went on to create the first game fishing organization, the Tuna Club of Avalon, which in turn created a template for the creation of other such clubs around the country. In the beginning, sportfishing, especially for big-game species, was accessible only to men—and they were almost exclusively men—with expensive boats, expensive specialized fishing tackle, and the financial means to travel. As sportfishing grew in popularity, it remained largely the purview of the wealthy, but that began to change in the years leading up to World War II and especially in the postwar boom. Nowhere was this more pronounced in the United States than in the New York Bight and especially on the eastern tip of Long Island, in a sleepy fishing village turned sportfishing hub called Montauk.

Montauk's rise as a sportfishing capital was due, at least indirectly, to the vision of Carl Fisher, who, after developing Miami Beach, Florida, set his sights in the mid-1920s on creating what he hoped would become the "Miami Beach of the North." Part of the beauty of the eastern extent of Long Island was the fact that this virtual wilderness of prime acreage was easily accessible by train from New York City. Once Fisher breached the land that separated a large freshwater lake from the sea beyond, the area became one of the best harbors in the region, with easy access to Block Island Sound and the Atlantic beyond. Fisher's dreams were ultimately dashed by the bursting of the Florida real estate bubble, the Wall Street crash, and the ensuing Great Depression, but the area's appeal persisted, especially to anglers who, in 1933, could purchase a round-trip ticket from New York City to Montauk for $1.50. "Fishermen Attention!" the Long Island Rail Road brochure implored. "A Great Opportunity for a Day of Sport and Pleasure. Boats ready upon arrival of train. $2.00 per person including bait."

By 1936 the train was dubbed "The Fisherman's Special," and the railroad boasted additional amenities while ticket prices remained $1.50 (by comparison, a World Series ticket in 1936 cost $5.50). The railroad advertised there would be "Prizes for the Biggest Fish"—tuna, marlin, striped bass, bluefish, blackfish, sea bass, or porgie—landed by Montauk anglers. In addition, anglers were assured their catch would be as fresh at journey's end as when they caught it because fish were iced at no charge during the return trip to the city. "Go by train to the fishing grounds—cheaper, safer, quicker!" was the call, and anglers from all walks of life answered it in droves.

When World War II ended, interest in recreational fishing boomed, and the democratization of the sport was well under way. Even a man of modest means could catch the Fisherman's Special to Montauk and then buy on to a head boat, a boat on which each person on board was paying for themselves, as opposed to a private charter. One of the first successful fishing businesses catering to a new type of angler was Fishangri-la, which, according to *New York Times* outdoor editor Raymond Camp, was "giving Montauk a new lease on life." Camp notes in August 1950 that "the dock is crowded to capacity with private, party, and charter boats, and the operators have opened a number of restaurants and snack bars." Most noteworthy, Camp says, "For the angler of moderate means, this set-up provides everything that could be asked, and at a reasonable price." With business booming, the operators of Fishangri-la cast their proverbial net far and wide to hire more charter boat captains. One of the men who answered the call was Frank Mundus, and it was Frank Mundus who would soon put shark fishing on the map.

Frank had been operating a charter business in Brielle, New Jersey, since 1945, when he decided to move his custom-built, forty-three-foot fishing boat named *Cricket II* to Fishangri-la in 1951. Mundus was a larger-than-life character—surly, irreverent, and eccentric, he often greeted his clients wearing trademark hoop earrings and an Australian slouch hat. His toenails on opposing feet were frequently painted red and green (for port and starboard), and he was a brazen daytime drinker, who would berate his clients . . . much to their delight. In time he would become an inspiration for the character of Quint, the shark hunter in *Jaws*, but in 1951 nobody in Montauk fished for sharks on purpose, not even Frank Mundus. That was about to change, however, and it would affect Jack Casey's career and, a bit later, my career. Both scientists—first Jack and his team, then Jack with me as part of his team, and then me with some of Jack's team—would become common fixtures at the shark tournaments that would surge in popularity from Montauk to Martha's Vineyard—all thanks to Frank Mundus.

How I Became a Fisherman

The first episode of *The Undersea World of Jacques Cousteau* aired in January 1968, when the sixteenth green behind our house on Lookout Drive was covered with snow. Come spring, my mom was more than ready to get us kids out of the house at any opportunity, and there was no better place to go than "the pond" for a little frogging and fishing. At twelve years old, my brother Bernie was deemed old enough to lead expeditions with Burt and me in tow, out the back of the house and down the hill of the sixteenth fairway to the pond. Trips to the pond were always an adventure, not least of which because the pond was a hazard on the par-three fifteenth hole. Golfers had to go from the tee to the green directly over the pond, and a trio of muddy boys fishing on the country club's private property was not advertised in the golf brochure. For me, the fact that we weren't supposed to be on the golf course only added to the adventure of pond trips, but I also had a growing fascination with all things aquatic. Tapping into my vivid imagination, I pretended to be an explorer in the mold of Jacques Cousteau, and the pond may as well have been the most remote reaches of the world's oceans. Rather than sharks, marine mammals, sea turtles, and coral, we found sunfish, polliwogs, frogs, the occasional turtle, and, much to my mother's consternation, plenty of odiferous pond muck.

It was during these excursions to the pond that I started to fish. Admittedly, it was a world apart from the type of fishing that would later earn me a spot in the Martha's Vineyard Striped Bass and Bluefish Derby Hall of Fame, but it was fishing nonetheless. A typical fishing trip to the pond entailed scavenging some of Dad's old rods and mismatched tackle from the basement, digging up some earthworms (or finding hot dogs in the fridge or bread in the pantry) for bait, and then traipsing down the hill behind the house along the edge of the impeccably manicured fairway. Buckets clacked against rods, and nets dragged on the ground behind us—a Norman Rockwell illustration from *Adventures of Huckleberry Finn*. We would spend the day getting dirty and sunburnt and come home smelling of fish and pond water. Those "expeditions" to the pond remain some of my favorite childhood memories.

I enjoyed fishing because life below the pond's surface was a mystery to me and you never really knew what you were going to catch. When we would catch a fish or frog, I was more interested in studying it. The pursuit was fun, but the observation was just as important to me, if not more so. In time, this led to my brother and me repurposing an old plastic pool that Mom had bought for the dog. We dug a hole out behind the garage among some cedar trees and sank the pool into the dirt so that the water was level with the surrounding ground. We added mud and plants and a few rocks to create our own artificial pond. In subsequent trips to the pond, we would collect frogs and minnows—except for the sunnies, we called every fish a "minnow"—to bring home and add to the pond. Of course, without filtration, the fish didn't last long, and we ended up running them back to the pond. The whole exercise, however, was a harbinger of what was to become a lifelong obsession with fish and aquariums.

As I grew up, and as the golf course became stricter about enforcing the no-trespassing rule, we would fish other freshwater spots around town, and that continued right through middle school and into high school. Initially I learned to fish from Bernie, who had learned from my father's father. My grandfather died when I was still quite young, but my brother was close to him, and Bernie's love of both baseball and fishing was a testament to that relationship. That grandfather lived in Nebraska, so all of Bernie's fishing experience was in fresh water, which was probably part of the reason we never fished in salt water even though Long Island Sound was only a few miles away. In addition, the Sound in those days was pretty ugly, certainly nothing like the waters I saw in Cousteau documentaries. To me the Sound was dark and cold, and, quite frankly, it scared me a bit. When I went off to college at the University of Rhode Island in 1979, I joined the Saltwater Fishing Club but quickly learned it was more geared toward talking about fishing during parties than actually fishing.

It wasn't until 1987, when I took the job as a regional fisheries biologist for the Massachusetts Division of Marine Fisheries on Martha's Vineyard, that I became reacquainted with my passion for fishing, and like Jack Casey, fishing then became central to my job.

Turning the "Shark Problem" on Its Head—Monster Fishing

When Frank Mundus, soon to become the most famous shark fisherman of his time, brought the *Cricket II* to Long Island in 1951, no charter captain was specializing in shark fishing. Instead it was common to hear anglers talk of "the shark problem." Most anglers still considered sharks to be, at best, a nuisance that damaged gear and chased the same species they were after. At worst, they were a dangerous menace—man-eaters, vicious brutes, and, in especially hyperbolic moments, "the very incarnation of evil." To the big-game sport angler who envisioned himself in the mold of Ernest Hemingway or Zane Grey, the shark was not the noble animal that the tuna, marlin, or swordfish was. In both Hemingway's novel *The Old Man and the Sea* and Grey's film *White Death*, shark fishing occurs, but the shark is not cast in the image of a revered and worthy opponent, like a magnificent billfish or tuna. Although Grey did popularize fishing for mako sharks on rod and reel, the shark in general remained something malevolent and sinister that must be killed.

Even for the everyman angler, sharks just weren't worth it. As Raymond Camp wrote in his *New York Times* outdoors column in 1952, "A few anglers have been taking along larger rigs and getting in some time on sharks, but the average angler who invests in a charter and a lot of chum likes to get bluefish for his money and effort." During Camp's long career as the outdoor editor for the paper, it's rare to see him mention a shark in the "Wood, Field, and Stream" column without the accompanying descriptor "menace" or "problem."

"It was considered snake hunting," Frank Mundus said of shark fishing when he arrived on Long Island. "It was considered collecting garbage." But it wouldn't remain that way for long. As Frank tells the story, he was out fishing for bluefish one night with clients when they got into some sharks. One of his clients referred to the sharks as "monsters," and something clicked. Frank tells Russell Drumm in the book *In the Slick of the Cricket*, "You couldn't say shark fishin'—we needed a fancy name. Three idiots come walkin' down the dock, and I tell 'em we're goin' MONSTER fishin.' They said okay, like that. They was shark fishin' and didn't know it. It worked. The only thing we caught was sharks. The next time, they said, 'Let's go shark fishin' . . . I was chummin' people with the twist of a woyd, from sportfishin' to monster fishin'."

"It was poor man's big-game fishing," Frank said, and that's exactly the way he set about marketing it.

The meteoric growth of Frank's "monster fishing" business was not mirrored in the sportfishing community at large, where sharks remained "scavengers" and a "menace." To be clear, Frank wasn't the first angler to target sharks; there were others before him, like William Young, who at the end of the nineteenth century, about the same time Charles Frederick Holder landed that tuna off Catalina, set out from San Diego to hunt sharks. "I wanted to go hunting real sea scavengers in their homes, learn their habits and track them down," writes Young in his 1933 autobiographical book *Shark! Shark!* "They were a challenge, they were dragons of seagoing legend, possessed of infinite strength and savagery." There was nothing noble about hunting sharks, though. As the book's reviewer points out, "Catching sharks for any other than the vindictive sailor's purpose of throwing back into the sea the hacked remains of a traditional enemy, was something unheard of."

There are isolated examples dating back to the 1930s of local sportsmen's clubs hosting shark fishing tournaments, but those "tournaments" were unlike other sportfishing tournaments. The objective of a shark tournament was almost always to kill as many sharks as possible. It was population control. In terms of actual "sport," a shark of just about any species—with the possible exception of the mako shark—was considered drastically inferior to any of the tunas or billfishes. The shark often found itself as the butt of the joke, as it did in the late summer of 1949, when the US Atlantic Tuna Tournament, one of the most revered sportfishing tournaments in the country, went "tunaless" for two consecutive days despite a fleet of seventy-eight boats (and two thousand disappointed spectators). On the first day of the 1949 tournament, the only fishes landed were sharks, and one participant sarcastically quipped, "Why not call it the US Atlantic Shark Tournament?" As if there could be nothing more disgraceful.

Often, however, it was far worse for the shark than being a punch line. In the 1950s, shark tournaments increasingly became reactionary events following reports of sharks swimming close to shore or, in far rarer cases, actual shark attacks. A late July headline in the *New York Times* in 1951 proclaimed an "Open War on Sharks," when Long Island fishermen embarked on a "rifle and harpoon campaign to rid nearby waters" of sharks based on sightings alone. Upon landing a seven-and-one-half-foot "shovel nose shark," which, the *Times* reported, "is not a maneater," the fishermen strung it up "[t]o caution swimmers." "[L]ocal residents want to take no chances with their children at the beaches," the reporter wrote, and a "look at the fish's three rows of sharp teeth, it was thought, would be adequate warning."

By the late 1950s, shark tournaments were more common, but they remained almost always billed as an effort to control shark populations. In contrast to the shark tournament, it would be anachronistic to find a tuna fisherman intent on ridding the ocean of tuna. "The only good shark was a dead shark," sums up the sentiment of most people. When an eight-year-old Florida boy lost his leg to a shark bite in August 1958, it kicked off a seven-day "tournament" in which more than fifty sharks were killed. The largest weighed close to three hundred pounds. A fatal shark attack in Hawaii the following January led to a "year-round hunt for sharks" organized by the nephew of shark hunter and author William Young. The younger Young remarked in the local paper, "We just don't like sharks." Young went on to tell the press that he was trying to interest more sportfishermen in "fighting the shark menace." "Anyone who fishes these waters," he said, "has lost good fish to the sharks."

Back in New Jersey and New York, even as Frank Mundus's "monster fishing" business thrived, serious sport anglers were not sold on most shark species as game fish. By 1953, Mundus's fishing clients were setting records for sharks landed, and Frank was having no trouble booking clients, often weeks in advance. "Mundus has made a very successful specialty of catching sharks, any and all kind he can find, on rod and real," wrote Dick Cornish, hunting and fishing editor for the New York *Daily News* in 1955. Cornish had gotten to know Mundus and had himself experienced "monster fishing," writing in 1956, "As all big game anglers know, if you really want to catch a shark, the man to take you is Capt. Frank Mundus, Montauk's 'Monster Fisherman.' This year alone, Frank has hung five world record sharks on the gin pole of the Cricket II." Even with this kind of press, shark fishing remained at the fringes of the sportfishing community. It wasn't until 1958, when Clark Bellows and his boat the *Mutiny II* started advertising "monster fishing charters," that Mundus had any real competition at all.

When forty-three sharks were hooked during the 1959 US Atlantic Tuna Tournament, the tournament's participants echoed what fellow anglers had quipped a decade before—so many of the "scavengers" were hooked that it should have been named "the US Atlantic Shark Tournament." Again, as if nothing could be worse. In late 1959, a group of Palm Beach, Florida, anglers bucked the trend and started a shark fishing club with the purpose of "developing the sports-angling potential of sharks."

The Sharkers, as they called themselves, organized a six-month shark fishing contest.

"Results surprised everybody except perhaps the Sharkers themselves," wrote sportswriter Vic Dunaway in the *Miami Herald* about the group's first tournament. It gave locals "an idea of what monstrous critters swim about in our coastal waters." Surely not the type of tournament news one would hear about a billfish or tuna tournament. "Though these catches undoubtedly failed to scratch the shark population as a whole," he continued, "Palm Beach bathers are probably just as happy that the buck-toothed denizens of the surf have been assassinated." For Dunaway and the Sharkers, shark fishing was a win-win and should be pursued more commonly by anglers because it both provided a public service and a relatively easy way to get into big-game fishing. "[A] fisherman's one best chance of tying into a really gigantic prize is to go after sharks. You can tackle the big ones with any sort of heavy gear and whatever kind of fish carcass or chunks you can pick up for bait."

As Frank Mundus had said a decade before, "It's poor man's big-game fishing," and that is what really began to finally turn the tide. Shark fishing was nothing if not accessible, something that could not be said of billfishing or even tuna fishing. "Developed to a popular degree," wrote outdoor writer Richard Knight of shark fishing in 1958, "it allows a man who has never taken a large fish on any type of tackle to be fairly well assured of hooking one offshore."

There was one group, however, that consistently opposed any form of shark tournament or advertised shark fishing—the Chambers of Commerce. How do you promote coastal tourism to tourists who have no interest in fishing for sharks but who are terrified of meeting one in the surf? Some business owners argued that advertising shark tournaments would hurt coastal tourism, not help it. Wouldn't tourists think the risk of being attacked by a shark while on their beach vacations was greater if there were enough sharks around to support shark tournaments? Wouldn't pictures of sharks hung up on the scales at local marinas scare the average tourist away and even keep residents away from the sea? Little did I know it back in the 1980s, but shark fishing and assessing how the public would respond to the presence of sharks would later become a central part of my career—and I would become intimately acquainted with how slow things are to change.

A Shark Attack That Changed Everything

When Jack Casey began working for the Bureau of Sport Fisheries and Wildlife at the Sandy Hook Marine Laboratory in the summer of 1960, sharks were not top of mind for him and his colleagues as they set about studying the life history of *important* marine sport fishes. In announcing the lab, the director of the bureau specifically mentioned "bluefish, bluefin tuna, striped bass, flounder, weakfish, and mackerel." Yes, by 1960, sharks were gaining popularity as sportfish, especially the mako with its fast runs and dramatic acrobatics, but recreational fishing for sharks remained a small fraction of the overall sportfishing industry, which the new marine game fish research center was charged with studying. Shark research was a low priority.

That changed in August 1960, when John Brodeur joined his fiancée at the New Jersey shore. What transpired over the next two weeks would come to shape Jack's career in a way he couldn't have imagined. It probably was the single most important event in terms of accelerating shark research in the northwest Atlantic in the second half of the twentieth century.

John Brodeur liked to bodysurf, and on a bright Sunday afternoon in August, he was preparing to ride a wave into the beach at Sea Girt, New Jersey, about thirty miles south of Jack's office at Sandy Hook. John was twenty-four and a junior accountant who had spent too many of the hottest days of the summer in a midtown Manhattan office. But things were looking up. In one month, he would be marrying his twenty-two-year-old fiancée, Jean Filoramo. "I was as happy as any twenty-four-year-old guy could be," John recalled.

The Stockton Hotel in Sea Girt opened in August 1921, which was five years after a string of shark attacks left four people dead and another seriously injured on the Jersey Shore. One of the attacks had occurred less than two miles north of Sea Girt, in the town of Spring Lake. In 1960, thirty-nine years after the opening of the Stockton, the hotel was showing its age, but it was still a popular summer destination, and few people had a vivid enough memory of the well-chronicled shark attacks of 1916 to fear swimming in the ocean. In fact, ocean swimming was more popular than ever.

Jean, who was a Jersey City first-grade teacher during the school year, was working as a waitress at the Stockton that summer. The evening before, the couple had walked the boardwalk hoping to see Project Echo, the world's first passive communication satellite. It was a new era. The sea was rough that night. Wind-shorn whitecaps cut stark, ragged lines against the dark night beneath the last wink of a waning moon. Earlier in the week, a major storm had developed off the Bahamas and strengthened into a hurricane. It churned its way north, and although it remained out to sea, its presence could be felt lashing the coast. It made Jean uneasy, and she remarked how cruel the ocean and Mother Nature could be. "To me," she told John, "the waves look like little fingers reaching out to grab you."

On Sunday morning, John attended church in Spring Lake. It was a muggy, overcast morning, but the rain kept at bay. By the time the priest was greeting the parishioners as they left St. Catherine's, the temperature had spiked to near ninety degrees—but at least the sun was out. It was a perfect beach day. John met Jean's parents, and they headed to the Stockton Hotel's employee beach, where John changed into his new blue-and-red-checked swimming trunks. Jean was planning to meet them later in the day. As they waited, John chatted with Jean's parents, lamenting the fact that he was leaving on a business trip to Pittsburgh the next day and wouldn't be back for three weeks, nearly all the intervening time before the wedding. He confided to Jean's parents that the wedding band he had purchased for Jean was a yellow gold that matched her hair. He already had the tickets for their honeymoon to Bermuda purchased and safely locked away in his apartment.

When Jean arrived, she was wearing a fashionable, horizontally striped beige and white tank suit. Her blond hair was cut pixie-style, and she was deeply tanned. They kissed in greeting and immediately went for a swim. The water was warm, just shy of 70°F, and murky, owing to the heavy rains. The surf was unruly and disordered—nothing like the picture postcards of the stately Stockton Hotel trimmed with neatly manicured green grass and fronting a white sand beach lapped by predictable sets of azure-green late summer swell.

Despite the conditions, John caught several good waves as Jean stood in the shallow surf and watched him. The couple eventually left the water, dried off, and settled onto their beach blankets. Jean produced a sample of the thank-you cards she intended to use after the wedding. In small, neat block type was the physical manifestation of their union: MRS. JOHN BRODEUR.

"It gives me such a warm, secure feeling whenever I see the printed name," she told him.

At a little before three thirty, John and Jean decided on one last dip. The tide was out, and they held hands as they entered the water. John commented as to how it felt like they had the whole Stockton employee beach to themselves. The water was barely thigh-deep, and the swells broke around them, pushing shoreward, and then exerting a steady pull, drawing their feet deeper into the sand. John saw a good, shoulder-high wave, and he prepared to launch himself a beat ahead of the wave's face as it slanted up and started to curl in on itself, white foam cresting its top. At the last minute, however, he hesitated, seeing a better wave behind it.

That's when he saw it out of the corner of his eye. It was a long, dark object about thirty feet away, and it seemed to John as if it were propelled by the wave. At first John thought it was a person, but it was too big. Then he remembered they were the only people in the surf. A telephone pole or some other debris washed into the sea by the rains? Perhaps. John's curiosity was short-lived, as the next wave closed. This would be his last wave of the day.

An unseen force slammed him from behind and then had him by the right leg, jerking him violently back. Disoriented with the initial shock, his mind raced to the dark Massachusetts ponds of his youth and the fear of giant snapping turtles that always lingered in his night terrors. But there was no pain. He slapped the water as the force doggedly shook his leg. He became hysterical, pounding the surf, driving at the unseen menace beneath the surface. He kicked wildly with his free leg, and his foot met a rough, heavy-bodied mass. With his hands, he found his assailant underwater and hit it with all his strength, cutting his fingers in the process.

That's when he saw the blood—his own blood, and so much of it. The ocean drew back upon itself, widening the brilliant red smudge in which ragged chunks of torn flesh bobbed to the surface like chum. This was it, he thought. This was the end. *God, save me!* He was aware of Jean screaming for the first time—screaming for help, screaming in fear. John found his own voice as if seeing the light from the bottom of a very deep well. "Jean! Jean!" He needed to warn her—to tell her to get out of the water, away from the thing that was going to kill him. But all he could do was shout her name. "Jean!"

As the large wave closed in around John, he momentarily blacked out. When he came to seconds later, Jean was there at his side. She was attempting to hold him up by the arm as she called frantically to the beach for help. The blood spread with the tide and completely encircled them. Before he passed out again, John saw three men running into the surf.

When John regained consciousness, he was sprawled on the beach. The lifeguard was there, but he was white as a sheet—immobile in his own shock at seeing the trauma to John's leg. Blood pooled in the sand, its acrid smell like the copper top of the hotel's bar. The back of John's calf was tattered pulp—flesh and muscle torn from ankle to knee and flecked with beach sand and tendrils of seaweed. The protruding bone end was crushed. Bystanders slid in and out of John's vision like shadows on a screen. He was nauseous. Jean was crying. Other bathers rushed from the guest beach. A man's face came into focus. It leaned in at him. Unlike the others, who either recoiled in horror or simply looked on, slack-jawed, with a morbid curiosity, this face was determined. The man fashioned a tourniquet from a leather belt and pulled it tight around John's thigh. John winced from the pain. It was the first sensation of pain he had experienced since the incident.

The pain awakened his senses. "Take my pulse," he pleaded repeatedly, recalling something he had learned in a medical class. A man offered him whiskey from a flask, but John knew alcohol was bad for someone in shock. He was in shock, right? Was he going to die? His future father-in-law leaned over him repeating, "The ambulance is on its way. You'll be all right, John. You'll be all right." Jean's mother clutched his hand. "Oh, John! Oh, dear John!" John slid back and forth between bouts of conscious clarity, where the pain scalded and he pleaded for someone to take his pulse, and then the shadowy world where reality became as dark as the previous night, when Jean had remarked how cruel the ocean could be.

A priest came running from the guest beach and muscled his way to the forefront. "Is this boy a Catholic?"

"Yes," Jean's father said.

"Please stand back, everyone," the priest said. "I would like to administer the last rites."

John, in his delirious state, responded by insisting "it must have been a snapping turtle." The last thing he remembered before passing out again was a confident voice that said, "No snapping turtle did that. It's a shark bite!"

CHAPTER 3

UNLIKELY PARTNERSHIPS

Civilization has not yet so refined the male animal as to breed out his zest for hunting the killers of his own kind.

—1960 editorial in the *Dayton Daily News*

To try to make the facts as we know them conform to the "rogue shark" theory is stretching sensationalism and credibility beyond reasonable limits.

—Richard Ellis

You know, people get heated up. In the past it's been a big macho thing to harpoon a shark. Now it seems that may change.

—Jack Casey

AUGUST 5, 1961—40°20'36.081"N, 73°57'33.3936"W

A quarter mile off the busy beach at Sea Bright, New Jersey, the water is barely twenty-five feet deep. The young shark is uninterested in the sounds of swimmers coming from the beach, instead focused on hunting fishes and squid. He began his life with embryonic teeth, but those were shed for teeth covered by a thin membrane and not yet fully erect. Soon those teeth also gave way to ones that are more formidable-looking—longer and narrower. They are perfect for the prey items on which he now relies.

As the shark patrols south off Monmouth Beach, he locates a school of menhaden, a small forage fish in the herring family with deep bodies, black spots, and deeply forked tails. Several strong sweeps of the shark's caudal fin brings him within the margins of the school, where he turns with astounding quickness and agility. His piscivorous teeth grasp a flash of bright silver as the school shifts violently away from the impending attack. There's a plume of carnage, and the shark bites again, really a gulp, and consumes the fleshy body of the menhaden nearly whole.

He has fed primarily on sea robins and menhaden throughout the summer, although he isn't averse to hakes and flounders, as well as skates and even a few crabs. He even ate a couple of dogfish, a small and plentiful shark in these waters. Later in life, he'll develop the white shark's telltale broad, serrated teeth that are designed to rip through blubber and mammalian flesh, but for now, fish are at the top of his menu.

The shark has remained in this same general area for months now. Unlike older white sharks that may travel great distances over the course of a summer, the young shark has taken up residence, perfecting his hunting skills. Plenty of food is present, including discards from the charter fishing boats returning from bluefishing trips out to Shrewsbury Rocks. The shark has grown attuned to the sound of the boats' engines and learned to associate the sound with an easy meal of bluefish heads and entrails discarded over the stern as the fishermen clean their catch on the way back to the dock.

In a couple of months the water temperature will begin to drop, and the shark will need to make his first foray beyond the area he inhabited all summer. He will begin a southern journey, moving through the Mid-Atlantic Bight without lingering anywhere for any length of time. It is an instinctual drive prodded by fewer hours of daylight and cooling water temperatures. Like young white sharks the world over, this shark seems to have a low tolerance for cooler water.

Unlike larger white sharks that also migrate down the east coast of North America, the young shark will not venture eastward into deeper water. Instead he will hug the shallow margins of the coastline, feeding on other species that are also making a similar migration south. He'll continue south to Cape Hatteras, North Carolina, but, also unlike larger white sharks, he'll go no farther. Not this year. This will be where he spends the colder months, and, if he survives, he will head north again next summer.

Before I started working for the Apex Predators Program in 1982, so much of what I thought I knew about shark research proved to be a misconception. In my mind, getting a job studying sharks was going to be next to impossible because, first, everybody wanted that job and, second, we must already know a lot about these animals. I also figured most sharks were in warm tropical waters and that was the place to study them. After all, shark documentaries at the time were filmed in the crystal-clear waters of the tropics. Nobody thought of New England when it came to sharks.

By working alongside Jack Casey and his team, I learned that these impressions were simply not true. For one thing, we didn't know very much about the basic natural history of almost all shark species. Second, there were plenty of sharks off the coast of New England, and many were being targeted by recreational fishermen. Last, I never imagined that, by immersing himself in the sportfishing community, a scientist like Jack could become central to unraveling so many of the mysteries surrounding these animals. I never considered that a shark scientist could learn at least as much at a shark fishing tournament as he could on an official government research cruise working with other scientists. While Matt Hooper came across to me as an extremely cool character with lots of toys to study sharks, and I had this notion that he was the kind of scientist I wanted to be, I don't think I actually believed that building a network of fishermen and performing dockside dissections could be just as valuable as networking with the top names in the field of shark science or looking at shark tissue under a microscope.

Looking back over Jack's career, it became apparent to me that he succeeded more often than not by thinking outside the box. In part, I suppose, that's because there was no template for studying sharks in the western North Atlantic when he began. He had to make it up as he went along, borrowing from other fields of inquiry and being creative. The alliances and partnerships he formed with recreational anglers, commercial fishermen, and even television all seemed unlikely, but each advanced the science in ways I don't think even Jack initially imagined.

About the time I was graduating with a bachelor of science degree from the University of Rhode Island, I realized that what I really wanted to be was a fisheries biologist like Jack Casey. It dawned on me how challenging that was going to be, as I was nothing like Jack. He was so outgoing, while I remained shy. He was a leader, and I was more inclined to drift into the background. But the one area in which I put myself on par with Jack was my passion for sharks. When I saw Jack begin a dissection, I saw the same excitement on his face that I felt in my gut every time I even thought about a shark.

In addition to studying hard, I made a point of learning the history of shark science, and what I learned surprised me. Most shark "science" up until the twentieth century was focused on naming and describing various species of sharks based on their physical appearance. While some of the species that regularly showed up in fisheries were fairly easy to classify, most remained virtually unknown and, in many cases, undescribed. In the late 1930s in the United States, commercial shark fisheries emerged on both the east and west coasts, and these fisheries provided an opportunity for scientific study. Although these shark fisheries were relatively small, they produced a reliable number of dead animals that could be examined. The fisheries data also provided an opportunity to get a sense of local population trends and seasonal variation. But very little scientific study was actually undertaken, with the exception of two species—the soupfin shark (*Galeorhinus galeus*) on the west coast and the sandbar shark (*Carcharhinus plumbeus*) on the east coast. On the east coast, a commercial fisherman named Stewart Springer, who working for Shark Industries Inc., did collect data on sandbar sharks, but the focus was on the animal's utility to humans, namely shark-liver oil. Nonetheless, in 1952, Springer summarized the data in a paper titled "The effect of fluctuations in the availability of sharks on a shark fishery."

One of the most significant advances in shark science occurred in 1954 when Eugenie Clark, who would later become widely known as "the Shark Lady," cofounded the Florida-based Cape Haze Marine Laboratory, which was the precursor to the Mote Marine Laboratory. In 1950, at the age of twenty-eight, Clark had earned her doctorate of zoology from New York University, having conducted research at such venerable institutions as the Scripps Institution of Oceanography, the American Museum of Natural History, and the Marine Biological Laboratory, located in Woods Hole. After graduation, she continued her marine research in the Pacific islands of Micronesia before receiving a Fulbright scholarship, which took her to the Red Sea. In 1953 she published a book about her life, which attracted the attention of Anne and William H. Vanderbilt, who built the Cape Haze Marine Laboratory in order to bring Clark's research to Florida.

One of the lab's first clients was the New England Institute for Medical Research, which needed fresh shark livers for research. The Vanderbilts had recommended a local fisherman named Beryl Chadwick to Clark as an assistant at the lab. As it so happened, Chadwick was a capable shark fisherman. Chadwick easily caught twelve sharks for the New England Institute for Medical Research, and Clark dissected them. The lab's reputation for shark research was soon established, and it wasn't long before Clark and Chadwick were regularly fishing for sharks and conducting behavioral studies on live animals kept in sea pens at the lab. The more Clark studied shark behavior, the more she bristled at the common perception that sharks were little more than "dumb eating machines," who were generally considered to be dangerous. "After some study," she later recalled, "I began to realize that these 'gangsters' of the deep had gotten a bad rap."

Regardless of Clark's magnanimous view of sharks, most of the shark research in the middle of the twentieth century was firmly focused on how to keep people safe from shark attacks. Nearly all the pioneering shark researchers were associated with that work, and most—including Eugenie Clark, Albert Tester, Richard Backus, and Perry Gilbert—were present at a 1958 conference titled "Basic Research Approaches to the Development of Shark Repellents," which was funded by the US Navy's Office of Naval Research. Following the conference, the American Institute of Biological Sciences established the Shark Research Panel, which, in addition to "expediting and activating" recommendations originating from the conference, would also become the clearinghouse for "all information related to the field of elasmobranch (the subclass of fishes that includes sharks and their relatives) biology in general and to the shark hazard problem in particular." Perry Gilbert, whose name would come to dominate shark research, chaired the committee.

In 1960, Stewart Springer, the former employee at Shark Industries Inc., summed up the state of shark research in a paper titled "Natural history of the sandbar shark *Eulamia milberti*." Springer, who was then working for the US Bureau of Commercial Fisheries, wrote: "Sharks have been studied because they are occasionally dangerous to man, often a nuisance to fishermen and, in the past at least, have been valuable as a source for food, leather, vitamin A, fish meal, and some specialty products." He went on to explain why, in 1960, information on the natural history of many shark species was, at best, "fragmentary." "This is to be expected," he wrote, "because large species not only are difficult to catch and handle, but also are far-ranging and require observation over a wide geographical area."

From these humble beginnings, shark research in the US advanced relatively quickly over the next two decades thanks in large part to government funding, including military funding. The military was interested in sharks as both weapons and templates for submarine technology based on sharks' effective predatory strategies. Increased global trade with China also created a market for sharks and especially shark fins. Finally, there was the increasing interest in recreational shark fishing. As a result, what was known about sharks when I began my work at the Apex Predators Program in 1982 was leaps and bounds beyond what was known when Eugenie Clark founded her lab twenty-eight years earlier.

Ever since seeing *Jaws*, I was firmly in the camp of Eugenie Clark—sharks had gotten, and continued to get, a bad rap. I was shocked by how quickly the old and often hyperbolized tropes emerged whenever an incident involving sharks occurred. I remember, for example, reading about two sailors in October 1982 who were attacked and killed by sharks after abandoning their yacht off North Carolina. The men were reportedly hallucinating after drinking seawater and jumped overboard into what some newspapers reported was "shark-infested seas." The incident would later inspire the natural horror/survival film *Capsized: Blood in the Water*. The film aired as part of the Discovery Channel's Shark Week in 2019 with the following blurb: "After a yacht bound for Florida capsizes during an unexpected storm, its crew is left to drift for days in the chilling waters of the Atlantic where they become prey to a group of tiger sharks." While the entire ordeal was certainly terrifying, it struck me that the unexpected storm, the forty-foot waves, and the sinking of the yacht were at least as fear-inducing as the tiger sharks, but in the minds of the media and popular culture, the sharks clearly made a more corporeal antagonist.

Aftermath of a Shark Attack

As I would come to learn later in my career, when a shark attacks a person, the response is often predictable. The difference between the middle of the twentieth century, when John Brodeur was attacked, and the second decade of the twenty-first century, when I was at the epicenter of a new white shark phenomenon on Cape Cod, was the scientific knowledge about sharks. The incident with Brodeur occurred at a time when little was known and much was assumed about sharks in the Atlantic. Many of the experts of the day confounded the situation when they made statements that were unsupported by the scant science, and the misinformation was received and amplified by a general public already primed to fear sharks. I don't think I realized until much later that World War II undoubtedly did at least as much to villainize sharks to an older generation as *Jaws* would later do to my generation.

John Brodeur had been attacked by a shark in less-than-waist-deep water just twenty-five yards from the beach that late August day in 1960. It was the first serious shark attack in New Jersey in more than forty years. Thankfully, the quick response by US Marine Corps veteran Norman Porter, who fastened a tourniquet around John's leg, meant that the last rites were superfluous—but the wound was severe. John was taken to the hospital, where he underwent surgery for over four hours, during which doctors tried to save his limb. Gangrene set in, and more than a week later, John's leg was amputated beneath the knee.

The aftermath of the attack ran the gamut from irrational fear to unwarranted bravado. The messages were mixed, and the headlines were stark and contradictory. The Sea Girt mayor said, "It was just one in a million. It probably won't happen again for a long time." At the same time, the Sea Girt lifeguards were shuttering the beaches. When a fourteen-year-old boy reported being attacked by a "a long brown fish about four feet long" in waist-deep surf the next day, every shadow in the water became a shark. The media homed in, amplifying even the most marginally credible reports.

It was news when two New York Yankees players claimed they saw a group of at least six sharks swimming off Sea Girt. It was news when a Sea Girt lifeguard reported three sharks less than a mile from the beach. Long Branch lifeguards claimed they sighted twenty sharks. The papers reported it all. Coast Guard crews patrolled the water, while Navy helicopters watched from the air, and the newspapers published pictures of Coast Guardsmen from the Shark River Station armed with rifles and dragging raw meat as bait. A blimp from Lakehurst Naval Air Station was assigned shark patrol. William Gray, the director of collections and exhibits at the Miami Seaquarium, stoked fears when he told the press that the shark that attacked Brodeur was a "renegade shark."

"Once in a while there is a renegade shark that gets a taste of a human being," Gray stated. "Then he will keep on looking and when he finds a human being, he'll attack him." Gray said the shark that attacked Brodeur had "gone crazy," and that it would probably return to seek another victim. This concept and the term *rogue shark* had originally been introduced into the shark lexicon by an Australian surgeon named Victor Coppleson in the 1930s in a paper published in the *Medical Journal of Australia*. "A rogue shark," he wrote, "if the theory is correct, and evidence appears to prove it to the hilt—like the man-eating tiger—is a killer which, having experienced the deadly sport of killing or mauling a human, goes in search of similar game. The theory is supported by the pattern and frequency of many attacks." The data did not actually support the theory, but it nonetheless fed into the public's already oversize and largely irrational fear of sharks. Two years before the 1960 attack on John Brodeur, Coppleson reiterated his rogue shark theory in a book titled *Shark Attacks*, which may have led to so-called shark experts of the day talking to the press about a sick or renegade shark being responsible—despite no real evidence to support their claims.

Fishermen soon joined the melee. In an article in the *Asbury Park Press* titled "Shark Attack on Man Sparks New Interest Among Fishermen," the reporter wrote:

> The shark bite incident earlier in the week has kindled a sudden interest in sharks and shark fishing. Ordinarily sharks are not thought of as desirable game fish and for that matter are usually detested by the majority of anglers. But when other big fish are scarce and a fisherman wants a change of pace from bluefishing on the Shrewsbury Rocks, sharks can fill the gap.

A member of the Brunswick Surf Club wondered about a tournament:

> Why not a shark derby with a substantial prize for the biggest shark caught during a specific period of the summer and a bounty for all sharks caught over four feet long? Most of the shore communities face a

> loss of revenue every time there is a shark scare and these communities should be more than willing to finance a shark derby to help rid the surf of areas of the hazard the sharks create. . . . Such a derby might not rid the ocean of all sharks but it would certainly reduce the danger to a large degree.

A Most Serious and Vital Problem

Eight days after the attack, John Brodeur's leg was amputated. The next day, a twenty-five-year-old medical student from South Korea was attacked by a shark while swimming in thirty feet of water roughly a quarter mile off the beach at Ocean City, New Jersey. Richard Chung was a strong swimmer preparing for a competition at the time of the attack. A boardwalk policeman saw the commotion and raised the alarm. Two lifeguards were able to reach Chung in a boat, apply a tourniquet, and get him to the beach. His right leg was "bitten through to the bone," but doctors said he would not lose his leg. He reported the shark was ten feet long and had followed him before attacking.

After this second confirmed attack in 1960, the specter of the shark attacks that took place in the summer of 1916 loomed large. "Once an attack has occurred," Perry Gilbert said, following a California shark attack the previous May, "chances are high that more will take place in the same area." Gilbert was the nation's undisputed leading authority on sharks. Gray's rogue shark theory was, for many, no longer theoretical, despite no real evidence. Today the rogue shark theory has been roundly debunked, but in 1960, state officials and local politicians felt compelled to act, especially as the next weekend was Labor Day weekend. A decrease in tourists would be economically devastating. The New Jersey director of public safety, D. Allen Stretch, announced he would "confer with individual boat owners and flotilla officials in the hope of setting up a volunteer picket line of boats to man bathing areas." He wanted "a ring of boats formed around the swimmers to form a protective shell." A Coast Guard cutter was stationed off Ocean City with Coast Guardsmen standing watch with .30-30 rifles. The owner of the Stockton Hotel purchased and installed a "bubble fence" to protect the hotel's beach.

There were no additional attacks, but state assemblyman Clifton T. Barkalow reiterated his call for a conference of resort area officials in mid-September "to draft plans for fighting the shark menace." The politician said he'd petition the State Department of Conservation and Economic Development for assistance. He told one reporter that the shark threat is "the most serious and vital problem to the entire state of New Jersey, particularly the resort counties," and he demanded to know why proven technologies like the shark fences used in other parts of the world were not employed in the Garden State.

The shark attacks were national news. Even in the heartland, more than six hundred miles from the New Jersey beaches, people saw headlines like "Predators Terrorize US Beaches." Readers in Dayton, Ohio, kept close tabs on the shark news. After the second attack, the editors of the *Dayton Daily News* went so far as to dedicate editorial space to the issue. "Turn sportsmen loose on sharks," the landlocked editors urged. "As more and more bathers and skindivers take to US coastal waters, fatal or maiming shark attacks increase." The solution? "The most natural method of shark control, sportfishing, remains untried. . . . Sport fishermen now prowling US shore waters by the several hundred thousand poke fun at the angler who hooks a shark with the heavy equipment he rigged for tuna, marlin, or some similar prize. Sagacious promotion by boatmakers and resort interests could change their jeers to cheers in a few seasons and stir their sporting blood to new, more exciting endeavors. Civilization has not yet so refined the male animal as to breed out his zest for hunting the killers of his own kind."

The Cooperative Shark Tagging Program Is Born

By late 1960, and largely because of what one paper called the "full-blown shark scare—real or imagined" that was put into motion by the attack on John Brodeur, Jack Casey and the other scientists at Sandy Hook Marine Laboratory were now very much engaged in the "shark issue." The public seemed to think the Sandy Hook Lab could, as one reporter put it, "reduce the odds of a possible recurrence of shark attacks," although it was unclear how they'd accomplish that. From a government standpoint, sharks were a threat to shipwrecked sailors and airmen, an economic liability to coastal communities, and a public safety concern. Sharks had no positive economic value, and they were a detriment to fishermen and fishing . . . or were they? The Bureau of Sport Fisheries and Wildlife was charged with maintaining and increasing public opportunities for recreational use of fish and wildlife resources, and the recreational use of sharks had not been part of the equation—but should it be?

Should the bureau promote recreational shark fishing?

The next summer, Jack, along with other staff from the Sandy Hook Lab, conducted their first set of longline surveys from Jones Inlet, New York, to Cape Henlopen, Delaware. The surveys, which were conducted aboard a commercial trawler donated by a local fishing company, yielded more than three hundred sharks, including two considered a "public safety concern"—the tiger shark and the white shark. What wasn't immediately revealed to the public was the number of white sharks caught. "In order to avoid panic amongst tourists," recalled Al Ristori, a sportswriter who became good friends with Jack and a big supporter of his shark research, "Casey asked outdoor writers not to publicize his catch of ten young great white sharks on a half-mile, thirty-two-hook longline set a quarter mile off a bathing beach in False Hook Channel." Those were not the only white sharks caught in the survey. "Casey also caught juvenile white sharks within a mile or so of beaches at Rockaway and Coney Island, New York, but again kept that information quiet as it appeared the New York Bight might be a pupping ground for whites, and adult females were a much greater threat than their offspring."

Even with some of the specifics pertaining to white sharks kept quiet, many anglers were excited by the survey results and the bureau's newfound interest in possibly promoting recreational shark fishing. The idea of pursuing sharks as big game was starting to go mainstream. Jack was excited as well. Already they'd learned so much, and that was just from one season of longlining. "The more I learned about them," Jack said of the sharks, "the more I wanted to know." Shark fishing in the New York Bight, Jack believed, could be a "welcome addition to sport-fishing," while at the same time providing invaluable data about these little-understood animals that had piqued his intellectual curiosity as a scientist. Over the next year, more than seventy-five anglers and angling clubs partnered with Jack and the Sandy Hook Lab, agreeing to collect data and provide sharks for dissection and study.

A major advancement in shark research on the east coast occurred in September 1961, when Jack partnered with the Bay Shore Tuna Club on its second annual Mako Shark Tournament off Long Island. What may be viewed as controversial today—scientists collaborating with a tournament that promoted killing sharks of all species—was viewed in a very different light in 1961, given the dearth of scientific knowledge regarding sharks. During the shark tournament, Jack and his team from Sandy Hook were on hand to study and dissect the fourteen sharks that were landed by the forty-nine participating boats during the two-day event. Given that there was little appetite for eating most shark species besides mako, the utilization of the dead sharks for science made sense to Jack and spurred a cooperative relationship between scientists and sportsmen that would benefit both shark research and shark fishing for decades to come.

As useful as dead sharks were to the scientists, dissections did not provide a lot of detail about the animals' movements, and given that most of these sharks were migrating, Jack wanted to know where they went. Jack was familiar with the Cooperative Game Fish Tagging Program, started by Frank Mather out of the Woods Hole Oceanographic Institution in 1954, and he wondered about initiating a similar tagging program with sharks. Tagging fishes was nothing new. By 1961, people had been tagging fishes for well over one hundred years, but the reasons for tagging had changed. The first fishes tagged were likely salmon and trout in the middle of the nineteenth century in Scotland. Fishing for salmon had been regulated as early as 1030, and recreational salmon angling with rod and reel had been a popular pursuit since the second decade of the nineteenth century. In Scotland, landowners owned the streams running through their land, and salmon fishing rights were a heritable property. As such, it was understandable why landowners wanted to know where these anadromous fishes went when they migrated from stream to ocean and back again. Tagging was a way to track their movements.

The first successful fish tagging study in the United States was in Maine's Penobscot River in 1873. The goal of the study was "to obtain evidence bearing on the frequency and duration of the salmon's migrations and its rate of growth." To do this, the salmon "must be distinctly and durably marked, yet in such a way as to do them no injury." The researchers working on the Penobscot salmon tried a variety of marking strategies, including branding the fish, but, as the researchers wrote, "the serious mutilation that befell the first fish operated on . . . caused that method to be abandoned." The next effort used a metallic tag attached to the fish's tail by way of a rubber band, but that method was also abandoned because no salmon marked that way was recovered. "It is probable that most of the bands slipped off," the researchers surmised, "and that those which were tight enough to stay on cut through the skin and produced wounds that destroyed the fish." Next the researchers tried attaching an aluminum tag with a platinum wire to the first dorsal fin. They thought this location was ideal because a tag attached near the middle of the fish would be subject to less lateral motion when the fish was swimming. Also, being in the wake of the first dorsal fin would protect the tag from contact with foreign objects. To attach the tag, the researchers removed the fish from the water, strapped it to a table, and pierced the fin with a needle attached to the wire already connected to the tag. The third time was the charm and tagging the salmon in the first dorsal fin became the de facto methodology for most fish tagging moving forward.

While the general concept was the same—mark the animal with a durable and distinct tag that did not injure the animal and which could provide data on its migrations and its rate of growth—Jack knew that tagging a shark was a lot different than tagging a salmon. Jack was not the first scientist to tag a shark—that distinction probably goes to Paul Hansen of Greenland Fisheries Investigations, who had tagged Greenland sharks as early as 1936. To tag those sharks, Hansen ended up using so-called Petersen disc tags (named for the Danish biologist who had invented them in 1894), which were just a little over half an inch in diameter and attached to the shark through either a pectoral fin or through the cheek with 1mm silver wire.

The Petersen tag became the most widely used and successful tag through the early 1960s, but Jack knew it wouldn't serve his vision well. The beauty of Frank Mather's Cooperative Game Fish Tagging Program was that it harnessed the power and reach of recreational anglers to tag many more fishes than a handful of highly trained scientists could. A cooperative shark tagging program modeled on Mather's program could, Jack figured, harness the growing popularity of recreational shark fishing. The challenge, however, was that volunteer anglers would be doing the tagging, and handling a shark to attach a tag is far riskier than handling a gamefish like a tuna. The solution, Jack concluded, was to devise a shark-tagging method that kept the shark in the water, and he thought a dart tag, attached to the end of a ten-foot pole, would be ideal. The angler could simply jab the tag into the shark's back through the base of the first dorsal fin, which is supported by a series of rodlike, cartilaginous structures called *ceratotrichia*. Pushing the dart through the ceratotrichia would anchor it long-term. In November 1962, Jack participated in a research cruise with Frank Mather, during which they tested dart tags on sharks. That November they tagged thirty-eight sharks, including thirty-three blue sharks, four mako sharks, and one hammerhead shark. Jack had proof of concept.

With both the technology and the "mark-recapture" methodology in hand, Jack set out to implement his shark tagging program in the summer of 1963. His volunteer recreational anglers would hook a shark on rod and reel, tag it, and then release it. The tagger would record important information, including date, location, species, length, and sex, and send that information to Jack's program. Then Jack would wait for that same shark to be caught again and for the tag to be retrieved and returned to him. These data would be fed into a growing database that would provide not only data on shark movements, distribution, age, and growth, but also estimates of the size of the population.

The next summer, Jack selected sixty "cooperative sportsmen" from the hundreds of applications he received from enthusiastic anglers and distributed tags to the volunteer shark anglers. After training these anglers in proper tagging technique, the anglers went on to tag two hundred sharks from Delaware to Cape Cod. During that first year, three of the sharks with tags were recaptured and reported back to Jack. While both the Canadians and the Norwegians had attempted some limited shark tagging in the North Atlantic previously, Casey's program, which he named the Cooperative Shark Tagging Program, soon became the most successful shark tagging program to date. According to one reporter, Jack's shark work was "perhaps the most dramatic project currently underway" at the Sandy Hook Lab, and Jack planned to expand the Cooperative Shark Tagging Program to encompass the entire eastern seaboard of the United States within a couple of years.

A Modern Ahab

Jack's Cooperative Shark Tagging Program exploded. In one year alone, 1978, 4,504 sharks from thirty species were tagged and released, and 216 tagged fishes, representing fourteen species of sharks (including one white shark), were recaptured. Ninety percent of these sharks were tagged by volunteers, proving just how valuable collaboration with recreational anglers could be to science. It was a record year for the program, given that the number of sharks released, the number recaptured, and the recapture rate of just under 5 percent were all higher than in any year since the program began.

The next year, 1979—the year I graduated from high school—Jack mobilized a team that included Frank Carey from the Woods Hole Oceanographic Institution to try to study white sharks. Montauk-based sportfishing captain Carl Darenberg Sr., with whom Jack had a close relationship, was also part of the team, as was Wes Pratt. The work was funded in large part by ABC television, which sent a film crew to shoot footage for their popular *American Sportsman* show.

The plan was to chum up white sharks to tag and film them, and the team had been at it unsuccessfully for a couple weeks when a dead fin whale showed up eleven miles off Long Island at the end of June. The dead whale, Jack learned from fishermen out of Moriches, Long Island, had attracted at least seven large white sharks. The event was remarkable for Jack, as previously white sharks had only been sighted as solitary individuals or pairs. Before Jack and his team could get there, one of the white sharks feeding on the fin whale, a 2,075-pound male, was harpooned and landed by a charter boat. They allowed Jack and his team to perform a necropsy on it.

At Jack's request, the National Marine Fisheries Service "gave" the whale to Jack and his team for research. The government didn't want the whale going onto the beach, for obvious reasons, and so they were more than happy to have Jack tow it out to sea so long as he kept it out of the shipping lanes. The sharks followed, and on July 1, Captain Gene Kelly tagged one of the white sharks with a special tag Frank had built in his lab at Woods Hole. What ensued was the biggest breakthrough in white shark research to date. Unlike the "conventional" tags Jack was using in the Cooperative Shark Tagging Program, which simply contained printed information that could be retrieved by anyone who recaptured the shark, the tag Frank devised possessed three transmitters and allowed the scientists to follow the shark by boat using acoustic telemetry. It was a first in white shark science, and newspapers called Frank a "modern-day Ahab."

The tag emitted a high-frequency acoustic ping, which allowed the scientists to follow the shark using a receiver connected to a listening device, or hydrophone, that was mounted on the hull of the boat. In addition to producing a track of the shark, the tag also transmitted depth, water temperature, and, because the dart was inserted into the shark, its muscle temperature. Over the next three and a half days, they followed the shark over 118 miles from southeast of Montauk Point to the Hudson Canyon roughly fifty miles south of Fire Island. It ranged between twenty-two and thirty-four miles offshore and, on average, swam at a speed of between 1.2 and 2.5 miles per hour. Most notably, while the shark did come to the surface and dive to the bottom, it generally remained in the thermocline, the thin layer between the warmer, upper layer of ocean water and the deeper, colder water, where the temperature remained fairly constant. Frank guessed that staying in the thermocline helped the shark to stay in a place where it could sample the mixed layer above and the cold water below.

By comparing the shark's muscle temperature to the ambient water temperature, Frank was able to show that the muscle of white sharks is significantly warmer than the water. Most fishes are "cold-blooded," or ectothermic, meaning their body temperature equalizes with the temperature of the surrounding seawater. Showing that white sharks were warm-blooded, or endothermic, meant they were capable of retaining the heat generated by their own metabolism. While the white shark's muscle temperature changed slowly with water temperature, it remained as much as 12°F warmer than the surrounding water. "It is indeed a warm-blooded animal," Carey told the press upon returning to shore. This one fact had huge implications. An endothermic shark could not only migrate into colder water at higher latitudes and greater depths than many shark species, but it could also hunt there.

The trip was unfortunately cut short due to an electronic malfunction on the boat, and although Frank wished they could have tracked the shark at least one more day, he was pleased with the results. It was a rare opportunity to observe a living white shark, about which, he reminded people, very little was known. The shark was alternatively called the "Montauk Monster" and "Big Ernie," and it was estimated to be fifteen feet in length and 2,500 pounds.

The data themselves were groundbreaking, as was the paper that Frank, along with Jack and Wes as coauthors, published in 1982 in the journal *Copeia*. It was not only a testament to Frank as a pioneer in acoustic telemetry—he built this tag in his lab long before commercial manufacturers did so for me—but those three and a half days provided the very first glimpse into the fine-scale behavior, ecology, and physiology of the species. Based on his temperature measurements, the paper provided the first rough estimate of metabolic rate, which indicated that a single meal of seventy pounds of whale blubber would sustain the shark for up to one and a half months—although these estimates have been subsequently refined to be much shorter.

The contents of that paper also reflected the state of white shark ecological research in the Atlantic at the time and, I suspect unknowingly, the conclusions provided a harbinger of what was to come. Frank and his colleagues hypothesized that the behavior of this shark was typical of white sharks in the Atlantic. They stated that "the seals, sea lions, and elephant seals which are common items in the diet of white sharks in other regions are not available [in the Atlantic]" and "dead whales may be more common than the number of sightings would indicate and might represent an important source of food for white sharks." The implication was that, given the decimation of seal populations, white sharks in the Atlantic were forced to rely heavily on the scavenging of dead whales, which occurred farther from shore and generally out of sight.

This, of course, would change in the years to come as seal populations responded to protection.

Over the next several decades, the acoustic methods developed and deployed by Frank would become standard techniques used by shark researchers around the globe. They would, in fact, become central to my own work studying sharks and, specifically, white sharks off Cape Cod.

Reading this paper still puts a giant smile on my face, and I truly feel honored to have known and worked with Frank during my early career.

There was another aspect of the project that was particularly innovative—its funding. In large part, the research was made possible because of funding by ABC, which believed white shark research, or at least underwater footage of a "man-eater," was something that would appeal to their TV audience. In other words, the investment in the research on behalf of the studio would produce a financial return based on the popularity of the programming. It was a win-win scenario and one that would also become an important part of my own career more than thirty years later.

"I think everyone pretty much got what they wanted out of the whale and sharks," Jack concluded. "ABC got a good show. Carey was elated and probably advanced the knowledge of white sharks by a couple giant steps with the work he did in just a few days. . . . I know I got something," he continued. "[T]agging two great white sharks is something that has never been done before." Of the two sharks that Jack tagged, he noted that both had harpoon marks in them, indicating previous attempts by anglers to hang the animals from the gin poles. On a more philosophical and even progressive note, Jack concluded that "the Moriches guys got their shot at a great white and a lot of publicity with it. It was a shame that there weren't any scientists there with them at the time, but this has been an unusual event for them. You know, people get heated up. In the past it's been a big macho thing to harpoon a shark. Now it seems that may change. I suspect if the whale stayed in place where they could have gotten at them they probably would have stuck the entire lot, all five of the whites."

A Class-A Tournament for Biologists

In the spring of 1983, twenty years after Jack's first successful tagging season, I was finishing my bachelor of science degree in zoology at the University of Rhode Island and working as a research assistant under Wes Pratt at the Apex Predators Program. Jack and his Cooperative Shark Tagging Program had moved from the Sandy Hook Lab to the Narragansett Lab in Rhode Island while I was still exploring the pond on the fifteenth hole of Brooklawn Country Club. Soon after, in 1970, the Bureau of Sport Fisheries and Wildlife was dissolved and the Cooperative Shark Tagging Program became part of the new National Marine Fisheries Service, itself under a new federal agency called the National Oceanic and Atmospheric Administration (NOAA).

When I graduated in May 1983, I continued working at the Narragansett Lab, as I was thinking about the next steps in my career. I was not yet ready to move on to grad school, and the job at the Apex Predators Program was a perfect way for me to gain some experience that could inform my decisions. A little over a month into my first summer with the program, I had the opportunity to see firsthand the secret of Jack's success over the last two decades. The data, of course, speak for themselves, but what I observed made an indelible impression that would come to inform my own career. In 1982 the Cooperative Shark Tagging Program tagged more than 4,500 fishes, including thirty-six species of sharks. Forty-four percent of those fishes were tagged by volunteer recreational rod-and-reel anglers, who were also responsible for 41 percent of the 139 tags returned. Looking at the numbers, I knew that Jack's relationship with recreational anglers and the sportfishing community at large was the key to the tagging program's success, but to see those relationships play out on the docks at a shark-fishing tournament was another thing altogether.

The Bay Shore Mako Tournament, the same tournament with which Jack had collaborated back in 1962, was still going strong in 1983, and Jack and his crew, now including me, were still on the docks performing necropsies on the sharks landed. The tournament was conceived by Charles Entenmann (of baked-goods fame), who wanted to create a big-game angling tournament that wouldn't be limited to the wealthy the way billfish tournaments were. His shark tournament was so successful that many other clubs copied the model. By the early 1980s, shark tournaments were a mainstay of the sportfishing world and Jack was well connected to all the major ones in the Northeast from New Jersey to Massachusetts.

On the first morning of the 1983 Bay Shore Mako Tournament, the weather was less than ideal when the boats left the marina at 6 A.M. A half-gale was blowing, with winds gusting to thirty-five miles per hour out of the northwest. The weather didn't dampen Jack's mood as we crossed the parking lot at the Bay Shore Marina to where the scale was set up for weigh-in at the end of the day. Jack seemed to know everyone he passed on the docks, greeting them warmly and engaging in conversations about everything from bait and fishing techniques to what fishermen had been catching. While I'd become comfortable with Jack and Wes and everyone else at the Lab, I was still shy, and I certainly didn't feel comfortable engaging folks I didn't know. I was in awe of how easily Jack fit in. He was not a man out of his element, as one might expect of a scientist at a fishing tournament. Rather, it appeared to me that Jack was a fisherman who happened to also be a scientist.

"This right here," Jack said to me, "is a class-A tournament for biologists."

At the end of the day, as the anglers returned to the marina, the sharks were identified and weighed to determine the leaderboard, and then we got to work. Jack had us working with the efficiency of a pit crew at the Indianapolis 500. We approached the animal from outside in—first taking measurements and making observations, then cutting the animals open. As the work continued, it moved progressively from researcher to researcher. Wes, who studied reproduction in sharks, was generally first to open the shark, taking reproductive organ measurements before collecting tissues. In the case of pregnant females, we collected, weighed, and measured unborn shark pups. Working directly for Wes, I recorded data for him and then packaged, labeled, and preserved the tissue samples. Wes was followed by Chuck Stillwell, who, working with his student Nancy Kohler, examined food habits by sampling stomach contents. After Chuck, a section of backbone was removed for age and growth work, while other tissues and parasites were sampled and collected for other researchers.

As low man on the totem pole, I did whatever was asked of me as the team made quick work of each animal. I was also charged with the task of documenting the event, the scientists, and the sharks as the team photographer. Like my father, I was fascinated by cameras and loved this part of the job as I tried to enhance my skills as a photographer. As it is for many people who are not gregarious and outgoing by nature, the camera also provided me with a reason to be directly involved in the action, while remaining behind the lens and out of the spotlight.

Like most shark enthusiasts, I was still obsessed with white sharks, although I'd yet to see one in the flesh, either alive or dead. Increasingly, I'd resigned myself to the fact that I was probably not going to have the professional opportunity to work directly with white sharks unless I relocated to the west coast, where there was a white shark hotspot—a place where white sharks predictably aggregate and therefore can be studied, off the central coast in the Farallon Islands. I had hoped this tournament might provide an opportunity to at least see a white shark, as historically it was the Bay Shore Mako Tournament where more white sharks were landed than in any tournament in the Atlantic. But as Jack said, "A marine biologist might go most of his life without having a white shark to examine." I had wanted to be a marine biologist who worked on white sharks since I saw *Jaws* in 1975, and, at times, I'd seriously considered moving to California, but I wasn't willing to give up working with Jack, Wes, and the rest of the team at the Apex Predators Program. At least not yet.

Even though no white sharks were landed at the tournament in 1983, I was thrilled to hear the crew of one boat excitedly report that they'd hooked a white shark about thirty miles southeast of Fire Island Inlet in about 150 feet of water. The *Lady Pat* was fishing with her engines off, drifting with the tide and throwing half-frozen, ground menhaden over the stern. This was the typical shark-fishing procedure: create a large oily slick from the chum that would bring the sharks into range. On a typical day, a shark-fishing boat might use seventy-five pounds of chum. Once the chum slick was established, the fishermen rigged up several rods with mackerel and herring and, using Styrofoam chunks or balloons like bobbers, suspended the baitfish at different depths in the chum slick. Bluefish were often the nemesis of shark anglers fishing a chum slick. It was not uncommon to see bluefish surface feeding in schools as far as the eye could see, and they would readily take the bait intended for sharks. When one of the floats went under, the fisherman at the stern of the *Lady Pat* reeled in the line and, disappointed, removed a bluefish. Instead of throwing it back, he inserted a shark hook through the fish's back, partially severed its tail to hinder its swimming ability, and then tossed it over the side, fishing it live.

The shark fishermen drifted a while before the live bluefish bait was hit by something big. The closest angler grabbed the rod and let the fish run close to thirty yards. As the fish slowed, the angler gave two sharp tugs to set the hook, at which point the fish turned and swam straight back toward the stern of the boat. As it passed under the transom, all five fishermen said they clearly saw that it was a small white shark of maybe 150 pounds. Unfortunately for the anglers (and for me), it broke free, but it was still a good story for Popeye's Bar after the tournament ended, and it gave me hope that I would one day get to see a white shark in the northwest Atlantic. Admittedly, however, as I nursed a beer at the bar, I was beginning to feel a bit like a Sasquatch tracker whose faith in seeing the mythical beast was kept alive by little more than the occasional anecdote.

In addition to the 115 sharks landed during the two days of the tournament, another forty were tagged and released. Some scientists, including Jack, were beginning to think this was the future—tournaments that tagged sharks instead of killing them. It was, however, a tough sell to many of the anglers and tournament organizers, who relied on the scenes at the dock during weigh-in to generate excitement and to confirm the awarding of prize money. The largest shark caught during the 1983 Bay Shore Mako Tournament was a 663-pound female dusky shark. It was the largest shark landed in tournament history, and it was also the largest shark I had observed in the flesh. It was an impressive sight. In writing about the tournament for the *New York Times*, Nelson Bryant, who was aboard the *Lady Pat* when the white shark was hooked, captured it well:

> Inert and flaccid on a dock, its skin dried and wrinkled by the sun, a shark loses the almost spectral quality it possesses in water. Alive in its milieu, it is utter grace and undulating power, an ancient creature with a boneless skeleton seeming to flow rather than swim.

Like the sunnies, minnows, and tadpoles from the pond, I was fascinated by each and every shark brought to the dock. The big sharks were impressive, but it wasn't so much the size as it was the opportunity to see these animals up close—to study them and take note of the differences between individuals. Like any person who steps out of a classroom and into a job, there is the moment when those things about which one has only read are experienced, and the opportunity to observe so many species in such a short period of time was everything I'd dreamed it would be. Sure, book knowledge is important and working in the lab is invaluable, but being on the dock and making firsthand observations of the work millions of years of evolution put into creating these near-perfect predators was without comparison. This was the stuff of Matt Hooper, and I was starting to realize that this was the stuff on which I, like Jack before me, wanted to ground my career.

CHAPTER 4

LIVING IN THE KEY OF G

After *Jaws*, we got a different kind of customer—the majority were killers, not sportsmen. . . . Benchley created an image of the shark that was to the detriment of the shark population. He's been a thorn in the side of the shark industry.

—Ernie Celotto, charter captain who harpooned a 2,800-pound white shark in August 1983

[A]s much as our scientific curiosity demands that we take advantage of every opportunity to examine a white shark, we do not recommend that white sharks be indiscriminately harpooned. They are dangerous when provoked, and they are rather rare individuals that deserve to be left alone.

—Jack Casey and Wes Pratt

JUNE 29, 1979—40°31'30.432"N, 72°48'51.4548"W
(SOUTH OF LONG ISLAND, NEW YORK)

The fifteen-foot male white shark swims purposefully through the Atlantic Ocean, heading north and then north-northeast along the east coast of the United States in its annual migration from the waters off the southeastern US to the warming waters of New England. After paralleling the mid-Atlantic beaches for several days, he starts to cross the New York Bight. As he traverses the Bight, he enters the Northeast Shelf Ecosystem, a region defined by its cold, nutrient-rich water and high productivity.

From above, he is a slate gray—dark and brutishly broad, with a girth of some eight and a half feet. Just behind his head and in front of his dorsal fin, his pectoral fins protrude from each side of his body, arcing back like snub wings on a fighter jet and capable of precision adjustments to the shark's direction and depth. The body narrows toward the tail, where the rhythmic sweep of the caudal fin propels the shark forward at a deliberate speed of about two miles per hour. He's about twenty-five miles off Long Island's beaches and moving easterly, his senses adjusting to the changing ecosystem. Instinctually he knows he is entering an area where seals may be present, although the seal population is nothing compared to what his ancestors knew—before the humans hunted the seals to near extirpation. Still, the effort to find one can be worth it. An adult male gray seal can weigh more than eight hundred pounds. With its fatty blubber, even a smaller seal means far more energetic payoff for the exertion required to find, hunt, and kill it than chasing squid or tuna.

The shark is hungry, but he's patient.

At over two thousand pounds, the shark has encountered few animals larger than himself, except for several whales to which he paid little attention during his long trip north. He may well double in weight over the next couple of decades, provided he lives that long. His scientific name, *Carcharodon carcharias*, is derived from the Greek: *Sharpen teeth type of shark*. It's an appropriate name given the impressive rows of piercing, broad, serrated teeth that are well adapted to rending chunks of fatty blubber from his prey.

As the shark swims east along Long Island toward Montauk Point, all its senses are working together. It is his excellent sense of smell, though, that first hints of a meal not far away. The shark processes smell by taking water in through nostril-like openings called nares, which, unlike a human nose, are not attached to its respiratory system. Instead the shark's nares are highly specialized structures located at the tip of its snout and divided by a flap of skin into separate channels for incoming and exiting seawater. Once the water enters the shark's nares, it moves over a series of plates covered with olfactory receptors, which in turn communicate information to the olfactory bulb in the shark's brain. Of all the sharks, the white shark possesses one of the largest olfactory bulbs, weighing close to a fifth of the entire brain weight. That's how critical smell is to the shark's ability to successfully hunt and feed.

The particular scent in the water today is both familiar and welcome to the shark, and he increases his speed and straightens his course as he makes his way to the source. Feeding doesn't always involve hunting, at least not hunting live prey. In this case, the scent is originating with a decomposing fin whale, and the shark excitedly follows the slick of malodorous oil seeping from the putrefying flesh that stretches miles downstream.

The summer of 1983 changed my life.

I'd just graduated from URI with a degree in marine science, and I was continuing to work as a research assistant with the Apex Predators Program out of the NMFS Narragansett Lab in Rhode Island. I was working directly for Wes Pratt, but under Jack Casey's overall direction. My duties ranged from field sampling to photography to lab work. Wes had trained me on the histological techniques of tissue preparation in the lab. This involved taking fresh or preserved tissues collected in the field and running them through the process of embedding them in paraffin, slicing them very thin, staining them to highlight cells, and mounting them on slides for analysis. I did this with reproductive tissues to examine if sharks were mature as well as with shark vertebrae to age them. I also gained a lot of photographic experience. This was long before the days of digital photography, so I became accustomed to shooting Kodak slide film, like Kodachrome and Ektachrome, as well as black-and-white film like Tri-X. I used a darkroom at the lab to develop and print the black-and-white photos, which were sent out to media outlets and used for publications.

Like most college graduates, I thought I knew a lot, and while I wasn't entirely certain what my next move would be beyond working at the lab that summer, I felt confident that a promising career as a marine scientist was in my future. The hard work of the past four years was paying off, and I was finally able to focus on the animals with which I'd been obsessed since first seeing them in *The Undersea World of Jacques Cousteau.*

By 1983, Jack had studied sharks for more than two decades, and the Cooperative Shark Tagging Program was a huge success by any measure. Researchers were learning a lot about the sharks that inhabit the northwest Atlantic, but, not unlike the college graduate, the more they learned, the more they realized how much they didn't know. This was especially the case when it came to the sharks they didn't see as often—the ones the Apex Predators Program scientists called "rare-event species." Species like the white shark. In the Atlantic, researchers saw these animals so infrequently that their knowledge of them was like a seventeenth-century map with a lot of blank areas occupied by fanciful drawings of sea monsters. To me this was amazing. I distinctly remember aspiring to be Matt Hooper after first seeing *Jaws* in 1975, but then thinking to myself that by the time I finished school, there wouldn't be much left to learn about them.

How wrong I was.

The more I spent time around these scientists studying sharks, the more I came to appreciate how much work there was to do. Jack and his team were fisheries biologists, and although I'd taken a fisheries course or two, I wasn't really in tune with what it meant to conduct biological studies focused on fisheries. In general terms, a fisheries biologist collects data and conducts biological research that will directly assist with the sustainable management of a species. The fisheries biologist is concerned with information like how fast a fish grows, what size it is at maturity, how often it reproduces, and how many young it produces. These data are all essential for natural resource management, whether we are talking deer, codfish, or sharks. This was all new to me.

One thing that surprised me was how much of the work being done by the Apex Predators Program was dependent on access to dead sharks. The shark tagging program was telling us a lot about where and when the most common species of sharks migrated, about population structure, and how much a shark grew between tagging and recapture, but it didn't tell us a lot about the basic biology of sharks. To understand the natural history of these magnificently evolved animals, traditional methods involved using necropsies to sample a variety of tissues and organs so you could study food habits, age and growth, and reproduction—all essential for the management of shark fisheries.

When I think back to the summer of 1983, I think most about the dissections. Sometimes people ask me if dissecting a shark is sad or difficult for me, especially sharks that haven't died of natural causes. Is it sad working with sharks caught as bycatch in fishing gear or ones that are killed by recreational anglers? When asked that question, I think back to July and August of 1983. During those two months, I dissected many sharks landed by recreational anglers, but what I remember most is the work, not the emotion. Doing science is hard, often messy work, but if you're passionate about it, it's the kind of work you feel lucky to be able to do.

In terms of emotion, yes, I was excited. Dissecting sharks is fascinating. It's one thing to read about a shark's internal anatomy in a book, but it's quite another to actually see that anatomy in a real shark. To feel the weight of its liver—see the color of its muscle and look in its stomach. I would have loved to study live sharks in their environment, but I was completely sold on the value of what Jack and his team at the Apex Predators Program were doing. Their work was important to science. To be frank, I was too busy to be sad, and this information was much needed for shark management, which was largely lacking in those days.

I guess I always believed in the back of my mind that the opportunity to study live animals would eventually materialize.

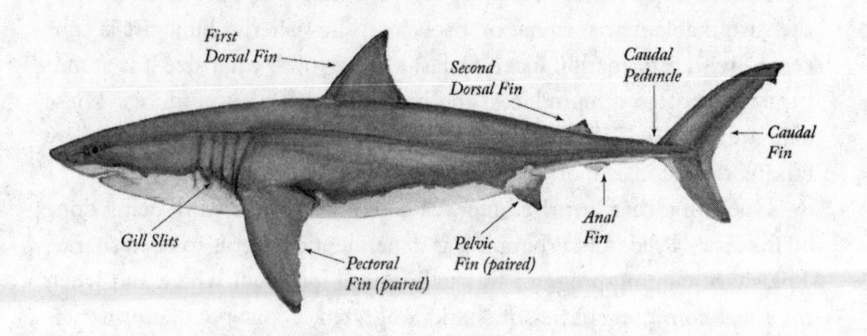

White Sharks: Rare and Vulnerable

The summer of 1983 was hot. By Labor Day, the northern plains were begging for federal aid during the worst drought since Dust Bowl days. "We're asking Uncle Sam to help where Mother Nature has cruelly neglected her responsibilities," pleaded Missouri's governor. Record-breaking heat also scorched the Northeast. July was the hottest month ever recorded in Boston, where the temperature exceeded ninety-five degrees seven times. The Baltimore Orioles were having a great season, and the Police and Michael Jackson topped the pop charts. A gallon of gas cost $0.96. The first American woman and the first black astronaut went into space aboard the space shuttle, and Motorola released the first mobile phone. The third installment in the *Jaws* franchise, *Jaws 3-D,* turned out to be a disappointment—*Return of the Jedi* was the film to beat.

Despite the performance of *Jaws 3-D*—the best the *New York Times* could muster was to say it's "probably no worse than *Jaws II*"—the summer of 1983 was a great one for me. I liked the movie in the way anyone passionate about a thing has a vested interest in seeing a movie about it, but there was no denying that, when it came to sharks, fact was better than fiction. As my father was fond of saying, I was "living in the key of G"—marching to the beat of my own drum—as I worked daily at the Narragansett Lab alongside scientists on the cutting edge of shark research in the northwest Atlantic.

In addition to tournaments, like the Bay Shore tournament I worked in July, the lab also got access to shark specimens by way of commercial fishermen or recreational anglers who knew to call Jack if they landed a "rare-event species." It's what I loved most about my job—the fact that although there was the monotony of data entry and other jobs around the lab, each day also held the promise of something completely unexpected. I was continuing to hold out hope that one of those rare-event species calls would be about the shark that inspired it all for me: the white shark.

I knew white sharks were present in New England waters, but the data confirmed the anecdote that they were not common. The reasons they were rare are unclear, as little was known about the western Atlantic population. Since the beginning of the Cooperative Shark Tagging Program in the early 1960s, just twelve white sharks had been tagged—one in 1968, one in 1969, one in 1973, one in 1977, two in 1979, one in 1980, three in 1981, and two in 1982. Compared with sharks like the blue shark—with more than one thousand blue sharks tagged some years—the white shark remained elusive.

In other parts of the world where white sharks were more common, some scientists were starting to express concern. In 1982, a commercial fisherman landed four white sharks near California's Farallon Islands, an established white shark hotspot. Scientists closely studying that population observed a marked decrease in predation events in the weeks that followed. This surprised them because seal and sea lion populations were skyrocketing following the 1972 Marine Mammal Protection Act, which made it illegal to kill, harm, or harass the white shark's favorite prey item. There should be more sharks. Instead the number of observed predation events was cut almost in half, and it took several years to return to the levels previously observed. Could the removal of just four individuals have so significantly affected an otherwise healthy population with a plentiful food source?

In other places where recreational fishing for white sharks was popular, researchers reported similar trends. One of the most dramatic examples was a location in Australia known as Dangerous Reef. In the late 1960s and early '70s, it was a reliable place to find white sharks. Scenes from both *Jaws* and *Blue Water, White Death*, my favorite film about white sharks, were filmed there, and it's where Zane Grey traveled in 1939 to fish for white sharks. By the 1980s, scientists reported that increased fishing pressure was coinciding with very few white sharks observed. Leading shark advocates and researchers like Valerie Taylor and Jacques Cousteau expressed concern about the species and its ability to withstand fishing pressure.

It wasn't just white sharks about which there was concern. The more scientists learned about the biology and life history of sharks, the more they realized how different sharks are from other fishes. Sharks grow slowly, mature later in life, produce low numbers of well-developed young, and live a long time. In many ways, they are more like mammals, and one need only look at whales to understand the vulnerability of this evolutionary strategy in the face of a predator as effective as humans. Despite these concerning signs, by 1983 there was still no widespread effort to protect sharks, and in the western North Atlantic, sharks remained entirely unmanaged.

The focus for so long had been "to rid the seas of sharks" and "end the shark menace" that few people were asking the question of whether sharks might need protection. Jack, who was part of the federal government's effort to promote shark fishing in the 1960s, was one of those people asking a new question. In 1976, sportswriter and shark fisherman Nelson Bryant wrote in the *New York Times* that Jack Casey was pondering the question of whether sharks needed protection:

> The motion picture *Jaws* and the book that spawned it accelerated the interest in sharks and shark fishing and may also, says shark specialist John Casey at the Narragansett Marine Laboratory, have contributed to increasing concern for the well being of the species. Not very much is known about the various sharks but it is clear that they are vulnerable to overfishing. . . .

To be clear, Jack was not against shark fishing. He continued to actively collaborate with shark anglers and shark tournaments because he believed they remained the best way to access specimens for study. Expressing concern about overfishing and maintaining strong relationships with shark tournament organizers were not mutually exclusive in Jack's eyes, and he was not alone in that position.

Just a year after *Jaws* was released, an article in the *Naples Daily News* about the ninth annual Florida Lake Worth US Open Shark Tournament posited that shark tournaments might soon be a thing of the past. It's "a contest whose days may be numbered," wrote the reporter. Tournament organizer Gerald Mickley was quoted saying that initially the purpose of the tournament was indeed to reduce shark populations, but that may be changing. While some types of sharks are "absolutely dangerous," he said, he believed others "may be becoming endangered." The tournament's scientist, Regina Skocik, a shark researcher affiliated with the Smithsonian Institution and the author of *The Sharks Around Us*, was more direct, arguing that there are reasons for not killing sharks. "Sharks are not all wild meat-eaters. Some have specialized dietary habits. By destroying them, we're destroying the ecosystem of the ocean."

Like Jack, Skocik believed there was also scientific value in shark tournaments even as she encouraged caution. "The only reason we know as much as we do about these species is the fishing public," she said, "and it's the people who are fishing for sharks and who are willing to fill out questionnaires who are helping the most." She was not opposed to shark fishing tournaments, she said. "We do need the vital data."

By 1983, I hadn't given too much thought to shark conservation. Working for Jack, I'd seen the fishing public's value to shark science, and I found the work fascinating. There was no coherent opposition to shark tournaments in the early 1980s (that would come later) and shark tournaments were growing in popularity. From my perspective, there were going to be dead sharks on the docks, and not utilizing them for science was a waste. In addition to the data gleaned from shark dissections, I also felt a more visceral connection to opening up a shark on the dock. There was a powerful narrative that aligned this messy, bloody work with the archetype who inspired my own quest to become a marine scientist: Matt Hooper. During the summer of 1983, I thought a lot about the scene in the original *Jaws* movie, where Police Chief Brody (Roy Scheider), absentmindedly fidgeting with the foil on the neck of a bottle of Beaujolais, says to Hooper, "Why don't we have one more drink and then go down and cut that shark open?"

In the movie, Brody and Hooper had watched a posse of local fishermen triumphantly land a large shark at the town dock earlier in the day. The fishermen are intent on slaying the man-eater that killed a young boy. The mayor, resolute in his decree to keep the beaches open and the tourist dollars flowing through the summer holiday, proclaims as fact the story sweeping the dock. *This is the shark that killed the boy! It's safe to go back in the water!* A reporter and photographer are in the process of consecrating the narrative in what they hope will be a nationwide media blitz, because the shark attacks on Amity, like most shark attacks, will be national news. But Hooper remains rationally unconvinced. While other characters are motivated by political pressure, ego, and fear, Hooper, the scientist, focuses on objective truths—what he can observe, measure, and, ultimately, dissect.

The scene in the movie is rowdy. The crowd alternates between jeers and cheers—celebratory bedlam as the animal is hoisted aloft. It's a mob wielding fishing rods and harpoons instead of pitchforks. One man has a rifle. Salty fishermen speculate on what species the shark is, as they peer into the blood-slathered, toothy mouth of a monster. Two arrows dangle from its flank. A large hook pierces its snout, causing the mouth to gape. It's a lynching—a hasty conviction at the gallows without due process. A witch at the pillory. No evidence. No data. Yet at this point in the movie, most every person in the theater is rooting with the crowd on the dock. They are hoping this is the man-eater (even though they know their optimism will be short-lived—such is Spielberg's magic). The audience is caught up in the frenzy of the hunt. The revenge kill. The vindication. Man triumphing over the monster and making the wilderness safe again.

It's primal.

But that's not the way I saw it sitting in the movie theater in Fairfield, Connecticut. I saw something entirely different—something that would ultimately change my life and lead to me dissecting sharks alongside Jack and Wes up and down the New England coast in the summer of 1983. I saw Matt Hooper approach the shark steadfast and determined. He's unemotional against a backdrop of hysteria. He observes the animal as a scientist. He's rational. Thoughtful. Deliberate. Objective. Blocking out the noise, he acts methodically. He measures the shark's mouth. He calmly identifies it and uses terms like *bite radius* as he makes mental notes of the animal's measurements. He does not think this is the shark that killed the boy, but nobody wants to hear his opinion, however well informed it may be. The desire to slay the beast is so strong that even Brody buys into it . . . until the awkward scene with the Beaujolais and the ensuing dockside dissection by flashlight, which proves this is not the man-eater.

Brody holds the flashlight. Hooper wields the knife. I was on the edge of my seat. It was so cool.

Sharks Go Mainstream

As the summer of 1983 progressed, I saw similarities between the scene on the Amity dock and the shark tournaments. There is something elemental about sharks—something hardwired into the human brain that makes hunting them different than hunting striped bass or tuna or marlin. The scene repeated itself from New York to Maine. Sharks hoisted from boats and hauled up onto the scale for weigh-in. The crowd cheered, as the announcer emceed the event over a loudspeaker with all the emotions of a sports commentator giving a play-by-play, complete with color commentary. The best announcers built suspense and manipulated the audience's expectation to ensure an electrifying spectacle: "Behold this behemoth from the depths!" While all this was going on, the scientists, usually just a few meters away from the scale, measured and cut and recorded. They were seemingly immune to the raucous backdrop. They were eager to answer questions from the children fascinated by seeing the sharks' insides—the size of the liver, the color of the muscle. The jaws. Everyone wanted to see the jaws.

It was like a three-ring circus, with the ringmaster controlling the audience's response to these truly magnificent animals that, removed from their natural environment, became a prop used to fit a cultural narrative. Even a tournament favorite like the shortfin mako was turned into a man-eater when the snout was lifted to expose the jaws to the bystanders, who instinctually took a step back from the dead animal. "How would you like to see that coming at you in the ocean?" the announcer asked a woman in a bikini. It was no matter that the shortfin mako was implicated in fewer than ten unprovoked attacks on humans since 1850. This was theater at best, propaganda at worst.

By 1983, sportfishing tournaments in general had changed. They were no longer dominated by the exclusive, staid affairs put on by elite fishing clubs and populated by well-healed marlin and tuna anglers—anglers of means. Shark tournaments tended to go a step further, however. It was a morass of swagger, bravado, cheap beer, and diesel fumes—a little ragged around the edges, but immensely popular. Shark fishing, thanks in large part to pop culture, was the type of fishing where the angler could frame him- or herself as doing battle with a true monster at great personal risk. The annals of "provoked shark attacks" were full of minor injuries to anglers, but the risk of serious bodily injury—the loss of life or limb—was not borne out in the data.

Beyond *Jaws*, another reason shark tournaments were so popular is that shark fishing remains one of the most accessible forms of sportfishing for big-game fish in the Northeast. Whether you had a twenty-five-foot Boston Whaler or a forty-foot Hatteras, you could go catch a shark or take part in a shark tournament. By the early 1980s, billfishes and tunas were becoming harder to catch. Once plentiful in nearshore waters, these trophy gamefishes were now harder for anglers to find, forcing them to go farther offshore in bigger boats, with more fuel and plenty of time.

Sharks, on the other hand, appeared plentiful. Anglers and outdoor writers were no longer talking so much about "the shark problem" or "the shark menace." They were instead extolling the virtues of shark fishing, and shark meat had entered the culinary lexicon as a substitute for more expensive swordfish and tuna steaks. Shark was even being marketed aggressively to the stay-at-home mom preparing dinner for a family of four. During the summer of 1983, the grocery chain ShopRite tempted its shoppers with "uniquely delectable, broiled mako shark steak" at just $4.39 per pound. "Get even," the advertisement read. "Bite a shark!"

There is no question that, by 1983, sharks had gone mainstream, and I was in the thick of it.

Back to School

About midway through the summer of 1983, I was hit with a shocker when Wes told me that my position with the Apex Predators Program was dependent upon me being a student. I couldn't continue working for Wes unless I went back to school. The trouble was that it was already July, and the semester started in September. I'd missed all the deadlines. In somewhat of a panic, I scrambled and reached out to my undergraduate advisor, Bill Krueger.

Krueger was an ichthyologist in the URI zoology department, and he and I had not always had the most personable relationship. Bill was not one of those professors who made you feel all warm and fuzzy, and I was initially scared to death of him—of his seemingly stone-cold personality. As an undergraduate, it had taken me awhile to figure him out. Once, for example, Krueger boasted that he had the world's largest three-spined stickleback collection, and all I could think was, *Why?* All those jars, each with dozens of small dead fish, just didn't make sense to me. There was also the fact that whenever I went into his office, he expected me to do all the talking. He'd sit behind his desk and just look at me.

As strained and awkward as some of our meetings were, especially early on, I knew Professor Krueger saw some potential in me. It probably had a lot to do with the day that I walked into his office for one of those uncomfortable meetings. I was a junior. Krueger sat at his desk waiting for me to talk.

"I want to take your oceanic ichthyology course," I told him.

Krueger raised an eyebrow. "That's a high-level graduate course with only three people in it."

I summoned a confidence I didn't know I possessed. "I know."

Krueger agreed to let me join. I took the course and did exceptionally well. After that, our relationship was less stilted, and I came to appreciate how lucky I was to have such a phenomenal advisor. In addition to filling my schedule with graduate-level courses, I conducted an independent study with Bill, sampling more of those three-spine sticklebacks from a small pond in Narragansett and measuring all those pickled specimens in his collection. It was Bill who had told me about the job opportunities at the National Marine Fisheries Service Lab and wrote me an excellent recommendation. In my final year as an undergraduate, he nominated me for the URI President's Award in Academic Achievement, which I received at graduation.

So, when I walked into Krueger's office in July 1983 and asked straight-out, "Will you advise me as a graduate student?" he agreed without hesitation. When I added, "I know you're a stickleback, eel, and oceanic fishes biologist, and you don't really do anything with sharks, but would you allow me to do a shark project?" Again, Professor Krueger said, "Yes."

I hesitated.

"Yes?" Krueger prodded after an uncomfortable silence.

"Can you get me in this September?"

Krueger smiled.

"One of the Largest Ever Sighted off the Eastern US"

Everything was falling into place for me. The only thing that could make the summer even better was seeing a white shark. Some tantalizing anecdotes had drifted into the lab suggesting that New England's most infamous and elusive apex predator was in the area, but insofar as I was concerned, they were little more than tall tales told over beers at the bars closest to marinas and boat ramps. That all changed, however, when Captain Charles E. Donilon III reported tagging one of the largest white sharks ever seen in New England waters.

Charlie operated a thirty-five-foot charter boat named *Snappa* out of the Galilee Docks in Point Judith, Rhode Island, just twenty minutes from the lab. Charlie was also a well-known and highly respected shark fisherman, and as one might expect, a friend of Jack's. He'd been tagging sharks for the Cooperative Shark Tagging Program since 1976.

On July 18, Charlie and six clients were fishing twenty-five miles south of Point Judith when he saw the fish. He initially mistook it for a pilot whale, given the size. "It opened its mouth and came right up to the back of the boat," Charlie said later, "and when its mouth was open you could see his jaws and his teeth, and you knew what it was." It was the largest white shark Charlie had ever seen. "If he had a three-hundred-pound man in his mouth, he could easily have bitten him in half."

The shark appeared curious and was not at all aggressive. "He didn't bite the boat," Charlie said, "but he just opened his mouth and kind of mouthed our boat, just scraped down the side of the stern. From there he went under the boat and came along the side and did the same thing, bit at the side, scraping along it with his teeth."

Charlie's clients wanted him to harpoon the shark, but Charlie refused. For one thing, the fish was massive. "I tried to explain to them that it would be an impossible feat to harpoon it," Charlie said. "If we did get him mad, he could destroy something or somebody." Further, Charlie was curious. "I didn't want to kill that thing," he said. "I would like to find out some things about that shark, like how old is it and how many times has he cruised around the world."

It took Charlie three tries to tag it, but finally he was successful. The clients were disappointed but agreed it was for the best. "We wanted to take it, but the captain didn't," said Jerome Thompson, a United Parcel Warehouse employee who worked in the Hartford, Connecticut, warehouse. "I guess he would have been responsible. He probably made the right decision. He might have lost his boat and his life and us, too."

Charlie estimated the white shark at twenty feet long and weighing five thousand pounds. When Jack was later asked by a reporter about Charlie's estimate, Jack said without equivocation that Charlie is "an avid tagger of sharks" who knew "what he's talking about when it comes to sharks." I believed it, too, and for the next several days it was all I could think about—a twenty-foot-long white shark right here in New England!

Charlie's white shark was heralded in the press as "one of the largest ever sighted off the eastern US" and quickly became national news. The phone at the lab was ringing off the hook, and as I walked past Jack's office, I could hear him speaking with reporters. Jack was accustomed to the news cycle that followed a notable shark incident, and he knew how to respond. He told the press he was thrilled Charlie tagged the shark instead of killing it. He confirmed that the shark was one of the largest reported off the east coast. Even as he corroborated the superlatives to the eager reporters, who were already formulating *Jaws*-inspired puns and witty headlines, Jack was also careful to emphasize that seeing a shark that large didn't indicate anything abnormal. "It's not unusual for us every year to have white sharks reported off southern New England, usually twenty or thirty miles offshore or beyond. There are a lot more white sharks around than people think."

The final part of Jack's message to the media was intended to circumvent overreaction—to root the sighting in data and reality rather than Hollywood. "There hasn't been an unprovoked attack off the northeast coast north of Cape Hatteras in twenty years that I know of," Jack said. To another reporter, Jack said, "This does not mean there is no possibility of being bitten, but in my opinion, it is a nonproblem." I found it interesting how Jack responded: first confirm the sighting, second to say it's not abnormal, and then to play down any potential risk. Wasn't Jack the one who had concealed the fact that he'd discovered so many juvenile white sharks off popular New Jersey and New York beaches in the early 1960s because of not wanting to alarm the public? I realized that Jack was trying to thread a very small eye of a very small needle. He knew a lot more about white sharks now, and he was both trying to educate the public with honest information about the species while at the same time tamping down anxiety and fear related to public safety. I was oblivious at the time, but watching Jack respond to these events was an invaluable primer for what would become a major part of my career in the future.

Despite Jack's efforts, nuance didn't sell newspapers, and the more salacious headlines were written despite Jack's best efforts: "Reality Surpasses Legend in Mysterious World of Sharks" and the simple "Jaws IV." I was oblivious to how frustrated this must have made him, but in time, I would come to know that frustration firsthand.

Many Are Called, but Few Are Chosen

In the weeks that followed, Charlie Donilon's white shark was the talk of the coast, so anglers took note when a forty-foot fin whale carcass was reported off Block Island in early August. If a five-thousand-pound, twenty-foot white shark was in the vicinity, there was a good chance it would feed on the dead whale. Plenty of shark anglers readied their boats and set out into Block Island Sound in yet another scene reminiscent of *Jaws*. Ironically perhaps, the author of *Jaws*, Peter Benchley, spent his summers on a nearby island in Long Island Sound, and he, too, was following the shark news. Word reached him Friday evening, August 5, that two charter boat captains out of the neighboring town of Noank, Connecticut, were bringing in a massive white shark, and he felt a too-familiar twinge of guilt.

A lot had changed for the author over the past decade since he wrote *Jaws*, and in 1983 he was as thoughtful about the "*Jaws* effect" as anyone. Benchley created a work of fiction complete with a fictitious monster, but many people apparently couldn't separate fact from fiction, especially after the movie. The result was bad for sharks. Just the year before, in 1982, Benchley had published *The Girl of the Sea of Cortez*, a very different novel, inspired by John Steinbeck's *The Log from the Sea of Cortez* and Benchley's own experience diving there while filming an ABC *American Sportsman* segment on hammerhead sharks. *The Girl of the Sea of Cortez* was a seminal book for Benchley and marked a public shift toward an environmental consciousness with tendrils rooted in culpability over the *Jaws* effects. As a reviewer wrote in the *New York Times*, "this is not a novel for those who were buoyed by the fictional suggestion that we could not go back into the water."

Benchley got in his car Saturday morning and drove over to Mystic Aquarium to see the shark. The story of this shark's end had begun a few days earlier, when a forty-eight-year-old charter captain named Ernie Celotto steamed out of Noank aboard his thirty-two-foot sportfishing boat *Reelin*. While the fishing season had heated up early that spring due to a warm winter, the fishing hadn't been great as of late. On Thursday, however, there were plenty of makos and blue sharks twenty-five to thirty miles off Montauk Point. Ernie saw the dead whale on his way back to Noank, passing through the fourteen-mile straits between Montauk and Block Island. He recalled Donilon's encounter with the big white shark two weeks earlier. It had been a lifelong dream of Ernie's to catch a big white shark, and as he closed the distance to the dead whale, he saw a fin break the surface. He idled nearby as the shark circled and then rose out of the water to feed. It was a white shark, and it was the largest he'd ever seen. Ernie quickly rigged up a rod and attempted to catch the shark, but to no avail. Daylight was waning. Ernie figured he'd have no trouble finding the carcass again the next day, and he headed for home.

That night, back in Noank, Ernie called Greg Dubrule, captain of the charter boat *Seaweed Too.* Greg was younger than Ernie at thirty-two years old, but he'd been advertising "shark fishing" charters for a decade now. Back in 1976, Greg was fishing with Ernie and Ernie's son for blue sharks when he hooked a thirteen-foot, 1,200-pound tiger shark. "I wasn't prepared for a shark of this kind," he said. "If it wanted to it could have demolished the boat." He recounted to the press the ensuing battle, which ended with a shotgun blast "right between the eyes."

"When I shot him," Greg told the reporter, "he opened his mouth. You could put two men crosswise between his jaws. His head was as big as a fifty-five-gallon drum." The shark thrashed around for less than a minute, blood pouring from the gunshot wound and clouding the water as the animal sank to the bottom. The story that ran in the paper was better than any glossy brochure. Greg had no trouble booking charters from that point on.

On Friday at 4 A.M., Ernie, aboard *Reelin*, and Greg, aboard *Seaweed Too*, headed out in tandem to achieve what a couple of days earlier had just been a dream. Aboard their boats, they had six clients each. Over the years, Ernie said he'd seen changes in his clientele, but nowhere were these changes more obvious than in those who charter *Reelin* to go shark fishing. Ernie attributed it to the *Jaws* effect. "After *Jaws*," he said, "we got a different type of customer—the majority were killers, not sportsmen. I was aghast when the book came out. I knew what was going to happen. We lost customers because we didn't allow indiscriminate killing."

Ernie is not alone in that sentiment, as just a week before he spotted the white shark, an article in a New York newspaper described how the Montauk Boatmen and Captains Association was encouraging all their charter boat clients "to throw back all but the edible mako sharks." Although this sentiment may have aligned with Ernie's stated preference, he and Greg had no intention of tagging this white shark. They intended to kill it and bring it home.

They found the whale at 8 A.M., drifting about ten miles southeast of Block Island, and they set up adjacent to it, readying their gear. They waited seven hours for the shark to show up and begin feeding. Greg again used his fifty-five-gallon drum analogy, estimating the shark's mouth is three-quarters of a fifty-five-gallon drum. It sent chills up his spine, he'd later recall. Greg sank the first harpoon, and the battle began. The shark took three hundred feet of line over the side in a scene his clients likened to Quint harpooning the shark in *Jaws*. Like Quint, Ernie and Greg had rigged flotation balls to the line so they could track the shark as it ran. They watched in astonishment as the shark pulled first one, then another, and then a third flotation ball under the surface. It was like nothing they'd ever seen.

After two hours of battling the shark, they got close enough for Ernie to throw a second harpoon, but it missed the mark. Eventually it took a fourth floatation ball attached to the line to slow the animal down and give the shark hunters a chance to begin hauling the line in hand over hand. They made steady progress over the next hour, until the line abruptly went slack. The anglers leaned toward the stern, taking in the slack as fast as they could. From the wheel of the *Reelin*, Ernie saw the shark break the surface and, as Ernie put it, "lunge" toward the *Seaweed Too*'s transom, sending the anglers scrambling forward. Greg slammed the boat into drive.

It took another couple of hours for the anglers to work the exhausted fish to the boat, where Greg, using the same shotgun with which he had killed the tiger shark seven years earlier, fired two Magnum loads of buckshot into its head. "That took the life out of him," Greg said. They tied the shark to the side of the *Seaweed Too* and began a slow trip back to Noank.

On the way back in, Greg radioed his wife, Faith, who relayed a radio-telephone interview with the *Hartford Courant*. "This animal was taken within sight of the eastern shore of Block Island," Greg told the reporter. "The fish was well inside of its normal range and what kind of a danger that is to the general public I don't know, but it is certainly a killing machine." Ernie added, "This is one of the thrills of a lifetime. It's one of the highlights of my fishing career." When asked what they intended to do with the shark, Ernie and Greg said they didn't know yet.

When the boats entered Noank Harbor at around midnight, *Seaweed Too* was in the lead. Despite the hour, there was a crowd of close to seventy people waiting. "That's them!" someone yelled. "Here comes the shark!" Greg pulled the *Seaweed Too* up along the narrow pier at the Noank Shipyard. The onlookers gazed down at the silvery-white body of the shark, which appeared a ghostly apparition just beneath the dark water. It took a crane and ten men to lift it onto the dock. Greg and Ernie basked in the glow of their success. Ernie remarked that it felt like being Ronald Reagan.

Ernie and Greg needed to decide what to do with the shark. Ernie told a staff writer from the *Courant* that he'd like to commemorate the catch, and he was considering having the head mounted, like Frank Mundus did with his big white shark landed off Long Island. "It's a very, very deep feeling to be standing next to one of these animals and see them in their glory," he said as the shark was hoisted onto a flatbed truck for the short trip to the Mystic Aquarium, where the fishermen had arranged to have it placed in a refrigerated truck. "This is what we go fishing for," Greg said. "It's the name of the game. It's the chase.

"It's like being one of Christ's disciples," he said. "Many are called, but few are chosen."

I did not immediately hear the news of the huge white shark landed at Noank at midnight. I'd gone to bed early, as I needed to get up early Saturday to head down to Snug Harbor Marina for day two of the second annual Snug Harbor Shark Tournament. I had hoped Charlie's white shark boded well for the Snug Harbor tournament and my own chance to see a white shark. It's not that I wanted the fishermen to go out and kill a lot of white sharks, but a single specimen could reveal a wealth of much-needed information about this species. The first day of the tournament was a disappointment, though. While Ernie and Greg were fighting the white shark off Block Island, 150 boats competed in the tournament, but only two sharks were landed—a seventy-five-pounder and a 110-pounder, and neither was a white shark.

As I prepped the gear early Saturday morning for day two of the tournament, a local angler named Richard Pearce was readying his boat, *Adventure II*, at Kenport Marina in East Matunuck, Rhode Island. He'd heard about the dead whale and the Noank shark, and he figured if he could locate the whale, he'd find big sharks—maybe even a tournament-winning shark. Pearce smelled the whale before he saw it, and he tied off to the carcass. His adrenaline was pumping, but, God, it smelled bad. Picking up the radio, he hailed his friend Stephen MacDonald, who was also competing in the tournament aboard his twenty-four-foot boat *Yellow Jacket*. MacDonald said he was on his way.

Soon after MacDonald arrived, both anglers decided that, while the whale may well have been the ticket to a tournament-winning shark, the stench was just too much to bear. As Pearce shifted his boat into reverse to untie from the whale, he saw the shark rising from the murk between the whale and the transom of the *Adventure II*. He put the boat into neutral and stared as the shark came right up and appeared to "nibble" at his boat before disappearing beneath the hull. Pearce estimated the shark was at least twenty feet long. His hands trembling, he slammed the engine into reverse again and quickly disconnected from the whale. It was the largest shark he'd ever seen, and all he could think about was *Jaws*. As his boat started to drift away, he saw a second, smaller shark approach the whale.

"Hey," he called across the water to MacDonald aboard the *Yellow Jacket*. "Are you seeing this?!"

The second shark rose up and bit down on a hunk of blubber from the whale's flank. It shook its head from side to side—sawing through the dense fat. Pearce turned to see if MacDonald heard him, and that's when he saw another large shark come up right under the bow of the *Yellow Jacket*, lifting the boat several inches out of the water. He pointed, incredulous, but MacDonald already saw it, and the line "I think we're going to need a bigger boat" flashed through his mind. The big shark turned, the *Yellow Jacket* settled, and Pearce and MacDonald decided they'd had enough shark excitement. Pearce radioed the Coast Guard, which in turn issued a warning for everyone to stay clear of the carcass because of the sharks. When MacDonald got back to the marina, he was pretty sure most people would be skeptical of his story, but the eight-inch-long gouge marks in the fiberglass hull were clear as day.

Back in Mystic, Peter Benchley also contacted the Coast Guard to ask them to tow the whale out to sea and sink it. Benchley was not worried about the risk to humans. He was worried about the sharks. He feared more anglers would head out for their chance to catch a "monster" just like his own characters had done. It sickened him that he'd popularized the playbook that shark hunters like Ernie and Greg used to lethal effect.

When Benchley arrived at Mystic Aquarium, he was disgusted by what he saw. The author looked at the massive shark grotesquely laid out on the flatbed truck. The shark's head hung over the tailgate with its once-pristine white underside exposed. The mouth was open, the ragged teeth exposed in a wretched death grin. Like everyone gathered to see the animal, he was immediately taken by the size of the fish. Its sheer bulk was incomprehensively large. He crouched beside the massive head and reached out his right hand to touch the animal's snout. "Dear God," he muttered to himself. A wash of emotions. Sadness. Guilt. Anger.

The senselessness of it all.

Donated to Science

The night before, within an hour of the shark's arrival in Noank, John Buck, a marine microbiologist at the University of Connecticut's Marine Sciences Institute, requested permission to swab the shark's teeth. He was to become the first scientist to culture bacteria from the teeth of a white shark, and he immediately knew the value of this animal to other scientists. He made some calls, and UConn biology professor Carl Rettenmeyer drove down from Storrs on Saturday morning. Rettenmeyer moonlighted as the director of UConn's Museum of Natural History, which, at the time, lacked a building. The museum's couple million display items were instead scattered among basements and attics across the UConn campus. It was Rettenmeyer's dream to change that, but funds were lacking. When he heard about the white shark, he saw an opportunity. The white shark, he thought, could be the "showcase piece" to get the museum off the ground.

Rettenmeyer asked a parasitologist named George Benz to come to

Mystic with him. George was a tall, well-muscled man in jeans and a tank top, who didn't at all look the part of a fisheries biologist. George, who was in his twenties, had a passion for sharks, and Rettenmeyer thought the fishermen might respond better to George than to a middle-aged college professor who looked like a middle-aged college professor. Rettenmeyer also knew that George would want to collect samples of tiny shrimplike parasites, called copepods, that might be attached to the shark's skin and gills.

George, who was later to become a dear friend of mine, had studied parasitology at UConn and developed a fascination with the copepods found on sharks. To feed his fascination, he used the same methodology as Jack—attending shark tournaments to collect samples. That's how he first ran into Jack Casey and crew. To George they were like rock stars, and he mustered the courage to introduce himself during the 1977 New England chapter meeting of the American Fisheries Society. Wes recalled George as "this massive young guy in a muscle shirt, very out of place among the state fisheries guys," but Wes also remembered being immediately impressed. So was Jack.

Although George never actually worked for Jack or the National Marine Fisheries Service, he became an honorary member of Jack's team. George and Jack developed a strong friendship, and Jack, who looked out for young scientists who showed promise, opened some important doors for George, including an invitation to a 1978 research cruise aboard the F/V *Wieczno*. A trip aboard the *Wieczno* was like boot camp and a rite of passage for young shark biologists in New England. If you survived the long hours, cramped quarters, and rough weather, that was worth more than any degree, at least as far as Jack was concerned. For George, it was "the experience of a lifetime." The parasitic copepod samples he collected proved critical to his master's thesis, and Jack served on George's thesis committee in 1979.

After earning his master's, George headed south to the University of Maryland to work in the lab of the "Shark Lady," Eugenie Clark, but he returned to Connecticut to take a job with the Connecticut Department of the Environment's Fisheries Bureau, while flirting with the idea of undertaking a doctorate. Although he was mostly working with the state's freshwater fisheries, he satiated his passion for sharks and his continued interest in parasitic copepods by moonlighting as a parasitologist. That's how he came to be standing on the flatbed truck on that Saturday morning in Mystic looking incredulously at the biggest shark he'd ever seen.

George immediately called Jack and told him he was going to want to come down and see this shark. The trouble was, however, that the fishermen were not in agreement about donating the shark to science. A negotiation ensued. The fishermen said they'd had some lucrative offers from private individuals, including a $10,000 offer from a guy in Texas. George suggested the shark should stay in Connecticut, given how rare it is and the fact it was landed in the state. "Better yet," George said, "donate it to science." He knew Rettenmeyer wanted the shark for the museum, and he knew Jack would want to dissect it, but Ernie was adamant that the shark was not to be dissected. A dissection, Ernie said, would destroy his chances of having it mounted as a full fish. It was one thing for John Buck to scrape the shark's teeth and for George to pick off some external parasites, but nobody was going to cut his fish up. George had his work cut out for him.

As George negotiated, the press focused on an animated and emotional Peter Benchley. "It's like going out and shooting dogs," Benchley said to the gaggle of reporters. "The shame of it is, you get a whale out there and the white sharks come around, and every sportfishermen comes along to hunt for sharks." A reporter, her interest piqued by Benchley's criticism of the fishermen, started asking him questions. Benchley told her how little is known about the species and that nobody really knows how many there are. He told her that female white sharks only bear one offspring a year. "With that kind of reproduction rate," he said, "fishermen should not make a sport of killing them." The reporter scribbled furiously in her notebook. This was, after all, the author of *Jaws*.

"What about the risk to people?" the reporter asked, recalling Greg Dubrule calling the shark "a killing machine" and questioning the danger it posed to the public, given how close it was to Block Island beaches. Benchley shook his head dismissively. "The incidence of a great white shark attacking human beings is so small that it is statistically insignificant," Benchley said. "I can name you most of the people who have been attacked in the last fifteen years. Granted," he added, "when it happens, it's a terrible sight.

"This kind of wanton slaughter is madness," Benchley concluded. "The main problem with this is people are killing these animals wantonly as if it is sport, and there is no sport in harpooning a fish."

The press, eager for the fishermen's rebuttal, asked Ernie and Greg if it wouldn't have been better to let a shark like that live. Wouldn't tagging it instead of killing it be more valuable to science? "I think the scientific data we are going to be able to obtain from this shark is going to make its death worthwhile," Greg responded, referring mainly to the samples taken by John Buck and George Benz. "That's all I've got to say."

The Sunday newspapers were not kind to Ernie and Greg. Highlighting Benchley's quotations, they criticized the fishermen's decision to kill the shark. On Monday, the fishermen held a press conference to announce they'd be donating the shark to science. "All I can say to Mr. Benchley," Ernie said, "is that there are other great white sharks out there. I'd be happy to take Mr. Benchley out there and he can judge for himself if he thinks there is no sport involved [in harpooning a shark]." Ernie went on to justify killing the shark, saying "sharks of this size are rarely put under the microscope."

During a radio station talk show appearance later that day, nearly half of the callers disparaged the killing. Ernie again defended himself, claiming now that science was always his motivation. "I've had no second thoughts. We did it for the scientific and education community." He announced that a full "autopsy" would be performed but only after a fiberglass mold of the almost three-day-dead shark was made for UConn's natural history museum.

George called Jack back and told him they should be able to have access to the shark by Tuesday morning.

A Real White Shark!

I will always remember Tuesday, August 9, 1983. It was the day I saw my first white shark. Not a white shark in a movie or documentary. Not in the pages of a magazine or textbook. A real (albeit dead) white shark in the flesh.

It hadn't cooled off significantly overnight, and my sleep was restless as a result—such is August in New England. I was still tired when the alarm went off, but I was nonetheless quick to get out of bed, get dressed, and get to the lab. Although it wasn't unusual for the crew from the Apex Predators Program to take a field trip to document a "rare-event species," today was special. This was one of the largest white sharks any of them had ever seen, much less had the opportunity to study. The plan was simple: drive to Mystic Aquarium, where Ernie and Greg's shark had been refrigerated since early Saturday morning, take possession of the animal, and then drive it back to a national wildlife refuge in Rhode Island, where they would dissect it and bury what remains.

As the van exited Interstate 95 at Mystic, I caught a glimpse of Long Island Sound out the window and remembered seeing it out my bedroom window from the third floor of the house in Fairfield growing up. I thought back to watching those Jacques Cousteau specials and especially the ones about sharks. I remembered seeing *Jaws* on the big screen at the theater with the Art Deco marquee in Fairfield. The van pulled onto the access road leading behind the aquarium and eventually came to a stop near the loading dock. Everyone got out. While I was not exactly sure what I expected, I immediately sensed that what I saw before me was not it. The shark, laid out on the loading dock, was decomposing—which, in and of itself, was not unusual. Rare-event species were often not in the relatively pristine state of those animals landed at a shark tournament—animals swimming only hours earlier instead of days. What was strange about this scene was the flurry of activity. Usually, a rare species was washed up on a desolate beach or laid out on a commercial fishing wharf, but this shark was buzzing with activity. Several men were working frantically with sheets of fiberglass and buckets of resin, coating the shark in a skin that would harden into a mold. They looked tired and harried.

I recognized George Benz as being the parasitologist who was a regular at the shark tournaments. I had yet to really get to know George, but I was hoping to change that. When George saw Jack, he excitedly hopped down from the loading dock and the two men greeted each other in what appeared to me to be a perfect blending of several meaningful relationships—student-teacher, colleagues, and, ultimately, friends. After they shook hands, George gestured to the shark. He was explaining something to Jack with a pained expression on his face. For my part, I was performing my low-man-on-the-totem-pole duties of unloading the van, prepping the gear, and waiting to be told what to do next, and I was too far away to hear George, but I did hear Jack's booming response.

"Jesus Christ!" Jack bellowed. "You've got to be kidding me."

George had recounted the previous twelve hours to Jack. The taxidermy guys from Fish Unlimited arrived at the aquarium around 8 P.M. the night before and went straight to work creating the mold. The mold itself is a mixture of fiberglass and polyester resin, and it took them nearly six hours under the glare of artificial lights to get half the shark done. Then, without warning, the heavens opened, and a summer rain drenched their work. It was just a cloudburst, but the damage was done.

"They had to remove everything and start from scratch," George said.

Once the guys cleaned all the "goop" off the shark, they had to figure out how to dry the skin, which was continuing to decompose. "Someone brought out a heater," George told Jack. "Like the heater used in fast-food restaurants to keep the food warm, but it was much too slow." That's when someone else from the aquarium suggested they could try one of the large driers that aquarium scientists used to dry salamanders. "They got the shark dry enough to start over, and, well"—George gestured to the shark with nearly three-quarters of its flank encased in fiberglass—"here we are."

As the team from the lab settled in to wait, I decided to do some photography. It was an area where I felt confident in my skills, and more importantly, it was an area where I could take some initiative. My photographs were also an opportunity to make a lasting contribution to the lab's work and the advancement of shark research in general. Although I'd added a lot of "chicken scratch" to the Individual Fish Records (my penmanship left something to be desired), and I'd filled out more than a few "Morph Cards" (short for "morphometrics") that would be filed away to become part of the permanent record of the Apex Predators Program, I liked to think that it was my photography that would leave a more enduring mark.

I looked around, thinking about how best to document the scene. The Apex Predators Program is always trying to better educate cooperative shark taggers about proper shark identification. So, one of the standard shots is a lateral view of the entire shark to show coloration, fin shape and placement, and other unique attributes. This shark was huge, and I knew I'd need to photograph it from a distance to get the whole animal in the frame without foreshortening it. The goal was to document the animal in a way that reflected its actual proportions. That's when I noticed an I-beam projecting from the building above the loading dock. I suggested to the crew that if we got a pallet, we could rig a platform, and they could then hoist me up to get a bird's-eye view of the entire shark. With nothing else to do until the taxidermy guys finished, New England's brain trust on sharks turned to the task of rigging the platform and hauling me up. As I gained altitude, the pallet spun and swayed. In one direction, I could see cars zooming by on I-95, and in the other I could see Long Island Sound. Eventually the pallet steadied and found its equilibrium. I brought the camera up, framed the first shot through the viewfinder, and pressed the shutter.

I had photographed a lot of sharks that summer, but this one felt different. It *was* different. Even Jack, Wes, and George looked at this shark differently. Beneath me was an animal about which I'd only dreamed up until this point. I'd seen hundreds of photographs and watched and rewatched Peter Gimbel's documentary *Blue Water, White Death*, which included some of the first underwater footage of white sharks, and which inspired the writer Peter Matthiessen to describe the white shark as "a silent thing of merciless serenity." But to see this shark, this *great* white shark, laid out beneath me was awe-inspiring. This was it, I remember thinking to myself, as I released the shutter again. Here I was, Greg Skomal, with the leading shark researchers in the northwest Atlantic gathered beside a 2,779-pound white shark. This was not someone else's picture in a magazine or scientific journal. I was the photographer. I was one of them.

Click.

By midafternoon, the team from Fish Unlimited Taxidermy finished the mold. They were hesitant to begin work on the other side, given the state of decomposition. After consulting with George, they decided to truck the half mold back to Long Island, where they planned to create the other half as a mirror image, join the two together, and finally insert the actual jaws into the mold before shipping it to UConn. It was a process that would take more than a year to complete, and Carl Rettenmeyer still had no idea how they would pay for it.

As the taxidermy guys pulled away, Jack and Wes jumped into action coordinating the shark's final move. Tape measures came out and were stretched across the flank. Numbers were called out. I dutifully recorded each of Wes's measurements of the external reproductive organs on the data worksheet.

LEFT CLASPER LENGTH: 435 MM

RIGHT CLASPER LENGTH: 420 MM

It was a remarkable animal being reduced to a set of data points, but it was a set of data points that would be critical to better understanding the species.

SPERMATOPHORES: YES

RHIPIDION EXT.: NO

RHIPIDION ROTATES: YES

When we had finished with the initial observations, we set about transferring the shark for the drive to Rhode Island. We tried to lift it by the tail using the hoist, but the massive caudal fin started to tear away from the body due to decomposition.

"Shit," Jack muttered to no one in particular. Eventually we got the shark loaded and headed back to Rhode Island.

The dissection took place in the waning hours of light at the Ninigret National Wildlife Refuge in Charlestown, Rhode Island. The shark was broken down to extract what remains of value to the scientists. Samples were packaged and sent to various labs, where shark experts would each, depending on their fields, use them to build on the knowledge of the species. Among the team present for the dissection was Frank Carey, who a few years earlier had made white shark history in the western Atlantic when he tagged and tracked the Montauk Monster. Frank was looking at the heat retention system and the muscle in this massive sample. As the last step in the necropsy, Frank sliced the shark into one-foot sections, and I photographed each so he could calculate the ratio of red to white muscle. What remained of the shark was then buried as the sun set over the marsh grasses, accompanied by the strong smell of salt air, low tide, and decomposition.

⚓

It's remarkable how much the tide had turned in the last twenty years that Jack had been studying sharks. When he'd started at Sandy Hook, there were very few people who would have questioned the killing of a white shark. What transpired with the Noank shark showed that things had become more complex. The media ginned up the debate between the fishermen and the author, but they were also interested in what the scientists thought. It was delicate, a bit of a tightrope. The scientists said they were happy to have had access to the shark—Carl Rettenmeyer called it a "once-in-a-lifetime opportunity"—but they, like Benchley, were uneasy with the fact it was harpooned while feeding on a dead whale. Chuck Stillwell spoke for the Apex Predators Program, saying to a reporter that while he, like Rettenmeyer, believed the killing of the shark was "justified," his agency discouraged the killing of sharks, especially by harpooning them. "I can't really blame these fellows," Chuck said. "They're in the business of catching fish, but if we start wiping out the big fish, we're going to have problems."

Jack said he didn't condone harpooning sharks, but he added that the opportunity to study a white shark this large could be monumental. "There are a lot of people who know a little bit about great white sharks," he said. "We know they range the world, but no one really knows that much about them."

The encounter with the Noank white shark reaffirmed my fascination with the species. I was slated to start my master's program at URI just a few weeks after we buried that shark in the salt marsh, and I wondered if I should have instead gone to California, where I could have had a shot at reliably studying white sharks. But then I thought back over my short time with Jack and Wes and the rest of the crew from the Apex Predators Program. I realized how lucky I was to have become part of that team, and I was finally just starting to feel like I belonged. Even more important than the sense of belonging, however, was realizing what a phenomenal opportunity this was for my career. I'd always assumed I'd take some time off after getting my undergraduate degree, but here I was going full steam ahead into graduate work. As much as I would have liked the opportunity to work more with white sharks, I knew there was so much more to learn from Jack and Wes about sharks in general.

While it was true that white sharks had motivated me at an early age, I was beginning to realize that my passion for marine science extended well beyond this one magnificent species. Over the course of my time working with Jack and Wes, I observed, handled, and studied many species of sharks, and I realized that my passion extended to all of them. Sure, it was amazing to have the opportunity to see the Noank white shark, but as cool as the experience was, I knew now that I'd be just as enthusiastic to work with a blue, sandbar, or mako shark at the next tournament. The thing that I came to finally understand was that sharks, as a group of animals, were to me the most fascinating living creatures on earth. Likewise, I was convinced that there was nobody else from which I'd like to learn more about them than Jack and Wes.

CHAPTER 5

GOING TO SEA

Be a pirate. Steal time for yourself and those you desire to help, rather than wasting time on trivial bureaucracy and nonsense.

—George Benz

Study nature, not books.

—Louis Agassiz

JUNE 29, 1979—40°31'30.432"N, 72°48'51.4548"W (SOUTH OF LONG ISLAND, NEW YORK)

The fifteen-foot male shark closes in on the rotting fifty-foot whale carcass off the coast of Long Island. He senses the presence of other white sharks before he sees them. While he prefers a solitary existence, he's fed on whale carcasses with other white sharks before. He's cautious yet undeterred as he closes the distance, his dorsal fin breaking the shimmer of the slick. Above him, a hundred petrels dive-bomb the water, feeding on the fat droplets. There are no other sharks in the area besides the other whites.

He circles the decomposing whale, which floats more than four feet above the water's surface due to the gaseous tissue. Large holes mar the decaying blubber, where the other white sharks have fed. The head of one shark breaks the surface, thrusting upward. Reaching. Its teeth rip at the whale's flank, rending large pieces of blubber. A meal like this requires little effort but offers plenty of reward. In the absence of the abundant seal populations that once existed here, the dead whale has become an essential part of the white shark's diet in the Atlantic. Eventually the other shark slips back beneath the waves and turns away from the whale, exhibiting none of the displays white sharks often use to maintain their personal space. No tail slapping or tilting. Instead, just a sluggish descent into the ocean.

The shark moves in and begins to feed—repeatedly tearing at the carcass in half-halting, half-lunging motions. With each bite, he twists and turns, leveraging his massive body to sink his jaws and slice into the carcass. He swallows massive, five-gallon-bucket-sized chunks of blubber, moving his way down the whale's side. Sharks do not have tongues, and while they do have "taste bud" receptors in their mouths, they do not discern taste as humans do. Instead they perceive whatever they have bitten in a more binary fashion: food or not food. The blubber of this dead whale, bloated and decaying as it may be, is most certainly food for the shark, and it feeds with abandon. Sharks also don't have external ears, but they do have an inner ear sensitive enough to obtain a bearing on the low-frequency sounds of a wounded fish almost three football fields away. It is with this inner ear that the shark feeding on the whale senses a new sound in its environment. It's not an unfamiliar sound, and it is of no real concern given the task at hand. The shark continues feeding as the boats approach.

A new scent enters the water. Unlike the fatty blubber of the whale, this is fresher meat, and the shark turns to investigate. As he takes a bite of the bloody flesh, he is unaware (or unconcerned) of the man standing on the transom of the boat above him, harpoon in hand. The harpoon hits the shark in the back just behind the first dorsal fin. It pierces the V-shaped scales that are known as dermal denticles, a term that literally translates as "tiny skin teeth." Despite its size, the white shark possesses relatively tiny scales, each of which does, in fact, have a distinctive tooth-like structure. The top of each scale points backward toward the shark's tail, causing the skin to feel smooth when rubbed in one direction but like sandpaper in the other. While these scales are incredibly strong, providing protection against larger threats like the claws of a seal or the teeth of another shark, they also protect the shark from tiny parasites, as they move with the shark with the precision of a chain-mail suit.

As much protection as the shark's scales provide, they are no match for the thrust of the harpoon and the three-and-a-half-inch dart that is now buried in his muscle behind the dorsal fin, which will allow the boat and its scientists to follow the shark.

After the summer of 1983, I was clear on three things. First, I knew I wanted to make a career out of studying sharks, even if I couldn't study white sharks. Second, I wanted to make sure fieldwork always remained central to my career. Last, I wanted to commit to always thinking outside the box, as I'd seen my mentors do.

I had wanted to be a marine biologist for most of my life, but whenever I envisioned what that looked like, I saw Matt Hooper with a knife on the dock at night dissecting a shark, not doing data entry in the lab. Perhaps that's not unusual. Perhaps it's the way that most people envision their future career—the aspiring firefighter envisions rushing from a burning building carrying a child they've just rescued, not cleaning fire engines; an aspiring surgeon sees herself saving her patient on the operating room table, not filling out paperwork to submit to insurance companies. Unfortunately, I think most of us come to the realization that any job ends up falling far short of what our childhood minds imagined. We come to understand that the role models upon which we'd fashioned our dreams were often well-scripted characters in movies and television shows.

But working at the Apex Predators Program made me believe with an almost childlike enthusiasm that I could be the Matt Hooper kind of scientist. Of course, I needed to build a strong academic foundation, and I was doing that through school, but Jack was also teaching me that I could do a lot with that foundation. It wasn't all about school and textbooks and working in the lab. I saw how Jack's open-mindedness benefited his research. The fact that a recreational angler held no science degree didn't bother him, nor that a commercial fisherman had no classical training in fish identification. Jack understood that these people often possessed more firsthand working knowledge of fishes than the scientists did, and that's what mattered most to him. It wasn't that fishermen had all the answers; rather, it was that they, as collaborators, had valuable experiences, skills, and knowledge that opened new avenues of research and discovery.

Feeding Frenzy

In the first couple of years after earning my undergraduate degree, there were a couple of experiences that markedly influenced my approach to the study of sharks—experiences that affected the type of shark researcher I would become. While the dissections were invaluable, I increasingly had a strong desire to also observe living sharks. When I thought about being in the water with sharks, I didn't think about Hooper in the shark cage near the end of *Jaws*. Instead I thought about the footage of Jacques Cousteau, Peter Gimbel, Stan Waterman, and others working underwater with sharks. I wanted to experience that myself both as a scientist and as a human being. I'd seen nurse sharks in the Caribbean while diving, but they were mostly smaller, docile animals laying among coral heads with a plethora of other fishes and marine life creating a kaleidoscope of biodiversity and beauty. It was the sort of thing for which many divers traveled many hundreds, if not thousands, of miles and spent large sums of money to see. What I dreamed about, however, was open-ocean encounters with large pelagic sharks, and I imagined that would likely require a shark cage.

Jacques Cousteau developed the shark cage in the early 1950s, and his contraption debuted in a 1956 film called *The Silent World*, but it was *Blue Water, White Death* and *Jaws* in the 1970s that popularized it. My most vivid recollection of a shark cage was from *Blue Water, White Death* and a scene where Peter Gimbel and crew film, from within a cage, oceanic whitetip sharks feeding on a whale carcass. It sticks in my mind not only because of the awesome footage of one of the world's most notorious pelagic sharks swarming the water around the cage, but also because of what happens next. At Gimbel's suggestion, the divers choose to leave the cage and swim among the sharks unencumbered and unprotected. To the public who believe wholeheartedly in the popular notion of a "feeding frenzy," where sharks, aroused by prey, become indiscriminate, bloodthirsty killing machines, leaving the cage to swim among the whitetips was insanity.

Jaws, and in turn the white shark, get much of the credit for creating the deeply rooted societal fear of sharks, but the reality is that the seeds for a ubiquitous shark-related psychosis are more firmly rooted in wartime, when shipwrecked sailors and downed airmen found themselves in what were commonly described in news reports as "shark-infested" waters. There was surely some gruesome irony to the fact that a young man would survive his ship being torpedoed or his plane being shot down only to face the threat of shark attack in the water. As atrocious as warfare was, the notion that there was something even more ghastly and primal lurking just beneath the surface was enough to traumatize a generation. Capitalizing on these deep-seated terrors, "shark-infested" waters became a speculative (at best) set piece of many newspaper stories, drumming up a crescendo of irrational fear of sharks. It wasn't, however, all hyperbole. As *Jaws* pays homage during Quint's penultimate scene, perhaps the worst disaster of World War II for US servicemen was the sinking of the USS *Indianapolis*, where many sailors were indeed killed and some eaten by sharks.

In Gimbel's film, when he and the other divers leave the cage during the whitetip's "feeding frenzy," the audience is primed to expect the sharks will arbitrarily turn on them. They do not. It could be argued that the myth of the feeding frenzy is disproven in that moment as the audience observes the sharks get close to the divers, even bump the cameras, but stay focused on the whale carcass. I had always taken that scene as further proof that sharks were more selective and didn't have it in for humans—that *Jaws* was, as Benchley had intended, fiction. A horror story with an imagined monster. I liked to think that if I had been with Gimbel during the filming of that scene, I too would have readily left the cage.

The Shark Cage

With so much going on between school and work, I failed to realize that the contraption in the back of the lab was a shark cage. Everyone at the lab always had a side project going on, and this one was Wes's, but I hadn't paid much attention to it. Then one morning I came in early. I parked in the back of the lab looking directly at the back of the building, and it dawned on me—*That's a frickin' shark cage!* I was wide-eyed with excitement as I inspected it. It certainly didn't look like the shark cages I'd seen in television documentaries or in films. Instead, this one was a rudimentary boxy structure sheathed in chicken wire and spray-painted bright blue. Nonetheless, it was unmistakable. Wes had a shark cage!

When I asked Wes about it later that day, he told me he'd reached out to Charlie Donilon, the captain of *Snappa*. "I know you tag a lot of sharks for us," Wes had said to Charlie, "but I'm wondering if you'd be willing to take me out to do some cage diving?" Charlie immediately agreed, and so began a new chapter of shark research occurring out of the Narragansett Lab—the study and photography of live sharks observed underwater from the relative safety of a shark cage. Charlie didn't know it at the time, but it was also the start of a new business for the *Snappa*.

A few weeks later, Wes caught me once again inspecting the cage. "Hey," he said to me. "You want to jump on one of our shark-cage trips?"

"What?!" I was unable to contain my enthusiasm. "Are you kidding me? Absolutely!"

It was a foggy summer morning when Wes, fellow graduate student Nancy Kohler, Kevin McCarthy, and I pulled up in Wes's Volvo wagon to Galilee in Point Judith with the shark cage strapped to the roof. At the edge of the parking lot, outriggers and booms of fishing boats glided in and out of the miasma, the boats rocking methodically against an ebb tide. Gulls screeched and halyards slapped at masts. The smell of bait, salt, and seaweed permeated everything. We met Charlie and shuttled our dive gear and the cage down the dock to *Snappa*. At a little after 6 A.M., Charlie engaged the diesel engine and we pulled away from the dock. Everything was wet from the dense fog, so we found a dry spot inside the salon for the long ride offshore. Charlie negotiated *Snappa* slowly between fishing boats, eventually entering the channel and clearing the harbor.

The shark cage obtrusively dominated the aft deck of the thirty-five-foot boat, and I couldn't stop looking at it. Once we were outside the harbor and up to cruising speed with the bow pointed to the horizon, Wes came aft to speak with the team. He was understandably cautious. He'd only done a couple of dives with this cage so far, and so he went over the procedure in detail as the shoreline faded away in a wash of white wake and diesel exhaust. We would deploy the cage, Wes explained, and then drift a little way off. To access the cage, we'd use a small inflatable boat. I looked at the inflatable on the deck beside the cage. It really wasn't much more than the type one might have as a pool toy, I thought. My next thought was of all the sharks I'd observed at tournaments and, more specifically, all those shark teeth. I swallowed hard.

After more than three hours, we reached the location for the dive, and we prepared our dive gear. Once everything was ready, we pushed the cage into the green murk and allowed the *Snappa* to drift away with a tether running between the boat and cage. Wes put the inflatable over the side and tied it off to the *Snappa*.

Charlie started chumming using ground menhaden and sliced chunks of mackerel and, after a couple of hours, I saw the dark shadow of sharks moving into the chum slick. The sharks were blue sharks, as we expected. Their thin bodies, conical snouts, and long pectoral fins were unmistakable when viewed from above, and while the poor visibility somewhat dampened the deep sapphire blue of their backs, I thought how much different and beautiful they looked when alive—instead of lying dead on a dock after a tournament. Blue sharks also happened to be one of the most, if not the most, abundant oceanic shark species in the world. Unlike white sharks, which shark researchers were beginning to think migrated along a mostly north–south axis in the Atlantic, Jack's tagging data showed that blue sharks appeared to migrate throughout the entire North Atlantic, moving from the eastern side of the Atlantic, where they are born and live as juveniles, to the western side, where they mate. They were a fascinating species and had recently captured my interest, in large part because their sheer numbers made them a reliable research subject for New England shark researchers.

With the arrival of the blue sharks, Wes and Nancy climbed into the inflatable and the rest of us watched as they paddled over to the cage. The sea wasn't rough, but it was New England, and the inflatable seemed to bounce along on top of the swell like a toy. When they got close to the cage, Wes tied the inflatable off to it and then, donning his scuba tank, went backward over the side of the boat. Nancy followed. Unlike the cages I'd seen in documentaries, Wes's cage didn't have a large top entry hatch, so the divers needed to enter the water, swim to the cage, and then use a door on the side. In my excitement over the cage in the lab, this "design flaw" (as I now saw it) had eluded me.

When it was my turn, Kevin and I readied our dive gear, boarded the inflatable, and rowed over to the cage. By the time the inflatable was tied off and I had my mask in place and was ready to roll backward into the water, I was downright terrified. Why were we doing this in the first place? If the sharks weren't going to hurt us, why did we need the cage, and if we needed the cage, why were we in a tiny inflatable boat from which we were going to enter the ocean without any protection until we could enter the cage? I suddenly had so many questions, and the whole thing seemed like an absurd idea. *How the hell can I get out of this?* I wondered. But there we were, with no reason to delay any longer. No excuse. Resigned to my fate, I put the regulator in my mouth and rolled backward into the water in a burst of bubbles. The water visibility was horrible, and I found myself momentarily disoriented. I spun around looking for the cage. I was breathing far too fast, but none of my usual strategies for calming my hyperventilation were working. I started to panic. But I had my underwater camera, and I decided to focus on that, which calmed me down—a method I would use for many years to come whenever I ended up freaking out underwater. I snapped a picture of the cage and another of Kevin. Finally, I oriented myself in the water column and located the cage's door. Swimming toward it, I grabbed the door and swam inside. Kevin swam in beside me, and we closed the door.

Once I recovered control over both my breathing and my emotions, I realized just how small and cramped this cage was. It bobbed and rocked disconcertingly in the swell. We waited. Because the cage was tethered to the *Snappa*, every movement of the boat tugged violently on it, jerking us like rag dolls in the current. I hung on to the side, bracing myself. I looked over at Kevin, who wasn't looking at all well, then I turned my attention back to the empty, infinite gloom beyond the chicken wire, looking for the sharks. I recalled Peter Gimbel talking about a recurring dream. In the dream, he was alone in "the gray-blue blankness of the open ocean":

> I spun round and round, trying to look in all directions at once, sensing that something enormous was just out of sight. A form appeared, huge beyond imagining. It came rapidly toward me and materialized as the great white shark, the man-eater. It bore straight in with overwhelming speed, and as the jaws opened to swallow me, I would awaken and begin thinking of what desperate measures to take if the nightmare . . . came true.

Just then a large swell caught the *Snappa* at an angle, causing the tether to go slack momentarily before jerking hard on the cage. My tank slammed into the metal cage, making a hollow, otherworldly sound. I looked at Kevin and saw that his regulator was out of his mouth. Why was that? But before I could assess what was going on, Kevin vomited into the water, the detritus of his breakfast swirling inside the cage like chum. This really wasn't how I had imagined it.

Finally, the sharks arrived. One at a time. They glided in from the gloom, first appearing as little more than apparitions at the edge of visibility. Excitement surged through me. Adrenaline pumped, the beating of my heart amplified by the constriction of the wet suit. The sharks approached, and with nothing in the middle ground for scale, it was hard to assess their size. I was captivated by their movements—so graceful and lithe, at times appearing choreographed . . . and so quiet. It was incredible how their powerful bodies moved through the water seemingly without effort. If efficiency and precision of movement was the metric, this was an animal clearly better adapted to its environment than we humans were to ours.

Sharks represent a diverse group of species that have evolved into their modern forms over the course of 450 million years. Compare that to the roughly 200,000 years *Homo sapiens* has been trying to get it right. Even hammerheads, the youngest shark species in the fossil record, have been around for at least 35 million years. Once again, I'm reminded that reading a fact in a textbook is one thing, but to experience something—to be a cumbersome, air-breathing diver encased in a wet suit, restricted by gear and huddled inside a makeshift cage at the mercy of swell and currents (not to mention his dive buddy's vomit)—is a whole different ball of wax.

As time passed and the sharks circled and turned, my initial fear turned to focus; the excitement became wonder. It was amazing, I thought. The experience was nothing short of transformative and I wanted to keep doing it. Wes's cage trip became an annual event for many years to come. Over those years, our team expanded to include notable filmmakers and photographers like Nick Caloyianis and Brian Skerry. Eventually Charlie transformed his business from fishing for sharks to taking people out to dive with them—further bolstering the view that live sharks can be more lucrative than dead ones.

Going to Sea

As I worked on my master's at URI, the required coursework kept me busy. After a series of discussions with Wes, I decided to make age and growth of the blue shark in the North Atlantic the focus of my thesis. While some work had been done in this area, there were still many uncertainties. For an animal like a shark—which, when compared to other fishes, grows quite slowly and produces few young—understanding the growth rate, longevity, and various age-related milestones, such as sexual maturity, is critical to management and conservation.

With my thesis work determined, I immersed myself in learning all I could about age and growth methodologies. There was a great deal of good work already done on blue sharks by researchers like Gregor Cailliet, Sho Tanaka, and Hideki Nakano, but their work focused on the North Pacific. In the North Atlantic, early work on age and growth had been undertaken by Olav Aasen and more recently by John Stevens, H. M. Silva, and A. C. Henderson. I didn't know it at the time, but I would eventually meet most of these researchers, and they would become friends. I took the time to familiarize myself with each study and understand the assets and liabilities of the various methodologies employed. To my surprise, I found the work engaging and even exciting, although academic and requiring more time in the library than in the field. It motivated me to think in new ways and employ new technologies that would build on the established science and chart a course toward the future.

As I've said, age and growth studies on sharks represent some of the most important work undertaken from a management and conservation standpoint. If you don't know how fast a shark grows or at what age it matures, it's virtually impossible to put effective management measures in place. Many fishes can be aged by counting annual rings, or *annuli*, on their scales or their inner ear bones, called *otoliths*, but sharks lack both these structures, making it more difficult to age them. The only hard bony surface that can be used to age sharks is their backbone. Each vertebra develops growth bands much like a tree develops rings, but unfortunately, it's unclear how quickly sharks develop these rings or if they are even related to time. Before a scientist can count vertebral rings, he or she must validate how many years each ring represents, which is where a drug called oxytetracycline, best known to most as a broad-spectrum antibiotic, comes into play.

Among its other effects, oxytetracycline, OTC for short, incorporates into body tissues where there is active calcium deposition, like in the vertebral column. When a shark is injected with OTC, its vertebral column will possess a clear fluorescent mark when viewed under ultraviolet light, corresponding to when the shark was injected. When that shark is recaptured and donated to science, the researcher can remove a vertebra and count the rings, comparing the number of rings to the number of years since the shark was injected. While some scientists had used OTC with sharks in a laboratory setting, I wanted to use it in field experiments, which was at the time a somewhat radical step forward. But to make that happen, I needed the support of a shark tagging program. Luckily for me, I happened to be working for the largest one in the world.

On paper, the idea was simple: capture a shark, inject it with OTC, release the shark, and wait it for it to be recaptured. Then, using the elapsed time between capture and recapture, combined with an analysis of the number of new vertebral growth rings after the ring marked by the OTC, validate age estimates. The theory was sound but taking this work into the field brought with it a host of new challenges, along with some very long odds. For starters, a lot of sharks would need to be captured, tagged, and injected to increase the chances that at least a few of those sharks would be recaptured. I knew that in Jack's Cooperative Shark Tagging Program, of all the sharks tagged, only a fraction was recaptured. It was a numbers game, which is why Jack had been so successful partnering with recreational anglers and commercial fishermen and not just relying on fisheries scientists and research cruises to do all the tagging. The trouble with my work was that teaching a recreational angler how to properly handle a shark and attach or recover a tag is a lot easier than teaching that angler how to inject a drug into a shark. As a result, I figured I was going to have to rely on scientific research cruises to inject the sharks.

I didn't know a lot about research cruises besides what I'd heard from people like George Benz. As was often the case when Jack and the team gathered at a bar after a shark tournament, George, with his storytelling prowess, became the center of attention. I recalled one night when George was telling stories from his first research cruise aboard the

Polish fishing vessel F/V *Wieczno* in 1978. The use of the Polish-flagged *Wieczno* was the result of an agreement between Poland and the United States, whereby US scientists were allowed to use the vessel as a research platform, and Poland was allowed to fish for mackerel off the east coast. On his first research cruise, George was twenty-four years old, about the same age as I was when I was contemplating mine.

"It was like going home," is how George characterized the experience. "That trip molded who I am as a scientist more than almost anything else." I knew that George's friends often called him a pirate, but I didn't know why until I started hearing the stories come out about the pranks for which George was famous aboard the *Wieczno*. Most of all, however, I was impressed with how George spoke of a deep sense of camaraderie and mutual respect not only among the biologists but also with the Polish fishermen. It was like hearing about the way men with a common purpose bonded in difficult conditions—like Lewis and Clark, Shackleton and Worsley, or Mallory and Irvine.

After hearing George's stories, I was eager to go on a research cruise. The first opportunity came through the oceanography department at the Narragansett Lab, where oceanographer Bob Benway, another lunchtime volleyballer, managed the regular sampling of water temperature along an established path, or transect, between the coastline and Bermuda. The transect provided data on the temperature structure across the continental shelf to the Gulf Stream. Rather than relying on a research vessel to specifically go out and collect the data, he used a container ship, the M/V *Oleander*, that made a regularly scheduled run to Bermuda. Once a month, he would get someone to ride the ship with the job of dropping a temperature probe over the side once an hour, every hour, for twenty-four hours. The person would then get to spend a day in Bermuda before returning to the US. I saw it as a free trip to Bermuda, and I eagerly signed up. The experience was a great one, and it proved to me that I could go to sea, especially since we hit massive seas and I didn't get seasick.

But I wanted more.

In the spring of 1984, my opportunity came knocking. Jack was leading a team aboard the *Wieczno* for a research cruise, and I would be part of it. The catch was that I would have to miss my final exams for spring semester and take incompletes, which, given that it was only my second semester of graduate school, may have been a bad idea. I didn't think twice. As the great biologist Louis Agassiz said a century ago, "study nature, not books." I opted for nature.

My first cruise on the *Wieczno* was only two weeks, which, by comparison, was a short cruise. The goal was simple: catch and study sharks. To do this, the scientists used traditional pelagic longline gear. Longline fishing involves deploying a mainline that has shorter lines called *gangions* attached to it at regular intervals. Each gangion has a baited hook at the end. Once set, the longline, which may be miles long, is left to "soak." After a period determined by many factors, the line is recovered and brought on board. Some of the sharks we caught would be landed and killed for dissection and sample collection. Others would be tagged and released. Unlike shark tournaments, this type of sampling was considered "directed sampling," meaning that, while we had no control over what species we would catch on the longlines, we could usually make a choice about whether we wanted to tag and release or keep any given shark.

I knew my first cruise was going to be a rite of passage—a test of sorts. I was fairly confident that seasickness would not be an issue after the Bermuda cruise, but I knew the *Wieczno* would be at sea a lot longer, bringing with it a higher chance of bad weather and rough seas. I couldn't imagine anything more embarrassing than not being able to pull my weight because I was puking over the rail. In addition to lingering concerns about seasickness, I worried about the close quarters. The *Wieczno* was built for commercial fishing, not comfort, and I would spend more time than I'd ever spent with a small group of people with very little privacy. Half of those people were either my colleagues or my bosses, and the other half didn't speak much English. I was certainly more outgoing than I'd been, especially around people I'd come to know so well, but I had also become a get-along-with-everyone type of guy, which often meant avoiding conflict. I imagined that avoiding conflict during a couple of weeks aboard the *Wieczno* would not be as easy as it was on land, where I could simply walk away or avoid somebody.

The thing that worried me the least about the cruise was the work itself. My role on board was as a technician, meaning that I would help when and where I was needed. Primarily, however, it was understood I would be helping Wes, recording the specifics of his dissections of the reproductive system. The flip side of doing whatever was needed was needing to stay out of the way when I wasn't needed. This was a skill I had mastered already in the lab, and it's why Jack often called me "Mr. Frog."

"There goes Mr. Frog," Jack would bellow as I hopped clear of someone else's work.

In the days before departing on the first cruise, the work at the lab became frenetic. There were a million items on the to-do list, as well as checks and double checks to make sure all the equipment was ready and packed. It was stressful, but the workload helped temper my anxiety. Days before departure, we took vans and trucks up to Woods Hole, where the *Wieczno* was docked, to load fishing, tagging, and sampling gear. Before I knew it, departure day arrived. We loaded all our personal gear and made the drive to Woods Hole. When I first saw the *Wieczno*, I must admit I was a little disappointed. She was a large trawler, and her white hull was streaked with rust stains. While the others assured me she was seaworthy, there was no mistaking the fact that she was a work boat and a far cry from the National Marine Fisheries Service research vessels commonly docked at Woods Hole. I shared a cabin with Wes, Chuck, and George, and I watched everyone carefully and took my lead at every turn from those who were more experienced . . . which was everyone.

The *Wieczno* set sail in the late afternoon, and I stood on the deck waving to the Woods Hole dock crew, who always took a moment to celebrate the departure of a research cruise. The ship came around into the channel off Woods Hole at Cape Cod's southernmost point. As the deck rumbled beneath my feet, the Elizabeth Islands, specifically Nonamesset and Naushon Islands, drifted by off the starboard rail. (Little did I know that those unassuming islands would be the setting for one of the pivotal moments in my career more than two decades later.) We passed the Vineyard's western shoreline, and I thought about how Steven Spielberg had picked Martha's Vineyard as the location to shoot *Jaws*. Slowly the *Wieczno* gained speed and passed south of Nomans Land island, with the Vineyard's Gay Head Light barely visible in the distance. Beyond was the open ocean.

Be a Pirate

The research cruise was structured so that we would set the longline at night, allow it to "soak" overnight, and then haul it in in the morning. The entire day would be spent working on deck, either tagging and releasing sharks or dissecting them and taking samples. It was hard work, but I was completely enamored with it. Between Jack chiding me—"There goes Mr. Frog" (by the end of the trip, this changed to Greg-off, a spin-off of the crew's Polish name for me, "Grzegorz")—and Wes relying on me to assist with his work, I was feeling like part of the team in a way I had yet to experience. There was also a social component to it all that I hadn't expected—especially when the captain would invite the scientists to his cabin and a bottle of Polish vodka would come out. The unwritten rule was that once a bottle was opened, it was not put away until it was finished.

Alcohol aside, living and working alongside men like Jack, Wes, Chuck, and George—scientists that I admired tremendously—was fantastic, as I was getting to know them better as people. Tight quarters will do that. It wasn't all "living in the key of G," however. I was finding out that working on a boat in the open ocean during all kinds of weather conditions was exhausting, so there were times when I would bounce from the deck to my bunk, pop a cassette into my Walkman, and tune everything else out. Often, however, I tried to embrace the close quarters and lack of privacy. This wasn't an eight-hour workday followed by everyone going home. I was seeing a different side of people, and I'm sure they were seeing a different side of me. For example, when the fishing was poor, Jack would take it personally, clashing with the Polish captain over where and how to fish. Jack's commitment to his team was remarkable. If we weren't catching enough sharks, Jack would do everything in his power to make it right.

Of all the people on that first cruise, it was George who most impressed me. George's sense of duty to the science and to his research was unbelievable. Because he was interested in sampling the parasites, George was often the last in line to work with a specimen. Although he'd have an opportunity to collect external copepods from the gills, fins, and nasal cavities first, he would then have to wait until everyone else had collected their samples before finishing his work. There was one day I remember clearly when twenty sharks were on board, resulting in a significant backlog, as each scientist took their measurements and collected their samples. Nobody would have faulted George for skipping a few sharks as the sun slipped into the sea, dinner was served in the galley, and people settled into their evening routines. But George was meticulous. He would, and did, go without sleep to make sure that every shark that had been killed for the purpose of science was used to its fullest potential.

Beyond George-the-scientist, I quickly realized that George-the-person was someone I really enjoyed being around. He was a barrel-chested man with a huge personality, just the type of person you wanted on an extended trip at sea. He could regale any crowd with seemingly endless stories from his varied adventures, and his knowledge of scientific history was excellent. He loved old books and scientific illustrations, and he frequently illustrated his own work in a similar style, chiding his colleagues that they should be at least as good at the eighteenth- and nineteenth-century field naturalists. As we got to know each other better, George repeatedly told me to "be a pirate." What he meant by this was to "steal time for yourself and those you desire to help, rather than wasting time on trivial bureaucracy and nonsense." It was no wonder that Jack and George got on so well. Over the years to follow, George's advice would ring true in my ears many times—especially when dealing with the bureaucratic inefficiencies of government. When faced with a difficult decision, I would ask myself, *What would George do?* If I didn't know, I would pick up the phone and call him. Thirty years later, George died unexpectedly, leaving a hole in my heart. From the day I met him, I knew he was wise beyond his years, an old soul, and one of the funniest people on earth.

By the time the first cruise came to an end, I felt like it was only beginning. I had found my stride and felt the archetypal call back to the sea. We disembarked on a sunny spring morning at Woods Hole, and I followed everyone down the street to the Captain Kidd in a ritual that I learned followed every cruise. The Kidd is nestled between Woods Hole Oceanographic Institution and Eel Pond harbor. It's a fixture in Woods Hole, having offered a comfortable seat and a cold beer to nearly a century's worth of Woods Hole scientists, visiting scientists, Marine Biological Laboratory students, fishermen, and, of course, tourists.

When the members of the research cruise entered the Kidd that morning, the place was nearly empty. The scientists crowded around a table in the corner laughing and sharing stories. I allowed my mind to drift across the countless conversations I imagined taking place there over the years. Either sitting at the hand-carved, forty-foot mahogany bar or nestled conspiratorially at a table beneath the pirate mural, the giants of marine biology—men like Henry Bryant Bigelow and William C. Schroeder—shared a drink. There were no doubt enthusiastic discussions among aspiring scientists and probably more than one beer downed by a disillusioned academic. The Kidd was a repository of stories—an unwritten history that hung in the air as the smell of the sea blew in through the open door facing onto Water Street. It made me feel very much like I belonged.

As beers arrived at the table, I felt conflicted. I knew I was indeed going to miss the excitement of working daily with so many sharks, including many species like hammerhead, tiger, bignose, and night sharks, which I had never seen before. But I was excited about getting back to the lab and my studies and getting back to the daily routine. As I looked around the table, I thought about how my relationships with each member of the science team had strengthened over the course of the cruise, and although we could barely communicate with each other, I had developed friendships with many of the Polish crew members. I was learning, I realized, that doing good science was not always just about the science.

Jack paid for the round, and we all raised our glasses.

"To the *Wieczno*!"

Like Family

The next several years at the Narragansett Lab were some of the best years of my life. Working with Jack and Wes was a privilege I never took for granted, but others in the shark project became family and would remain valued colleagues and friends for life.

When I joined the program, Nancy Kohler was already there working with Chuck Stillwell on food habits of the blue shark for her PhD. Unlike other members of the team, Nancy was more reserved and somewhat quiet, but she shared a common passion for sharks that permeated the team. In that first year of working for the program, I quickly realized that behind that quiet demeanor was one of the hardest-working people I had ever met. Whether it was packing and loading gear, digging into the stomachs of blue sharks, or managing the tagging database, Nancy always seemed to be working. She was with me at my first dissection, my first white shark, my first cage dive, my first research cruise, and at almost every milestone thereafter. Ultimately, Nancy would take over the Apex Predators Program when Jack retired, but none of us knew that at the time.

In 1985, a couple of years after I started at the lab, Lisa Natanson joined the team. She'd grown up in California, just south of San Francisco, where she spent time observing fishes at the Steinhart Aquarium and hearing about white sharks in the nearby Farallon Islands. When she was seven, her mother taught her the word *ichthyologist*, saying, "Well, you love sharks and fish in general, why don't you study fish?" Lisa decided she was going to be an ichthyologist and, by thirteen, she refined her goal to becoming a shark biologist. Everything she did from that point on was undertaken with that one objective in mind.

Lisa met Jack in May 1983 at the White Shark Symposium at California State University, Fullerton. It was her first time meeting many of the legends in white shark research, and upon her first interaction with Jack and Wes, she mistook Wes for the head of the Apex Predators Program. It was at a party, and Jack was having a great time interacting with everyone—as Lisa recalls it, "just kind of being a goofball." Wes, on the other hand, was more reserved and hardly said a word all night. "He was just kind of taking it all in," Lisa remembers. "Very Wes." As a result, Lisa assumed Wes must be the head of the program because Jack was "just too much like a normal guy."

When Lisa finished her master's, studying the age and growth of angel sharks in Gregor Cailliet's lab at the Moss Landing Marine Laboratories, near Monterey, in 1984, she was ready to pursue her PhD. The Apex Predators Program was where she wanted to go. She reached out and set up a phone conversation with Jack, who was enthusiastic and encouraging about her coming east and working at the lab. She applied for and was accepted as a student at URI's Graduate School of Oceanography, and Jack brought her into the lab, where she, along with me, worked directly for Wes. Over the next couple of years, Lisa and I developed a relationship not terribly unlike a brother and sister, with everything that accompanies a sibling relationship.

Lisa entered the world of shark science when there were few female role models, but that didn't particularly bother her. She knew what she loved, and she worked hard to achieve success. Although Lisa had experiences early in her career with men who didn't think women could cut it when it came to serious fieldwork, she proved herself, and by the time she arrived at the Narraganset Lab, she was confident in her skills and abilities. It was a different time, and looking back, there were certainly instances of sexism—but Lisa had come up in a male-dominated field, and she was not afraid to speak up for herself and, as Lisa says, "give it right back." There is little doubt she had to work harder than many of her male colleagues, and I respected her greatly, even though we would clash from time to time.

Working on the docks at shark tournaments and in close quarters on research cruises was demanding, with its own potential for gender-related challenges, but Jack was the great equalizer when it came to work, and he expected that everyone bring their A-game regardless of gender. Like me, Lisa looked up to Jack as a mentor; he helped to shape her career in the same way he had helped George Benz and other students in whom he saw promise.

For me, the people with whom I worked closely at the lab were very much like a family. We worked together, played together, and partied together. We shared small hotel rooms during tournaments and close quarters on research cruises. Lisa once shot an entire roll of film of the things into which I molded my unwashed, salt-crusted hair during a month-long research cruise. We helped each other, got on each other's nerves, bickered, and ultimately supported one another no matter what challenges faced us. For me, a big part of wanting to remain on at the lab was how much I valued these relationships. I valued my shark science family.

Wallop-Breaux

Over the next several years, as I continued to work at the Narragansett Lab and chip away at my thesis, I participated in many more research cruises aboard the *Wieczno*. My love of the research cruises merged with my thesis work, as the cruises had indeed proved the most effective way to inject oxytetracycline into blue sharks. As the number of blue sharks injected grew, so too did my hopes that one would be recaptured. I finished my coursework for the master's degree, but was dragging my feet finishing the thesis. In part, I was hoping for more of the injected sharks to be recaptured, but there was also the looming knowledge that I was in a situation like the one in which I had found myself in the summer of 1983. As soon as I was no longer a student, my job at the Apex Predators Program would end. Bill Krueger told me I really needed to move along, and I knew I did, but the idea of not working at the lab wasn't one I wanted to face. My solution was to pay a few hundred dollars each semester to remain a student.

Who knows how long I would have dragged my feet if it hadn't been for my friend Brad Chase. Brad worked just across the hall from me at the lab, and one afternoon he stuck his head into my office. "Did you hear the Division of Marine Fisheries in Massachusetts is advertising for four new positions?" Brad asked. "I'm going to apply. What do you think? Do you want to apply, too?"

"Sure," I said, not thinking much of it. "I'll throw my hat in the ring."

The four new jobs were regional biologist positions funded through the 1984 Wallop-Breaux Amendments to the Federal Aid in Sport Fish Restoration Act. The act, commonly referred to as the Dingell-Johnson Act, had been passed in August 1950 and enabled the Sport Fish Restoration Program. Modeled after the 1937 Federal Aid in Wildlife Restoration Act, it provided federal funds for state fisheries programs. The funding came from excise tax collected from the manufacturers of fishing gear such as rods and reels, and it allowed states to better address fisheries management issues. By the late 1970s, however, the number of state fisheries programs was outpacing the funds available, so in 1984, Senator Malcolm Wallop of Wyoming and Congressman John Breaux of Louisiana proposed their amendments.

The so-called Wallop-Breaux Amendments increased revenues under the Federal Aid in Sport Fish Restoration Act by a factor of three, to $122 million, by tapping into untaxed sporting equipment items and collecting additional excise taxes. Most significantly, under Wallop-Breaux, federal motorboat fuel taxes and import duties on pleasure boats were paid into a trust fund called the Aquatic Resources Trust Fund, which funded both boater safety and sport fish restoration. Under Wallop-Breaux, each state's funding allocation was based on a calculus that considered the number of licensed anglers in the state and the state's total area, including water.

Most simply stated, Wallop-Breaux gave the states an influx of money earmarked for better managing sport fisheries. In Massachusetts, this translated to the state creating four new regional fisheries biologist positions: one in Boston, two on Cape Cod, and one on Martha's Vineyard. These regional biologists were charged with conducting "surveys, investigations, and studies" that would assist with sport fishery management and habitat protection, as well as aid in resolving user conflict. The thing that caught my eye, however, was that the fundamental nature of the job was to communicate with the sportfishing public. When it came to recreational angling in the United States, Wallop-Breaux was as significant as the creation of the Bureau of Sport Fisheries and Wildlife had been back in the late 1950s and early 1960s. Likewise, just as the creation of the bureau was a pivotal moment in Jack Casey's career, Wallop-Breaux was to become a decisive moment in mine.

Brad and I both applied for the job, and we both got interviews. Being good friends, we turned it into a road trip and headed up to Boston. We took turns interviewing for the job. I had mixed emotions about this job opportunity. On one hand, I knew that I had to finish the thesis and move on. On the other, I loved my current job and was hoping that Jack would find a way to keep me on—but that prospect wasn't looking too promising. Regardless, I wanted to keep working on sharks. So, during the interview, when asked by Deputy Director Randy Fairbanks about how I viewed my role in this new position, I didn't hold back and told him I would like to develop a shark research program for the state of Massachusetts. Based on the experience garnered working for Jack, I explained that shark fisheries off the Northeast, including Massachusetts, were primarily recreational, which would well qualify for Wallop-Breaux funding. To my surprise, he didn't push back, and told me that they didn't have anyone currently working with sharks. With new federal shark management looming, that might work out.

After the interviews, I was beginning to warm up to the idea of moving on. Brad and I went to the North End, had a couple of beers, and compared notes before heading back to Rhode Island. For a few months, I heard nothing and assumed I hadn't got the job. *Oh well, what is meant to be, will be.* Then I got the call. They offered me the Martha's Vineyard regional biologist position, a position in which I had expressed interest. I'd never lived on an island, but the idea excited me because I associated island living with my trips to the Caribbean. Unbeknownst to Brad or me, all four of the positions had been slated for women because of how male-dominated the Division of Marine Fisheries was, but none of the women who had applied would move to the Vineyard. Brad didn't get offered a position, but Randy was so impressed by Brad that he created a fifth position and hired Brad to work out of the agency's Cat Cove Marine Laboratory, in Salem.

Little did I know—despite my trepidation about moving out of my comfort zone—that this was the step that set me on a career path to working with my most admired predatory fish.

PART 2

PREDATOR OR CELEBRITY

There are a lot of nice fish underwater, but the shark's beauty is unique. It's the swimming motion—it's like a rigid snake. Sinuous. The muscular movement begins at the head. They swim slowly. They just ease their way through the water. When you see a shark dead on a concrete slab, this is nothing like it looks like in the water. When they're in the water, the beauty is incredible.

—Greg Skomal, as quoted in the *Boston Globe*, July 14, 1991

CHAPTER 6

THE VINEYARD

As other fisheries have been depleted, sharks are more and more attractive. That is what is causing the great mortality we see nowadays.

—José Castro, NOAA Fisheries shark biologist and author of *The Sharks of North American Waters*

They're vicious, they're aggressive, but they're also very vulnerable.

—Jack Casey, 1989

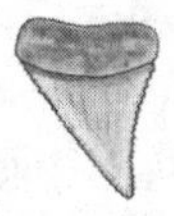

AUGUST 4, 1983—41°7'49.35"N, 71°33'9.05"W (SOUTH OF BLOCK ISLAND, RHODE ISLAND)

At twenty-two years of age, the large male white shark is approaching sexual maturity, but with a potential life span of seventy years, he is still a young shark by any measure. He is pushing north along the Atlantic seaboard off New York, Connecticut, and Rhode Island, following the warmth of the Gulf Stream to where it collides with the opposing Labrador Current, which brings cold water down from the Arctic. It's an annual migration to a place where the two currents meld and there is an abundant supply of food.

For many shark species making the trip north on the continental shelf, their northerly progress will slow as the warm, northeasterly trending Gulf Stream is diluted by the much colder, southwesterly flowing current. Most sharks, like most fishes in general, are cold-blooded or "ectothermic," and their body temperature is generally the same as the surrounding water. But the white shark is different—he is endothermic and can harness and retain metabolic heat to elevate his internal temperature above that of the surrounding water. He can continue north and east while some other sharks must stop when they hit the temperate waters of the Northeast. The white shark is not unique in this ability to elevate his body temperature, but endothermic fishes are certainly the exception, not the rule.

As he crosses from Montauk to the east of Block Island, the shark's power is awe-inspiring, owing to the fact that more than 60 percent of his body mass is muscle. Like all fishes, he has two types of swimming muscles, commonly referred to as white muscle and red muscle. The red muscle, which gets its color because it's loaded with blood vessels for oxygen delivery, is used for steady swimming or cruising at relatively low speeds—the type of swimming he's doing now. This muscle runs along the length of the body in a strip above the body cavity near the vertebral column. The white muscle, sometimes called the "fast-twitch" muscle, surrounds the red muscle and is what the shark uses for bursts of speed, such as when it's attacking prey. More than 90 percent of the white shark's muscle mass is white muscle.

Just like a human engaged in high-intensity activity, the shark's swimming generates metabolic heat. In most fishes, this heat is lost as it passes through the gills, where it comes up against colder seawater, but the white shark can retain this heat by way of specialized blood vessels called *rete mirabile*. This netlike structure of very fine capillaries facilitates countercurrent heat exchange as the warm blood leaving the muscle passes directly adjacent to the colder arteries, where the heat is transferred to the arterial blood and then pumped back into the muscle. In this way, the white shark can elevate his internal temperature, resulting in greater power, speed, and geographic and vertical range.

The shark is doing something that is instinctual—an annual migration that may take him beyond Block Island, Nantucket, and the Cape and into the rich feeding grounds of the Gulf of Maine, continuing as far north as Newfoundland. His ancestors evolved at a time when ocean temperatures were generally colder than that in which the earlier sharks had evolved. The first of the modern sharks showed up during the Jurassic Period, and when the dinosaurs disappeared, mammals, including marine mammals, flourished. These marine mammals, including many smaller whale species, provided ample fodder to fuel the development of larger predatory sharks like megalodon, which roamed the oceans from near pole to near pole.

As the ice caps grew and sea levels fell, new currents emerged, including the warm current traveling up the east coast of North America, fueling new seasons and seasonal migrations. Marine animals like the now-extinct, otterlike *Puijila darwini*—a webfooted, carnivorous precursor to the modern pinniped (that is, fin-footed animals)—began to exploit the Arctic and temperate waters. Behind them emerged predators like the white shark, which made semiannual migrations to temperate waters to take advantage of this new, energy-rich food source.

Although this white shark swimming past Block Island has the instinct to hunt pinnipeds in the inshore waters of New England and maritime Canada, he also knows from experience that he must rely on other food sources as well. His ancestors knew the seal migrations, especially the gray seals that can travel as much as thirty miles in a day transiting between Canadian and US waters. Unlike in other parts of the world—places like South Africa, California, and Mexico—where white sharks continue to follow the pinnipeds in seasonal migrations to white shark aggregation sites, white sharks in the northwest Atlantic have had to adapt. And so, when he picks up the scent of a dead forty-foot fin whale, the shark makes a large, arcing turn back toward Block Island Sound, putting him on a collision course with a fisherman named Ernie Celotto.

After New York's Long Island and Maine's Mount Desert Island, Martha's Vineyard is the third-largest island on the east coast. Like nearby Cape Cod, it was created during the last ice age by the Laurentide ice sheet as it plowed south and east across New England, a primordial dozer blade pushing a massive wall of detritus ahead of it. The ice, more than two miles thick in places, scraped the landscape and displaced most everything in its path. As the planet warmed, the ice slowed and eventually stopped its advance nearly twenty thousand years ago. The glacier started to melt and what began as freshwater rivulets turned to river torrents, winding like a serpent across the exposed continental shelf toward a sea more than three hundred feet below its current level.

Here, on the continental shelf, large animals like musk oxen, mastodons, and mammoths had taken refuge from the advancing ice, and the earliest New Englanders who hunted those animals followed. They arrived during the waning days of the Pleistocene only to find themselves on the precipice of a major extinction event. More than 80 percent of North America's large terrestrial mammals went extinct, which in turn caused a crash in populations of the terrestrial apex predators that hunted those animals. In the rising sea, however, marine apex predators like sharks continued to thrive. Unlike humans, sharks were not new to the scene. Evolutionarily speaking, even the newest sharks, the hammerheads, had been around for tens of millions of years and had adapted exquisitely to their environment. White sharks were *the* apex predator—so exquisitely adapted that they changed little in the intervening epochs between, when they evolved from broad-toothed mako sharks sometime in the Middle Eocene and the formation of Martha's Vineyard and Cape Cod.

As the sea rose, it covered the bones of the now-extinct mastodons and mammoths—the bones of which New England fishermen still occasionally find in their nets. The humans retreated west, escaping the rising water, and some came to inhabit heights composed of glacial debris—places today's residents of Martha's Vineyard know as prominent landmarks like Menemsha Hills and Gay Head Cliffs. Eventually these people became islanders, and they called the island Noepe, or "Land between the Streams."

On the southwestern peninsula of Martha's Vineyard, in a place they called Aquinnah ("Land Under the Hill"), members of the Wampanoag people ("People of the First Light") settled. In addition to harvesting shellfish by hand, they fished using nets, weirs, traps, hook and line, and spears. They also used dugout canoes from which Wampanoag hunters hurled spears into sperm whales, porpoises, and seals, all of which thrived close to shore, and all of which (except for perhaps the sperm whales) were also preyed upon by white sharks. The Wampanoag hunters' prowess with the spear was immortalized by Herman Melville in *Moby-Dick* in the character of Tashtego, an Aquinnah native described as the most skilled harpooner aboard the *Pequod*.

Aquinnah was renamed Gay Head by European colonists, and it was here, at a beach called Lobsterville, on a warm, moonlit summer night in 1987, that I became a saltwater fisherman. Although I'd caught sunnies and minnows at the pond on the golf course and then later fished for freshwater fishes with my brothers during high school, I wasn't much of a saltwater fisherman. My interest in recreational fishing had waned during my undergraduate years, and throughout my time at the Apex Predators Program, I'd come to see fishing largely as a sampling method, not a recreational pursuit. The only fishing I'd done since graduation was commercial longlining during research cruises aboard ships like the *Wieczno* and the *Delaware II*. Longlining may have earned me some street credit with commercial fisherman, but it was a long way from tactics that obsessed Vineyard anglers used with species like striped bass, bluefish, and false albacore.

I had arrived on the Vineyard a few months earlier in June 1987 as the newly minted Division of Marine Fisheries regional biologist for Martha's Vineyard and Nantucket Islands. Unlike most jobs I had worked in the past, this appointment came without much of a job description. I knew I was serving the sportfishing community, so I thought it best to become more personally involved with saltwater fishing, although it was clearly not a prerequisite of the job. Anyone observing me on Lobsterville Beach that night surely would not have mistaken me for the person charged by the state to assist with sport fisheries management. That guy was clearly not a fisherman. Twenty-five years old, I was barefoot and wearing shorts and an old T-shirt. I had a secondhand fishing rod rigged with a broken-back Rebel, a fishing lure made of two pieces joined in the middle to better imitate the swimming action of baitfish. In a Stop & Shop bag over my shoulder were a flashlight, a windbreaker, a couple of additional lures, some swivels, and a few wire leaders for the sharp teeth of bluefish. I was not carrying a hook removal tool, a net, a headlamp, or any of the other gear carried by most striped bass anglers, but I didn't worry too much. Needless to say, I didn't think I'd be catching any striped bass that day.

In my first month on Martha's Vineyard, I'd forced myself to become a regular at all of the island's tackle shops. At first I stopped in under the pretense of introducing myself. Then I returned with updated fishing regulations, recreational fishing guides, and other handouts from the Division of Marine Fisheries. It was the next visit that was harder, the visit for which I didn't have an obvious excuse. I'd become more socially confident while working at the Apex Predators Program, but the move to the Vineyard set me back. I hardly knew anyone on the island, and it wasn't my natural inclination to start up a conversation with a stranger. I knew, however, that if I was going to be successful at this job, that's exactly what I needed to do. As far as I could tell, this job required a close working relationship with the fishing public, and that required getting to know my constituents. I knew from observing Jack all those years that I needed to be on a first-name basis with the owners of the tackle shops, as well as the regulars who shopped there. I needed to be able to strike up a conversation with harbormasters and charter captains. I needed to know the commercial fishing vessels and who owned which boat. Jack's mentorship had strengthened my resolve, but it wasn't an aspect of the job that came naturally to me.

Coop's Bait and Tackle was one tackle shop where I felt I was making progress, but I still had to take a deep breath before opening the door and walking inside. The *easygoing smile* and *casual wave* I offered to Cooper Gilkes behind the counter were both well rehearsed.

"Hey, Greg," Coop said as I came through the door. "How's it going?"

"Great," I replied, and then hesitated, letting an awkward silence hang between us. "Really great," I said finally. "Thanks. You?"

"Same old, same old. It's summer on the Vineyard. Plenty of tourists, but the fishing's been good."

I smiled.

"You know these guys, right?" Coop asked, gesturing to two customers standing by the counter.

"No, I don't think so," I said, extending my hand.

"Greg's the new state fisheries guy out here," Coop said.

The men shook hands. "Oh yeah," a customer in a ball cap said. "I read about you in the paper. You're the NMFS shark guy, right?" He pronounced it *nimphs*.

"I was," I said. "I mean I spent the last five years studying sharks out of the Narragansett Lab with Jack Casey."

"But now you're here to write tickets for the state?" the one with the mustache asked.

"No," I said. I felt slightly taken aback, as someone who was still unaccustomed to some of the locals' perception of state employees. "My role isn't going to be enforcement. I'm more here to gather information, especially understanding conflicts between commercial and recreational fishermen. We want an accurate assessment of what's going on out here. I'm here to listen."

"Well then, have you heard about the situation at Chappaquiddick?" asked the guy in the ball cap.

"I'm not sure," I said. "Tell me."

"Those commercial guys are dragging their nets so close to East Beach that you could throw stones into the boats," he said. "They're catching up all the fluke, damn it."

I took out a little notebook I carried and wrote it down. "East Beach, you said?"

"Yep."

"That's exactly the kind of information I want to hear." I thanked him and put my notebook away. Another awkward silence hung in the air—only for a beat, but it seemed an eternity to me. I turned quickly and started examining a display of fishing lures. Coop and the two customers resumed their conversation.

"I got my brother and the nephews coming out for a week," the man wearing the baseball cap said to Coop. "They want to fish for stripers, but I don't think they've ever fished for anything that didn't involve a bobber and a worm."

"You should take 'em out to Lobsterville," Coop responded. "It's a perfect place to learn. The currents aren't bad at all out there, and the beach has a nice shallow slope to it. Plus, it faces north, so the wind will be at your back."

My ears perked up as they eavesdropped. I'd learned a few nights earlier about the prevailing southwest wind when I'd tried fishing with tackle that was too light for the conditions.

"I would imagine it's a good spot," said the one in the ball cap. "There must be plenty of baitfish packed into the shoal water after they leave Menemsha Pond."

"Yep, and the stripers know it," Coop said with a chuckle. "It also fishes well on both the rising and falling tide, so long as it's not slack. It makes it easier to work around 'family' schedules."

The three men laughed.

During each of these visits to the tackle shops, I made an effort to buy tackle, largely because I didn't have any, but also because I wanted them to know I was a fisherman. So, I purchased a couple of lures, said goodbye to Coop, and waved to the two guys. "Nice to meet you."

"See you out there maybe," the man with the ball cap said.

The evening after hearing about Lobsterville at Coop's, I headed to what is locally referred to as "Up-Island"—although it's actually in the southwestern part of Martha's Vineyard. In Chilmark I turned onto State Road and continued west around the southern end of Menemsha Pond, a large tidal salt pond that nearly severs Aquinnah from the rest of the island. About two miles past where State Road cut between Menemsha and Squibnocket Ponds, I turned north onto Lobsterville Road and followed it through woodlands until it gradually arced east before opening onto the beach. I parked in one of the roadside cutouts clearly used by many fishermen before me. As I stepped out of my truck in the waning light, I could hear a gentle surf, a sound that always made my heart flutter. From the path through the dunes, I could see the Elizabeth Islands on the far side of Vineyard Sound. I made a mental note that I needed to get over there at some point, as that island chain was technically part of my region.

On the beach, I instinctually scanned the ground looking for shark's teeth. I'd heard that large ones had been found all around Aquinnah, and I was always hopeful I might find one. The Wampanoag said the largest teeth were the remnants of meals consumed by Moshup, a supranatural being who lived in a cave in the Aquinnah Cliffs. Moshup was so large that he could lift a sperm whale or giant shark from the ocean by its tail and dash it against the cliffs, which is why the cliffs have a red hue to them. The shark teeth and whale bones were all that remained.

I'd heard that a few of the shark teeth found at Lobsterville were more than four inches in length, making them larger than the teeth of any living shark. Those teeth were from the megalodon (*Otodus megalodon*), a species that could exceed fifty feet in length and weigh close to fifty tons. The megalodon had gone extinct some 3.5 million years ago, but it had ruled the oceans for twenty million years, which seemed like a good run until one considers that the "new sharks"—called *neoselachians*—emerged with the dinosaurs during the Jurassic Period, some 200 million years ago. The "toothy" sharks of the order *Lamniformes*, the ones that better align with our popular conception of a shark, emerged later during the Jurassic. These sharks got their name from the Greek *lamna*, or "fish of prey," and included the mako, porbeagle, and white shark. It was once thought that the white shark was a close relative of the megalodon, just a scaled-down model, but the fossil record shows that the white shark is more closely related to mako sharks. By the time the Wampanoags were hunting Martha's Vineyard's nearshore waters in their dugout canoes, the megalodon was long since extinct—but white sharks were surely a common sight, hunting seals alongside the Wampanoags.

I found no shark teeth that evening, but I did catch fish. And it wasn't just a fluke, either. I caught six striped bass and a couple of bluefish. Coop told me to reel that broken-back lure in as slowly as possible and that is exactly what I did. Those initial strikes as the fish jumped on the lure were incredible and paled anything I experienced in fresh water—these were real fish and, although small by most saltwater standards, bigger than anything I'd caught before. "Oh my God," I said to myself, "I can actually do this!" Despite the difficulty of removing the treble hooks without a dehooking tool (note to self), despite the difficulty of dealing with a fish in the dark while trying to hold a flashlight (note to self), and despite a surprising battle with a marauding beach skunk (not sure how to prepare for that), I was literally and figuratively hooked.

A Mini Apex Predators Program

When I first arrived on the Vineyard, I wasn't completely sure what my days would look like. I had an office in the State Lobster Hatchery, which sat right on the shoreline of Lagoon Pond in Oak Bluffs. It was a beautiful location that I would come to love. The building itself was wood and sheathed in cedar shingles, and it was old, dating back to the middle of the twentieth century. I was told the location was the site of a hospital before the lobster hatchery, and there was a small graveyard in the woods adjacent to the building. A closer inspection of these revealed some to be gravestones of old mariners. As charming as it was from the outside, I was in for a surprise when I stepped through the door. The first floor was wide open and full of tanks for breeding lobsters. There was a front room open to the public five days a week, and people could come in and see the lobsters and read about lobster biology and how they were reared. The tanks were gravity fed from a holding tank, and the building was constantly filled with the sound of running water—which I actually kind of enjoyed.

Kevin Johnson, who was the state lobster culturist and who became my first true friend on the island, showed me to my "office." It was little more than a closet in the corner, which Kevin and I shared. The room was barely big enough for our two desks and the only window looked over the parking lot. Because of all the tanks, the office was always damp and moldy. It was clear that I had left a real lab behind and moved into a field station, which was a shock after working for so many years at the Narragansett Lab, with professional scientists in a professional lab. But the place grew on me, and I leaned into the vibe. It had the wild feel of a field station on an island, and I liked that.

My boss Randy had given me a long leash, which was daunting at first. The only real parameters were to get in tune with my region and develop research projects that would generate information for better management of recreational fisheries. Lastly, he added, "Have fun!" Of course, I wanted to follow through with starting a shark research program and Randy liked the idea because nobody in the agency was currently working on sharks. I envisioned a mini Apex Predators Program that would collaborate with Jack's program and other researchers. When I shared the idea with Jack, he was all for it.

Even though sharks were at the front of my mind, my job focused on all fish species, as sportfishermen were essentially the ones who paid my salary. Unfortunately, sportfishermen were generally suspicious of fisheries managers and scientists, given the conservation status of many species. Striped bass were recovering, but fluke and other stocks were in trouble, and I found myself working hard to earn trust while trying to avoid politics. I was learning that every fisherman had an opinion about what was happening with each species, and it varied depending on whether they were private vessels, charter captains, or shore anglers. If a species was in trouble, then it was always the other guy's fault, generally commercial fishermen. The state was either doing too little or too much. Some argued that the state should do nothing because fluctuations in species abundance were driven by natural cycles. I heard it all and enjoyed listening to all the opinions. Some of the folks were right, some might have been right, and some were simply wrong, but I thought it best to listen to them all. In the beginning, I wouldn't say much and would just nod my head. But as my own knowledge base grew and I became more versed in the science behind each species, I would offer my nuanced opinion. That seemed to garner me more respect and, eventually, some trust.

I had decided to give the new job five years. It would be a great experience, and it afforded me the opportunity to live on the island and, in many ways, be my own boss—but I always wanted to get back to sharks being at the center of my professional life. There were, however, two events that first summer on the Vineyard that not only convinced me that I was in exactly the right place, but also provided an opportunity to grow and to come into my own as a scientist.

Whether it was irony or fate is irrelevant, but Martha's Vineyard's first shark tournament occurred the same year I arrived on the island. The tournament, which would come to be called the Monster Shark Tournament and become one of the biggest shark tournaments in the country, was organized by the Boston Big Game Fishing Club. The club already hosted the Martha's Vineyard White Marlin Tournament, and it was the club's founder, Clint Allen, and his brother, Rick, who started the Monster Shark Tournament, although they also give credit to Arthur BenDavid, the Oak Bluffs harbormaster. The Martha's Vineyard shark tournament was one of three new shark tournaments in the Northeast in 1987. The other two were in Nantucket and West Yarmouth. Like elsewhere in the Northeast, shark tournaments were growing in popularity as other, more cherished species were becoming scarce.

The inaugural tournament was a two-day affair beginning on Saturday, August 8. Thirty-one boats competed, and the winning shark was a 452-pound tiger shark landed on the first day by Texan Tom O'Connell and crew aboard Tom's boat, *Renegade*. The *Renegade* left Oak Bluffs Harbor at 6:30 A.M., established a chum slick using menhaden, and suspended a mackerel-baited hook thirty feet below the surface. The line with the baited hook was attached to a balloon, which drifted away from the boat. O'Connell and his crew waited. The winning shark didn't bite unto 3 P.M., but the crew was ready and landed it in ten minutes. They shot the shark in the head with a pistol and then manhandled it through a hatch in the boat's transom. The *Renegade* headed for the harbor, where the shark was hauled onto the dock and weighed along with the dozen other sharks landed, which drew a crowd of onlookers.

The tournament was deemed a success.

For me, the inaugural Monster Shark Tournament coinciding with my arrival on Martha's Vineyard was almost too good to be true. Having a shark tournament in my backyard provided me the immediate opportunity to begin collecting data. Shark fishing was exploding in the late 1980s, so much so that it even made the front page of the *New York Times* when Donnie Braddick, fishing with Frank Mundus, landed a 3,450-pound white shark. While a debate ensued about whether it was a world record, Wes, who went with Jack to dissect the shark, told the *Times*, "It's the largest male great white shark taken in these waters." For his part, the reporter editorialized, "The great white shark—known to be a maneater—is one of the most savage creatures of the sea and the shark most often responsible for attacks on humans."

Jack included a note in the 1987 edition of the *Shark Tracker* newsletter saying: "[T]he increasing numbers of tournaments makes it impossible for our staff to attend all of them. However, we want to know about new tournaments and will do what we can to be of assistance."

Having me on the Vineyard was an advantage to the Apex Predators Program's efforts, and it was in fact Jack who first told me about the plans for the tournament. He gave me the name of Clint Allen and told me to go see him at the Wesley Hotel, across from Oak Bluffs Harbor, and which was tournament headquarters. That was easy for Jack to say—just go up to this tournament bigwig, a complete stranger, and ask if I could sample his event. As I climbed the steps to the hotel's big front porch, I was scared to death. I passed by people sitting in rocking chairs overlooking Oak Bluffs Harbor and taking in the view. Everyone looked like they were relaxed and having fun, and all I wanted to do was head in the other direction. There was an older woman selling tournament T-shirts—I would later find out she was Clint and Rick's mother, Loraine—and I hesitantly approached her and asked her if Clint Allen was available. She was polite and smiled but seemed suspicious. She told me to wait right there, and she went to find him.

Clint Allen was a big man both physically and in the way he talked and carried himself. He exuded confidence, and his voice commanded attention and respect. A self-made successful businessman, he had established the Boston Big Game Fishing Club to run some of the most successful fishing tournaments on the east coast, but this was his first foray into shark tournaments. Knowing little about shark tournaments, he called Jack.

Clint greeted me with a what-the-hell-do-you-want look, and I immediately dropped Jack's name and said I would like to sample his tournament. His face softened and he extended his hand.

"You need to meet my brother Rick," he said as he pointed to a smaller version of himself chatting it up with tournament participants on the other side of the porch.

Rick and I hit it off immediately, and he enthusiastically agreed to help me in any way he could. I told him I'd like to measure and dissect the sharks brought to the weigh station, and he slapped a hand on my back.

"Consider it done!" he said.

The Monster Shark Tournament grew quickly. By the second tournament in 1988, there were forty boats and 160 anglers, including "celebrity anglers" like the retired hockey player Bobby Orr. As the event grew in popularity, it also grew in utility to shark science. As the *Shark Tracker* put it in 1988, the Vineyard tournament, along with the other shark tournaments, provided "an important source of data and provide opportunities to examine sharks, distribute tags, and exchange information with fishermen who are indeed valuable windows to the sea." Like Coop's Bait and Tackle, it also provided a great place for me to meet, network, and ultimately work with the state's offshore fishing community. The Monster Shark Tournament started it for me, but I soon began sampling all the state's offshore fishing tournaments under the umbrella of the Massachusetts Sportfishing Tournament Monitoring Program, which I started in 1988. Walking in the footsteps of Jack, I got to know most of the private and charter vessels that routinely steamed offshore to target sharks, tunas, and marlins each summer.

I also became very familiar with all the tournament organizers, who began to reach out to me more frequently as time went on. In my mind, this gave me an opportunity not only to collect data at each event, but also to help them run a more conservation-oriented fishing tournament. For example, it was quite legal for every boat in the tournament to land a shark, but I saw no need to pile the docks with dead fish, especially those that were not palatable. So I worked with tournament organizers to eliminate species like blue sharks that were not good eating, to increase minimum sizes for makos and threshers, and to offer incentives like an award to vessels that released fish.

The other major fishing event I encountered my first year on the Vineyard was the annual fall Martha's Vineyard Striped Bass and Bluefish Derby. Unlike the Monster Shark Tournament, the Derby had been around since the 1940s, when it launched as a PR stunt for a struggling ferry service. Each year, the derby opened on the second Sunday of September and spanned the next five weeks. It was one of the oldest, largest, and most popular saltwater fishing tournaments in the country.

The derby had been around for decades and, by the time I arrived in 1987, it was itself experiencing a new beginning. A group of local fishermen had purchased the derby from the Chamber of Commerce for a dollar and rebranded it as a civic-minded nonprofit. I found that most of the anglers on the island were passionate about conservation, and while it was certainly true that some of the motivation for conservation was to ensure that the fishing would remain good, the net result was the same. In 1985, for example, the derby committee controversially removed striped bass, the most popular sportfish on the island, from the leaderboard because of conservation concerns.

Like the Monster Shark Tournament, I saw the derby as a perfect opportunity for research, and I was soon coordinating with the organizers and then with the anglers to collect data. During the five weeks of the derby, fish are weighed in daily at a centralized location called Derby Headquarters, meaning that there was an impressive database already in the making. In addition to the raw data regarding species and size, I could have my pick of fish to sample. Unlike the shark tournament, these were inshore species including striped bass, bluefish, Atlantic bonito, and false albacore, and this event allowed me not only to collect tissue samples, like stomachs, for life history work, but also to monitor the local abundance of these species.

As was the case with the shark tournament, I sought to coordinate with other researchers to collect samples that could further their research. At one point, I even collected samples of parasitic copepods for my friend George Benz. Although it wasn't quite collecting samples from white sharks, as I had seen George do back in 1983 with the Noank white shark, it was valuable research, and I made sure to maximize its potential.

In 1994, I was flattered when I was asked to join the Derby Committee, which organizes the annual event. I was down walking the wharves in Edgartown one day during the event and derby chairman Leslie Smith approached me.

"Hey, Greg, got a second?" she asked. I had known Leslie a few years at that point from Larry's Tackle Shop and knew she was chairperson of the Derby Committee.

"Sure," I responded. "What's up?"

"It's that time of year when we start talking about adding new members to the Derby Committee and your name keeps coming up from a lot of the current members."

Wow, I thought, *the committee is composed of some of the biggest fishing names on the island . . . and they want me to be part of it.*

"Would you consider joining?" she asked.

I tried to play it cool, but I don't think that worked because I blurted out, "Absolutely!"

I soon learned that everybody on the committee shared the love of fishing, but each also fulfilled a role. There were carpenters who set up the weigh station, salespeople who helped recruit sponsors, graphic artists who generated the Derby Booklet, and tackle shop owners who provided prizes. My role was to help keep the derby a conservation-based event. Using the data I collected, I helped guide the organizers when it came to conservation issues like minimum fish lengths and reducing overall mortality. As I had observed with shark tournaments, establishing size limits could drastically reduce the number of fish killed while still preserving the excitement of the competitive tournament. Using the data, I helped the derby adjust its minimum size limits for each species, which, in the case of false albacore, reduced the number of fish killed in the tournament by more than 50 percent.

The derby is a landmark event on the island, and most local anglers participate. As a regional biologist, working with the derby gave me a direct conduit to meeting and getting to know the local fishing community. But over the years, I also developed strong personal relationships with many islanders and consistent visitors. In doing so, I was overcoming my shyness and building my own self-confidence—both professionally and personally. To me the simple invitation to join the committee represented acceptance by the island community and my first sense of becoming a fellow islander. I was no longer just working on the island; I was now living on the island. My love for Martha's Vineyard and the community was growing.

A Sea Change: Protections for Sharks

By the late 1980s, the amount of data collected on sharks in the western Atlantic had grown exponentially since Jack started in the early 1960s. The bottom line was that the data increasingly showed sharks were in trouble, and the fact that they were almost wholly unregulated along the east coast of the United States was problematic. In response, in early 1989, US fisheries managers on the east coast began to get serious about developing a fishery management plan for sharks. In just a few years, sharks had gone from "an underutilized resource," for which fishing was encouraged, to becoming "overfished" according to federal fisheries data. At its January 1989 meeting, the Mid-Atlantic Fishery Management Council voted to submit a request to the National Marine Fisheries Service to begin collecting data on the shark fishery in order to facilitate the development of a fishery management plan.

The council emphasized the necessity of a management plan in the face of an expanding, often indiscriminate commercial fishery and a rapidly expanding recreational fishery. Regarding the recreational fishery, Nelson Bryant, the *New York Times* outdoors editor whom I had first met at the Bay Shore Mako Tournament in 1983, wrote in an article titled "Sharks Attract Protectors" that there were an estimated ten thousand anglers who focused on sharks in New Jersey alone. Jack Casey emphasized, however, that the recreational fishery was generally moving in the right direction when it came to shark conservation. The commercial fishery, on the other hand, was not.

Jack was serving on the task force charged with coming up with the proposed management plan, and he stood up for recreational anglers who so often were targeted as being a big part of the problem. Jack was clear to say that, while there had been a time when recreational shark anglers were guilty of overfishing, those same anglers were now leaders in both shark science and conservation.

"At the Bay Shore tournament," he said, "when they first started out almost thirty years ago, a good shark was a dead shark. They landed almost a thousand sharks in 1965. Now they're down to landing about thirty."

There is little doubt that many recreational anglers were supportive of regulations on the shark fishery because they'd observed precipitous declines firsthand. For example, shark anglers on Long Island, where modern shark fishing had begun with Frank Mundus, reported seeing the number of sharks caught decrease dramatically. The executive director of the New York Sportfishing Federation told a reporter, "I can remember going out and catching fourteen or fifteen sharks a day in a tournament. Now if you catch one shark a day—and I'm talking about per boat, not each angler—that's pretty good." He said that one day in June 1989, there were six hundred boats in the water at a shark tournament, and only five sharks were landed. In addition to fewer sharks, anglers were also reporting smaller sharks.

"There were years where we would go out and catch fish three hundred, four hundred pounds and it wasn't that big a thing," said the president of the Hudson Anglers, which hosted the self-proclaimed largest annual shark tournament in the western Atlantic. "Today it's almost inconceivable that you can catch a fish like that." He said that in 1989, he was lucky to catch a single shark on any given day, and only one shark caught all season was big enough to keep. "The rest," he said, "we tagged and put back."

Although it was true that shark angling and shark tournaments continued to grow in popularity, the reality was that the data showed these anglers were not the ones driving the decline in shark populations. Instead, it was really the commercial shark fisheries that were taking a toll. Federal fisheries data showed that the commercial catch had doubled each year for the last several years. "As other fisheries have been depleted, sharks are more and more attractive," explained José Castro, a shark biologist with NOAA Fisheries. "That is what is causing the great mortality we see nowadays."

In addition to directed shark fisheries, it was becoming apparent that sharks unintentionally killed as bycatch in other fisheries was a serious issue. One report by the National Marine Fisheries Service estimated that only 11 percent of sharks caught in nets and on longlines by commercial fishermen were landed. The rest were thrown overboard dead or dying as bycatch. To add insult to injury, the demand for shark fins in the export market was skyrocketing, and the record prices were further driving unsustainable practices.

Several case studies had clearly demonstrated that increased fishing pressure on specific shark species could quickly result in overfishing and population crashes. Fisheries managers pointed to the porbeagle shark as an excellent example of sharks' vulnerability to overfishing. After being overfished in the northeast Atlantic, porbeagles in the northwest Atlantic were targeted next by European commercial fishing boats. In the early 1960s, in a period of just four years, northwest Atlantic landings of porbeagle shark increased from 1,900 tons to over 9,000 by the Norwegian fishing fleet alone. The northwest Atlantic porbeagle population collapsed as a result. As if this was not bad enough, overfishing was magnified in the case of sharks because of their slow growth rates, late maturation, and few offspring. In the case of the porbeagle shark population in the northwest Atlantic, after the mid-1960s collapse, it took a quarter century to recover to only about 30 percent of the population prior to intensive targeted fishing by European fishing boats.

In the fall of 1989, the National Marine Fisheries Service finally proposed a fisheries management plan for thirty-eight of the roughly one hundred species of sharks known to inhabit the Atlantic coast. Based on the best available data, fisheries biologists determined that shark populations had been overfished for at least the past decade by both commercial fishing boats and recreational anglers. Growing markets for shark and shark fins, combined with growing interest in fishing for them recreationally, resulted in landings outpacing the maximum sustainable yield by several thousand metric tons annually. As Jack put it, "It's like fishing on interest in the bank. Once you get into the principal, there's going to be a crash, and it's going to take a long time to bring the population back."

Under the new management plan, the commercial catch would be limited to 14.1 million pounds and the recreational catch would be restricted to one shark per angler per day. It had been illegal for foreign fishing vessels to land sharks in US federal waters since 1979, but this was the federal government's first effort to protect sharks domestically, and it was met by a mixed response, and the opposition slowed the implementation of the new regulations. Despite the success of the Cooperative Shark Tagging Program, now entering its fourth decade, sharks, for the most part, remained a data-deficient species compared to other fishes, and there was much debate about the data on which the proposed regulations were based.

Recreational anglers criticized the plan as not going far enough, while commercial fishermen thought it went too far. The president of the Montauk Boatmen and Captains Association pointed out that shark-fishing tournaments already had more stringent rules than the proposed regulation.

"Conservation is good for our industry," he said. "Ten years ago, I could go ten miles offshore and catch plenty of sharks. Now I have to go out forty miles, and even then, I may not see any." Commercial fishermen and dealers said the new rules were too severe and were heavily influenced by the powerful sportfishing lobby. "I have seen plenty of sportfishermen who parade into the dock with big sharks hanging off the gin pole in the heat of the day," said a New York–based fish buyer who estimated that only about 5 percent of his business was in shark meat. "That's just wasting a food resource."

It wasn't until April 1993 that the Fishery Management Plan for Sharks of the Atlantic Ocean was finally implemented. The final plan regulated thirty-nine frequently caught species of Atlantic sharks and divided them into three groups: large coastal sharks, small coastal sharks, and pelagic sharks. It established a commercial permitting system, commercial quotas, and a framework for adjusting quotas. For recreational anglers, it put in place a trip limit of four sharks per vessel for large coastal sharks and pelagic sharks and a daily bag limit of five sharks per person for small coastal sharks. It also prohibited finning and the sale by recreational fishermen of sharks or shark products. While it was a long time in the making, sharks in the northwest Atlantic enjoyed widespread protections for the first time ever.

Shark Engagement: The 1990s

By 1989, the American public was probably more aware of sharks than it had ever been. Recreational angling for sharks was certainly as popular as ever, but sharks were also entering the public dialogue in myriad other ways, ranging from the debut of Discovery Channel's Shark Week in 1988 to shark being a more common menu item and shark finning becoming a rallying point for people concerned about the environment. On the last point, Jack Casey observed, "They're incensed far beyond what you would expect for a shark. People have been writing to us, to their congressman, saying this has got to stop. It has actually turned public opinion in favor of the sharks."

It was amid this intense interest in sharks that I formally established the Massachusetts Shark Research Program. The program, which I had modeled after the Apex Predators Program, was designed to help the state better understand the ecology, distribution, and relative abundance of the sharks for which there were fisheries in Massachusetts. Like the Apex Predators Program, I planned to conduct formal longline and angler surveys, but I also planned to rely on the relationships I was building with both commercial and recreational anglers throughout the state to opportunistically collect data.

Around this time, I was beginning to realize that people were often looking to me as the de facto local shark expert. From the stories of me working with local fishermen catching sharks from the beaches, coupled with my involvement with the shark tournament, I had developed a reputation as the shark guy. At no time was that clearer to me than when I got a call from the police in August 1989, asking if I would check out a potential shark bite.

I drove to the hospital to interview the patient. Ironically, it was the same hospital where Matt Hooper and Chief Brody had inspected the remains of a deceased swimmer in the movie *Jaws*, but this victim was in much better condition. He told me that his buddy was reeling in a sandbar shark from shore, and when it got close enough, he grabbed it by the tail. It turned and bit his calf in what those of us who assess shark attacks call a "provoked attack." After interviewing the man, I inspected the wound. It was not very large, but it was deep and sliced into the muscle of the calf in an arc-shaped gash. I recorded my observations in my notebook.

> Shark bite: Victim attempted to "see" or "get" shark in surf while the fish was being landed by an angler. With several fishermen ~1 mile north of Wasque brown [aka sandbar shark] shark fishing.

The incident had not been very serious (although I'm sure it was memorable for the guy who was bitten), but it was a big deal for me. To be called by the police to investigate a shark bite—isn't that pretty much exactly what Matt Hooper had been called to do? Brought in by the authorities to make an assessment about a shark attack? It didn't happen all at once, but soon I started getting calls from the media whenever a shark event would occur. Sometimes public safety officials would get in touch. I realized that I had a unique skill set—an area of expertise—and while I did not enjoy interviewing a shark bite victim, I felt a sense of accomplishment that brought with it a newfound confidence. I loved my job, and as it turned out, I was getting pretty good at it.

In May 1990, I finally earned my master's degree, based on the work I'd done with age and growth in blue sharks while at the Narragansett Lab. It was a tangible break with my past, and it was the first time in my life I hadn't been a formal student. But it didn't mean I was done learning—not by a long shot. My next teacher, however, would be unlike any professor I'd ever had.

By my third summer working with the state, I was making every effort to attend all the big-game fishing tournaments held in Massachusetts. The first of 1990 was the Cuttyhunk Invitational Tournament, which was held on the small island of Cuttyhunk, the outermost of the Elizabeth Islands, just west of Martha's Vineyard. Although the tournament focused on tunas and marlin, it also awarded points for mako sharks. Sixty boats were registered to fish that year, but poor weather plagued the event with rain, northeasterly winds blowing twenty-five knots, and six-foot seas. I rode along on one of the boats for the first day, and we took a pounding and caught nothing. We were not alone. By the end of the tournament, only fourteen small tunas were landed, by just a handful of boats. As I channeled my inner Jack Casey and walked the docks at the close of day three to chat with captains about what they were seeing on the water, someone called out to me.

"Hey, are you the marine biologist collecting data at this tournament?" he asked.

The voice came from a tall, tan man with a full head of blond hair tucked into a cowboy hat who was standing in the cockpit of an older Sea Ray and clutching a Bud Light. Stenciled on the transom was the name

Fish N Boat. I'd never seen this boat before, but it smelled of ground chum with a hint of fresh bluefish.

"Yes, I am," I responded hesitantly. The guy, who quite frankly looked like the Marlboro Man, had a broad smile and looked friendly enough, but I knew that biologists working for the state tended to get an earful of complaints, and I was tired.

He seemed to sense my anxiety and quickly replied, "Well, I've got a couple of tag-recaptures from blue sharks that you might be interested in."

That certainly got my attention. I made my way to the dock beside his boat and told him that I used to work for the Shark Tagging Program. "I'm definitely interested in those recaps," I said. "What do you have?"

He invited me aboard, and I climbed down into the cockpit, where I was greeted with a firm handshake, closely followed by a beer. The man introduced himself as Chuck Walker, and I learned that Chuck owned an excavating company in Belchertown, Massachusetts.

"I'm just a ditch digger," he said, "but I love shark fishing." He told me that he'd fished the first and last day of the tournament. On day one, he said, he trolled for tuna with no luck, so he switched to sharking on the third day, hoping to land a mako for points. Instead he caught and released thirty-five blue sharks between 8:20 in the morning and 2:54 that afternoon. Two of the thirty-five sharks were previously tagged. I was still trying to come to terms with the sheer amount of data this stranger was providing when he casually let slip that a sixteen-foot, two-thousand-pound white shark had swum right by the boat but wouldn't take a bait. *Is this guy for real?* I wondered.

A week later, I saw Chuck again at the Oak Bluffs Monster Shark Tournament. As the event was wrapping up, he approached me and invited me over to his house for dinner. I didn't know it when I first met him over on Cuttyhunk, but Chuck had a house on the Vineyard in Oak Bluffs, and he came out every weekend from western Massachusetts with his girlfriend, Linda, and his son Ben to fish for sharks. He was simply hooked on shark fishing. During the tournament, they'd landed a 250-pound mako, and he was planning to cook up the steaks. I readily accepted the invitation.

Chuck did everything on a grand scale, and the meal he prepared that night was no exception. We ate, drank beer, and he told me all about his shark-fishing experience.

"Most people think fishing for sharks is pretty mindless," he said. "They think you just head offshore, park, and chum. But that's not true." He sipped from his beer. "No. Fishing for sharks is like hunting. You need to strategize and think about where you fish and how you fish in order to catch big sharks, makos in particular." Chuck explained that it was all about fresh bait—he told me he used bluefish—and picking the right location based on water temperature, depth, wind direction, and tides.

I hung on every word Chuck was telling me. This perspective was invaluable, and it sure as heck wasn't the kind of thing you'd learn from a biologist in a lab. Chuck was obsessive the way some researchers I know are obsessive, and he never stopped thinking about how to do things better. As the empty beer cans accumulated and the stories continued, I knew I wanted nothing more than to go shark fishing with Chuck. Shortly before I left for home, my wish came true, and Chuck invited me to go with him on a future outing.

Chuck kept his Sea Ray, the *Fish N Boat*, on a mooring in Lagoon Pond close to my office at the lobster hatchery, and the next morning I met him there to head offshore and fish for sharks. Almost all of my angling experience since moving to the Vineyard had been from shore fishing for stripers, bluefish, and other inshore species. I was beginning to feel confident in my abilities, but I had no real offshore experience with big-game fish like billfishes, tunas, or sharks. I knew Chuck was knowledgeable, but I learned on that first trip that he was also patient and an excellent teacher. Plus, I just enjoyed his company. Working for Jack, we'd landed plenty of sharks on the *Wieczno*, but I quickly learned from Chuck that landing a shark taken on longline has very little in common with landing a shark on rod and reel. Some sharks go deep and require little more than brute force and time to bring them to the boat, but other sharks, like makos, take the bait with a ferocity that turns five hundred yards of fifty-pound test line into a smoking blur as the fish runs. Soon we started fishing together on many weekends, and with each successive trip, my confidence and knowledge grew. In addition to learning about shark fishing, I was also learning to love it. It was some of the most fun I'd had in years. Little did I know it when the Marlboro Man had called out to me on the dock at Cuttyhunk, but Chuck Walker would become a close friend who would teach me a lot about shark fishing, shark biology, and how to enjoy life to the fullest.

⚓

The 1990s turned out to be an interesting decade with regard to sharks. There was increased awareness that sharks were in trouble, but the *Jaws* effect was still hard to shake. An increasing number of shark anglers considered themselves conservationists, and they believed their engagement in shark fishing wasn't harmful to sharks so long as they followed the regulations—with some even believing that shark fishing was actually helpful for sharks. Many believed that by legally engaging in shark fishing, they were helping sharks by participating in surveys and tagging programs. I tended to agree—but at the same time, grassroots activism was on the rise and animal rights advocates increasingly viewed the pursuit of "sport" or "game" fish as nothing more than an outdated hobby. While I didn't agree with that sentiment, I did worry about perceptions. A 1991 article in the *Boston Globe* described shark heads lined up on the dock at the Monster Shark Tournament as "decorated with such cutesy adornments as hats, pipes, and cans of Budweiser in a kind of humiliation ritual of the fallen enemy." One couldn't really argue in good faith that these scenes were all about the science.

At the same time as the animal rights activists were condemning shark-fishing tournaments, high-profile issues like shark finning were being usurped to paint all shark fishing with a broad, ugly brush. It didn't help that fisheries data continued to show overfishing was still occurring in some shark fisheries even after the first federal regulations went into effect. In terms of white sharks specifically, scientists around the world expressed concern but agreed that the lack of data made it hard to know the species' actual conservation status. In 1991, South Africa became the first country to ban the intentional killing or sale of white sharks. Namibia followed suit in 1993, and a few months later California became the first US state to provide protection for white sharks. Since beginning my work on Martha's Vineyard, I had not personally encountered a white shark, although there had been exciting reports.

In January 1995, I had the opportunity to travel to Australia to cage dive with white sharks as part of a research project with Ian "Shark" Gordon, one of the many colleagues I had met at the annual meeting of the American Elasmobranch Society. Ian was starting a cage-diving business and wanted some help, so I seized the opportunity. On the way to Australia, I connected with west coast white shark researcher Ken Goldman. We flew together to Hawaii, where we stopped for a day and did some shark tagging with Chris Lowe, who was getting his doctorate degree under well-known, Casey-like biologist Kim Holland. After that we were on to Sydney, where Ken and I met with Ian and his crew, drove to Port Lincoln, loaded the boat, and steamed to the Neptune Islands, which is where Peter Gimbel had great success in filming *Blue Water, White Death*. To say I was excited would be an understatement.

We chummed for days, which turned into a week, and saw nothing. We were getting so bored that we decided to drop one of the cages to the bottom, about fifty feet below, to harvest abalone (a type of marine snail) for dinner. I volunteered to stay with the cage while the other divers searched the bottom for our quarry. There was an opening in the bottom of the cage through which I could "walk" the cage along the bottom to follow divers in case a shark showed up. For obvious reasons, I didn't need fins to do my job, so I left them on the surface.

Once on the bottom, the other divers left the cage and spread out in multiple directions. I was left behind in the cage, realizing there was no way to move it about, given the boulders and rocky outcroppings. Besides, everyone was too far away in too many directions to possibly keep up. So I decided to leave the cage and hunt for abalone myself—there was no point in sitting there like an idiot. Over the next thirty minutes, I wandered farther and farther from the cage, stuffing abalone into my buoyancy compensator. Suddenly I heard a muffled sound and looked up to see Ian swimming in my direction hooting through his regulator. He was holding his open hand straight up on the top of his head—the universal sign of a shark. Ian quickly turned and jetted back to the cage using his quick, efficient kicks with his fins. I had no fins. I'd never run so fast underwater, all the while thinking a white shark was going to snap me up from behind. Getting closer to the cage, I could see all the divers safely inside staring out at me and looking very scared. That didn't help my anxiety. Those last few feet took the longest, but I made it.

Safely inside the cage, we sent a small emergency float as a signal to the surface to be lifted from the bottom. On the way up, I saw nothing, just blue water. Once we were floating on the surface, I wanted answers.

"What the hell did you see, Ian?" I screamed.

He was breathing heavily but blurted out, "It was a big bronzey!"

The bronze whaler is a common shark in that part of Australia. It could be dangerous, but it's not a white shark. I was somewhat relieved and disappointed at the same time. As much as I didn't want to be bitten, I was really hoping that a white shark had finally found our slick. As we exited the cage and started to climb about the boat, a crewman on the other side screamed, "White shark, port side!" I thought he was joking, but the look on his face proved he was not. I didn't have time to think about the fact that we were indeed swimming about on the bottom with a white shark in the area. Instead I yelled out, "Back into the cage!"

Over a three-week period, we ultimately saw two white sharks, and I got to cage dive with both of them. The experience of being in the water with live white sharks was incredible. I was absolutely captivated by their size and, more specifically, their girth. It was incredible to me that something that large could move so gracefully in the water. The sheer beauty and behavior I observed were in sharp contrast to the marauding man-eating machines portrayed on television and in movies. It was a defining moment in my career, not only because I was seeing the animal that motivated me to be a marine biologist, but also because I realized that while amazing, I still had so much to learn about other lesser-known and less gregarious shark species on the Vineyard. In many ways, the trip to Australia provided a kind of closure for me. There was no doubt that the white shark still captured my imagination, and I was thrilled to have the opportunity to work with them in the wild, but I understood that at this point in my career, I was just as excited for the niche I'd created for myself back home.

My memories of cage diving with white sharks in Australia receded in deference to the day-to-day work upon returning to the Vineyard, but it all came rushing back in August 1996, when two commercial fishermen fishing out of Beverly, Massachusetts, caught an eighteen-foot, 2,500-pound female white shark twenty miles offshore. The fish drowned in their gill net, and I was eager to perform a necropsy, but I was unable to get there from the Vineyard. My friend Brad Chase ended up heading out from the Cat Cove Marine Laboratory to do the dissection. He called me later and told me he found a partially decomposed, thirty-one-pound porpoise, the remains of two dogfish, and a rock crab in the shark's stomach. Brad told me that a large crowd had gathered as the scientists worked. Many marveled that a white shark that size had been swimming just offshore. Over the years, fewer than ten adult white sharks from the northwest Atlantic had been sampled by scientists, and they remained a rare-event species, making this shark a big deal. It was, in fact, likely the largest female white shark ever sampled in the northwest Atlantic. Ken Sherman, the director of National Marine Fisheries Service, told the media, "It will likely be some time before we see a great white in these parts again."

By 1997, it became clear that the shark management plan put in place by the US government in 1993 was not working as intended. As a result, federal fisheries managers stepped in and tightened restrictions. In a statement, a spokesperson for the National Marine Fisheries Service said, "Atlantic shark populations are at a precarious state and fishing pressure needs to be reduced." Most notably, the new regulations reduced the commercial large coastal shark fishery quota by half. Recreational anglers were now limited to two sharks per vessel per trip for all but the coastal Atlantic sharpnose shark, of which each fisherman could take two. Fisheries regulators also banned direct fishing for five shark species considered extremely vulnerable: the whale shark, the basking shark, the sand tiger shark, the bigeye sand tiger, and the white shark.

CHAPTER 7

LIGHTS, CAMERA . . . WHITE SHARK!

The chances of encountering [a white shark] up [around Cape Cod] are markedly lower than in other parts of the country.

—George Burgess, director of the Florida Program for Shark Research at the Florida Museum of Natural History

We know there are great white sharks off South Australia, South Africa, and the coast of California. Those are the hotspots for great white sharks, where attacks are most frequent. In the Atlantic, the shark is enigmatic. We don't have access to the animal in predictable ways.

—Greg Skomal

AUGUST 17, 1984—51°22'01.2"N, 55°28'47.7"W (OFF NEWFOUNDLAND)

The adult white shark has traveled some 3,000 miles over the past several months, from the Gulf of Mexico to the northern tip of Newfoundland. He has made the trip for the same reason that French and Basque fishermen sailed 2,000 miles across the Atlantic in the early sixteenth century. The shark's ancestors understood the seasonal proliferation of marine life in Newfoundland's cold waters long before the Europeans did. White sharks were already here when the Norse constructed their settlement at L'Anse aux Meadows in the eleventh century. In fact, they were here long before that—thousands of years before the Norse arrived, when indigenous peoples occupied this land they call Ikkarumiklua and Ktaqamkuk—"enormous shoal" and "land across the water."

Wabinmek 'wa, as the shark is called in the Mi'kmaw First Nation language, instinctually comprehends—and has come to depend on—the summer abundance of the ocean. Seven thousand years ago, the shark's ancestors knew it, and they hunted here alongside Mi'kmaw hunters, often targeting the same prey. The human hunter, chasing a swordfish, poised with harpoon in hand aboard his seagoing canoe, knew the cut of the white shark's fin. He knew the rush of water around the shark's broad back, the explosive force of a breach, the decisiveness of the attack, and the devastating aftermath written in blood-stained surf. The Mi'kmaw hunter and the shark were both apex predators. Conspecifics by action, if not taxonomy. They shared an elemental understanding of the ecosystem in which they hunted. The human hunter admired and respected the shark's predatory prowess. He sought the predator's perspective. He sought to emulate the shark—even shifting between realms of existence. A shark tooth worn around his neck. Shamanistic transformation.

As the male shark passes the entrance to Hare Harbor, Newfoundland, he encounters a nearby boat. Like skates and rays, the shark possesses a unique sensory ability—the ability to detect electrical fields by way of small mucus-filled chambers loaded with sensory cells and peppering the animal's snout. Named for the seventeenth-century scientist who named them, each ampulla of *Lorenzini* is connected to the surface of the shark's skin by way of a tubelike canal. Collectively, these electroreceptors can detect electrical fields as low as one billionth of a volt—the tiny pulse of electricity generated by the muscle contractions of another animal, for example. Because of this, the shark can more efficiently hunt at night, find a nearby fish hiding in the sand, or, as is the case today, be attracted to the electrical pulse emanating from the slow trolling of the boat's motor. He follows the boat, curious.

While the presence of shark teeth, especially wabinmek 'wa teeth, in mortuary and ritual contexts, suggests the Mi'kmaw's respect for the shark, there is no doubt they also feared it. As Lkimu, a Mi'kmah shaman, said, "The bad fish that often infest these seas do not allow us to travel without worry and without fear.'"[*] An open water hunt into cold, perilous waters in bark canoes, pursuing animals like swordfish and seals, represented the zenith of the human hunters' competency. Entering the shark's realm and competing there in the most rudimentary contest for sustenance brought with it the possibility of great reward but also of total disaster. The shark's teeth, reads a firsthand account of one such voyage, are "made like gardeners' knives for cutting and boring, or like razors slightly bent."[†] Amidst the slate-gray swell of the North Atlantic, those teeth give the shark the upper hand, capable of rending bark canoes "open . . . so that they sink to the bottom." Yes, there is comradery and respect, but there is also fear.

Aboard the boat leaving Hare Harbor are fishermen armed with a cadre of tools that would be unrecognizable to the ancient Mi'kmaw hunters. The toggle harpoon, which twisted in the flesh to prevent a harpooned animal from escaping, represented the apex of technology at the time of the shark's ancestors, but it pales in comparison to liquid crystal display fish-finders, soft plastic lures containing water-soluble, fish-attractant compounds, and lightweight polyvinyl-injected reels with smooth drags and adequate gearing for big gamefish, even large sharks. These tools aboard a fiberglass hull powered by an engine represent how humans have leveled the playing field. How they have tipped it decisively in their favor.

The shark comes alongside the boat, interested in the pulses of electricity generated by the motor. The men race to the transom. The shark is still a thing to be feared, but the fear is largely conjectural. The shark is seen as a monster, but even more so, as a curiosity as it circles the boat. Should they attempt to catch it? What sport! Despite the advances in technology and shark science, these fishermen can't understand the shark in the same way as the people who buried a woman with thirty-six shark teeth not far from here at Port au Choix four thousand years ago. They can't comprehend it as the two humans interred with white shark

---

* "Lettre de M. l'abbé Maillard sur les missions de l'Acadie et particulièrement sur les missions micmaques," Les soirées canadiennes; recueil de littérature nationale (Québec), III (1863), 289–426.

† de Paul, Vincent, 1886, *Memoir of Father Vincent de Paul: Religious of La Trappe*. Translated by A.M. Pope. J. Coombs, Charlottetown.

teeth around their necks some two thousand years ago at Ministers Island on Passamaquoddy Bay, New Brunswick. There is knowledge and then there is *knowledge*.

The shark determines the boat is not food, and with a dismissive flick of his caudal fin, he disappears into the North Atlantic. He will hunt, grow, and reproduce, utilizing the same anatomical and behavioral traits that allowed his ancestors to dominate the ocean long before the first humans inhabited these shores. Provided that food remains plentiful and environmental conditions permit, his descendants will remain at the top of the food chain for many years to come, with few natural predators besides humans.

My first real experience with television came in 1988, when ESPN filmed the Oak Bluffs Monster Shark Tournament. As the second tournament approached, I started to have more frequent contact with tournament organizer Rick Allen to discuss logistics, including shark minimum sizes, scientific sampling, and the distribution of tags to participants. In late June, Rick called and said that ESPN was planning to produce a one-hour special on the tournament that year.

"Really?" I said, the excitement probably obvious in my voice.

"Yeah," he said, "they want to follow a few tournament participants while they're shark fishing, and I think they'll probably want to talk to you about local shark species and your scientific studies."

"That's fine with me, Rick," I said, "but who exactly will be reaching out?"

"The whole idea of covering the tournament was conceived and sold to ESPN by a California production company called Kerwin Communications," Rick said. "The owner, John Kerwin, or one of his people will probably be in touch."

I had never worked on any kind of film production and had no idea what that would entail, but I immediately thought back to what Jack had told me about ABC television sending a film crew in 1979, when Frank tracked the white shark off Long Island. The executives at ABC had calculated that investing in Jack's and Frank's white shark research would yield footage for their popular *American Sportsman* show, which in turn would bring more viewers to ABC. Could this potential relationship with ESPN yield a similar result?

Three weeks later, I got the call from John Kerwin. After telling me a bit about his production company, he elaborated on the concept for the show.

"I don't just want this prime-time special to show sharks being captured and landed by tournament boats," he said. "We'll get the tournament footage we need, but it would be great to get you or another biologist to discuss the biology of sharks, showcase them in the natural environment, and perhaps even show you diving with them."

Assuming Kerwin Communications on behalf of ESPN would have the boat, the cage, and the cameraman, this sounded really exciting to me. It would be a great opportunity for me to talk about sharks on camera and draw some attention to our research, which I was sure my bosses would like.

"Sounds great, John," I said. "I've got a lot of experience diving with local shark species. Just let me know when and where and I will jump on your boat."

"I wish it was that easy, Greg, but I was hoping you could actually help me find a boat, a shark cage, and an underwater cameraman."

What? My head was spinning. This was ESPN. Didn't they have people who did that stuff? I took a breath. I knew people, and this would be a great opportunity. "When are you thinking?" I asked.

"It would be great to shoot this the day after the tournament ends because my crew will already be in town."

It was already July 12, and the tournament started in less than a month. That didn't give me much time. Logistically, this was going to be incredibly challenging, especially since this was the late 1980s, when there were no commercial shark cage diving operations in New England, no cell phones, and home video cameras were a rarity—let alone ones that could film underwater. My first instinct was to call Wes Pratt. I knew Wes had a shark cage because I'd been diving in it. I then thought I could call Lance Stewart, a friend of Wes's who shot underwater film. It took me a day or two to reach them, but neither one of them was available. They suggested I reach out to Billy Campbell and Brian Skerry, my other cage-diving buddies, but they weren't available, either. My next stop was the local dive shop a few miles from my office, Vineyard Scuba. Since I'd moved to the island, I had developed a strong working relationship with the shop owner, Joe Leonardo, and his son, Joe Jr. They suggested I reach out to local cameraman and dive instructor Adam Geiger to see if he was available. Two days later, I was able to connect with Adam. He had the equipment and the experience filming sharks, but he also had a tight schedule. He asked about the cage. I had no answer, but I had an idea.

I could build one.

I called John Kerwin back. "I've got a cameraman, but he's only available a couple of days before the tournament."

"That might work," John told me. "Maybe we can shift the filming date for that sequence. Give me Adam's number and I will reach out directly. What about the cage?"

"There are not many shark cages in New England, John," I explained. "I can only think of one, and it's not available. But I'm pretty sure I can build a cage if you're willing to pay for it."

"Really?" he replied. "How much would that cost?"

I'd already done my homework, hoping that John would be amenable. "I think we can do it for about five hundred dollars," I said.

"Go for it!"

It was July 26, ten days from the start of the tournament and only nine days from the proposed date for filming. I had no time to waste. I planned the cage to be four feet by six feet and six feet tall. It would hold two people and have an aluminum frame covered in one-inch by two-inch steel turkey cage fencing with a side door and a top hatch. Flotation would come from four inflatable poly balls secured to the top four corners of the cage with nylon line. Most of the materials had to be ordered from off-island vendors, but they arrived in time for me to put it all together over the weekend. I recruited help from my colleague and office mate, state lobster culturist Kevin Johnson, who had all the right tools.

For a boat, I reached out to Bob Clausen, who fished his boat the *Normally Fun* in most of the big-game tournaments with his family. Bob agreed to provide the platform for filming with Kerwin's crew, Adam Geiger, me, and the cage. All the pieces were coming together, but juggling each of these parts of the shoot with my normal work responsibilities was stretching my nerves a bit. I kept my boss, Randy Fairbanks, in the loop. As a regulatory agency, the Massachusetts Division of Marine Fisheries was more often than not cast in a negative light, and so Randy supported anything that yielded positive publicity. He even offered to pay for the cage, but, just as Jack and Frank had done, I wanted to leverage the production company for that.

Finally, on August 3, we all boarded the *Normally Fun*, steamed out of Oak Bluffs Harbor, and headed south. The location I'd chosen was about thirty miles south of the Vineyard and was called "The Fingers" because the 180-foot depth contour is shaped like digits on a hand. To be honest, I was amazed that it had all come together, and I was really excited to finally be on the water. We chummed all day, and saw a lot of marine life—dolphins, whales, white marlin. We even saw a very large shark pass right under the boat, which rattled both Adam and me because we thought it might be a white shark, but no other sharks appeared all day. We decided to move back north to where Clint Allen was fishing. He had a small sandbar shark around his boat, so we jumped into the water and filmed it for about fifteen minutes with no need for a cage.

Our lack of success was reflected in the tournament standings. Very few boats that year caught the typically abundant blue and mako sharks. Most of the participants attributed this to warmer offshore water pushing these temperate sharks to other areas. Where we had been trying to film, the water temperature was pushing 80°F, which made it comfortable for us but not for the sharks. I was worried the whole thing was a failure, but the show did incredibly well. According to John Kerwin, it was one of the most viewed shows on ESPN that year.

In addition to being the point man for the production company, it was my first real experience on camera. I was so overwhelmed by the tasks at hand that the actual time in front of the camera didn't faze me. The trick, I learned, was to speak intelligently in short sound bites, and the whole experience gave me a taste of what it was like to work with a production company with unpredictable animals in an unpredictable environment.

. . . and I got a shark cage out of the deal.

The Boston Sea Rovers is one of the world's oldest dive clubs—past members include huge names in marine science such as Jacques Cousteau, Eugenie Clark, and Robert Ballard. Every year since 1955, the club sponsors a dive clinic, which includes an evening film festival showcasing some of the biggest names in underwater film production. Wes was the first to introduce me to the Rovers, and I began attending the clinic in the late 1980s. It was at one of these shows that I met the underwater cinematographer Nick Caloyianis. There is no shame in saying I was starstruck by Nick, who had a direct connection to *Blue Water, White Death*. Nick was an undergraduate working with Eugenie Clark at the University of Maryland during the filming of *Blue Water, White Death*, but he later formed an alliance with the film's director, Peter Gimbel, and it was Peter, another Sea Rover, who inspired Nick to pursue a career as an underwater cinematographer instead of as a shark biologist. Peter then hired Nick as a cameraman on his 1981 film expedition, *Andrea Doria: The Final Chapter*.

Nick went on to become well known for his underwater cinematography of natural history subjects, and in 1989 he was working on a National Geographic film called *Shark Trackers*. Through the Rovers, Nick was friends with Wes, and so he came to film some of the footage for *Shark Trackers* on one of Wes's cage-diving trips. I had already moved to the Vineyard, but I frequently returned to dive with Wes, and I sure as heck wasn't going to miss an opportunity to see Nick in action. Although I was only in the film for a moment, helping to load the cage on the boat, the excitement of being part of a production with a guy like Nick really whet my appetite for more. After my 1988 experience with ESPN and Kerwin Communications and then my brief work with Nick, I was hooked. The Boston Sea Rovers continued to be a great networking opportunity, and I was lucky to be involved in several more television shows throughout the 1990s, including another ESPN show about the Monster Shark Tournament in 1990, a show called *World's Most Dangerous Animals III* in 1997, Discovery Channel's *Sharks of the Deep Blue* in 1998, and *Fishing with Charlie Moore* on New England Sports Network in 1999. All of these experiences were exciting to me, and each helped build my network, but it wasn't until 1999 that my television career got a serious boost.

It was another February day at the office during my off-season. That meant data analysis, writing, and fielding a few phone calls. Most folks don't realize that the life of a shark biologist is not always teeming with dramatic acts of high adventure. No, on that February day, I was not hoisting a bull shark out of the water by its tail or diving on a basking shark to jam a satellite tag into its back. It was one of those days that consumed half my year—a day filled with the adventure of a different kind. While collecting the data is certainly the most fun and dangerous, making sense out of them is far more difficult. Key punching and crunching numbers, reading scientific publications, testing hypotheses—this is the science.

February on the Vineyard is not an exceptionally cold month. Temperatures generally range from the low twenties to the mid thirties, but the relentless northeast wind makes your bones brittle. I was compiling some behavioral data from a blue shark tracking study when the phone rang. I answered with my usual nonchalant "Marine Fisheries" greeting, only to hear a single word, uttered by a diminutive voice.

"Gampy?"

"Excuse me?" I said.

"Gampy, is that you?" was the response.

"No, I'm sorry, you have the wrong number."

"Will you take me fishing, Gampy?" asked the persistent caller.

Being the good civil servant, I had to maintain a level of professionalism. Hanging up was not an option. But something was not right about that voice—it sounded a bit contrived, and I sensed I was being had. I needed to tease the potential culprit out from behind the rock, like waiting in the bushes for kids that just egged your car. The best way to do that, I decided, was to play along.

"No, I'm sorry," I said, "but you do have the wrong number and your grandfather is not here." I couldn't push it too far. As a regional biologist for the agency, I get all kinds of phone calls, like the one from the mother who was afraid to let her kids swim at the beach or the one from a fisherman who found a one-eyed glob of flesh on the beach. Certainly, a child wanting to fish with his grandfather is not completely out of the question. What if I was wrong? But that voice, it rang with a certain familiarity.

"But Gampy, do you want to go to the Arctic Circle to study Greenland sharks?"

This response made absolutely no sense to me and my voice waffled from a weak *Huh? What the* . . . to a mix of anger, surprise, and embarrassment as I realized who it was. "Holy crap! Benz! You son of a bitch, you got me hook, line, and sinker. Why do I fall for your antics every time? You would think that after sixteen years of this I would know your voice and . . ." My voice trailed off to the sound of his laughter as I started to realize what George had actually said. Sharks in the Arctic Circle? That could only mean one thing: we got the funding.

"Hold on a second, George, are you saying what I think you're saying?"

"Yeah, we got the money" was his simple answer.

George was talking about going back to Baffin Island in the Canadian Arctic to study Greenland sharks. He had already been there once with Nick to examine their unusual parasites. That work appeared in an article Nick wrote for *National Geographic* about Greenland sharks, in which he showcased the first underwater images of the species ever taken. Since that trip, they had been working hard to sell the idea of going back to continue the research, and their efforts finally paid off. The new research would be funded by Discovery Channel for a Shark Week show called *Jurassic Shark.* Was I interested in a chance to study one of the most unstudied sharks on earth with George and Nick? "Heck, yeah!" I said without hesitation.

When George pitched Nick on the idea of bringing me, Nick wasn't entirely sure I was the right guy. Nick had met me on a couple of dive trips with Wes and his shark cage aboard the *Snappa,* and his recollection was that I was a shy and reserved kid. Was I really the right person for this type of expedition? George assured Nick I would be a great choice, and ultimately Nick agreed and called me. I had not worked one-on-one with Nick in the past, so these early conversations were pretty intense. Nick pays incredible attention to detail and rightly believes that by failing to prepare, you are preparing to fail—which could be deadly under Arctic conditions. What George hadn't mentioned was that Nick wanted me to dive with Greenland sharks under six feet of Arctic ice. As much as I liked George and wanted to work with Nick, if George had told me that I'd be diving under ice, I would have said, Absolutely not! Nick assured me that I would be ready for the expedition if I prepared properly. So, at Nick's direction, I quickly bought a dry suit, took a refresher diving course, and tested all my equipment with Wes and Nick at a deep-dive sight off Rhode Island. Then, in early May 1999, I was off to Arctic Canada.

Although it was just a short segment in a larger film, it was a thrill to see the final production during Shark Week. Besides providing some of the brute-force labor to burrow a hole through the sea ice, I'm seen piloting a remotely operated vehicle (ROV) to film a Greenland shark eating a seal and then diving beside a large female Greenland shark we had just tagged. My dialogue in the film is largely unscripted, but I can be heard discussing the depth of the ROV I'm piloting and observing the shark's location under the ice based on the acoustic signal the tag is transmitting. For those who know me, they say that my voice is unmistakable. In the final segment, there is only passing mention about the parasites that are attached to the shark's eyes, which is the primary reason for George's interest in the shark, but therein lies another lesson for me. Without the funding from Discovery Channel, the expedition would probably not have happened. Even though George's parasites play an even smaller role than me in the final cut, the expedition resulted in the first acoustic tracking of any animal under Arctic ice and the first insights into the behavior of this species. It was a win-win for science and television, and just like Frank Carey's tagging of the white shark back in 1979, resulted in the publication of an important paper. The whole segment including the Greenland shark expedition is less than five minutes in the film, but for me, it's a game changer.

The following summer, George was working on a film about parasites called *Body Snatchers* for National Geographic Explorer. This time he was working with a filmmaker named Nick Stringer, of Big Wave Productions. The film is narrated by George, who is introduced as a "parasite enthusiast" and takes a deep dive into several different parasites, including a copepod that lives in the nose of a blue shark. George once again reached out to me for help, and during the summer of 2000, the film crew came to Martha's Vineyard for two weeks and I helped them get the footage they needed. It was great working with George again, and getting to know Nick Stringer was another great opportunity for me. Both Nicks would prove central to my later television work, which would ultimately help to fund my research.

Basking Sharks

In late September 2001, the United States was in flux after the terror attacks on the World Trade Center and the Pentagon. I was living in Chuck Walker's house on the Vineyard at the time, and for a few days following the attacks, it seemed that time stood still, but slowly life resumed. For me this meant getting back out on the water. At this point in my career, the shark that most captured my attention and curiosity was the basking shark. It was a shark that showed up every summer off the coast of Massachusetts, and it was the species most frequently mistaken as a white shark. The cool thing about basking sharks is that just enough is known about them to intrigue even the most casual shark enthusiast. Given its massive size, which commonly exceeds twenty-five feet, and its planktonic diet, I always thought of a basking shark as a whale trapped inside the body of a shark. Behind the whale shark, it is the second-largest living shark species and one of only three known species to feed on plankton.

My goal at the time was to begin tagging basking sharks with pop-up satellite archival transmitting tags, commonly called PSAT tags. These high-tech tags were relatively new at the time. Once on the shark, each tag collects and archives water temperature, depth, and light-level data every ten seconds. With a built-in clock, they can be programmed to detach from the shark after periods of up to a year. Once they pop up (that is, pop off the fish and float to the surface), they transmit the archived data via satellite to an email address. The researcher can then rebuild the three-dimensional movements of the shark using the data. These tags revolutionized the science of movement ecology and, for me, they were a way to study the behavior of live sharks.

A couple of years before this, my friend Brad Chase and I had used PSAT tags to study bluefin tuna with another researcher, Molly Lutcavage. Now I really wanted to use them on basking sharks to try to answer a pressing question: Where did they go in the winter? Nick Caloyianis had also developed a fascination bordering on obsession with basking sharks, and he wanted to create a film about them. Following the trip to Baffin Island, Nick and I had become friends, and I saw an opportunity à la Frank Carey and ABC circa 1979—I could bring the research and Nick could bring the cinematography talent and funding.

To locate basking sharks to tag, we enlisted the help of Captain Billy Chaprales aboard his boat the F/V *Rueby* and a spotter pilot named Tim Voorheis, known to many as Wilderness Dave. I started working with Billy back in the mid-1990s when Brad Chase and I were heavily involved with studying the physiology and movements of bluefin tuna. We used a harpoon vessel because those are well designed to sneak up on fish. In the traditional harpoon fishery for bluefin, the harpooner stands on the pulpit well out in front of the noisy part of the boat, and the vessel is piloted from a tall tower, which allows the boat driver to see the tuna and maneuver the boat into position. In most cases, the vessel partners with a spotter plane who assists with initially locating and putting the vessel on fish.

Brad and I had found that working with Billy allowed us to tag free-swimming bluefin tuna without exposing them to capture stress. If you are going to study behavior, it's best not to stress out the fish. To shift from killing fish to tagging them, Billy adapted his harpoon to become a tagging pole. Instead of a standard harpoon dart, which is called a *lily* and is designed to kill fish, he used a small tagging needle so as to reduce the level of skin penetration and minimize tissue damage.

For this project, we wanted to tag basking sharks underwater so Nick could capture it on film. Billy's commercial harpoon fishing boat was called the *Ezyduzit*, but we elected to use his lobster fishing boat, the F/V *Rueby*, instead because it provided more room for dive and camera gear. In order to tag the shark underwater while diving, I had designed a small hand tagger that was light and could hold the PSAT tag. With the help of Wilderness, we'd located a group of ten to twelve basking sharks about forty-five miles east of Nantucket. Once Billy got us near them, we rolled off the back of the *Rueby* and I tagged a twenty-foot female while Nick filmed. It was the beginning of a project that would occupy much of my time over the next several years.

That tag was programmed to pop off the shark on February 14, but it detached prematurely in early December off the coast of North Carolina—regardless, it gave us fifteen good position estimates that allowed us to reconstruct the shark's journey. By early October, she'd traveled from the Gulf of Maine in a southwesterly direction to the Mid-Atlantic Bight, east of New Jersey. As continental shelf waters cooled, she continued her southern journey. On October 21 she was on the outer shelf southeast of Delaware Bay, and by the first of November she had reached the Outer Banks of North Carolina, where she stayed for the next month, up until the time the tag released.

While we failed to answer the question of exactly where basking sharks go for the winter, we did help to dispel a belief that basking sharks hibernated once they left northern waters. Although our results represented the movements of only a single shark, the novelty of the work allowed us to publish these findings in 2004, with Nick as a coauthor. More importantly, this pilot study allowed me to garner additional grant funding and we went on to tag more than fifty basking sharks over subsequent years.

A Fish as Rare as a Unicorn—2002

The F/V *Unicorn*, a seventy-five-foot steel fishing boat, had steamed out of Menemsha Harbor on Martha's Vineyard early on Monday, August 6, 2002. At the helm was fifty-six-year-old Captain Gregory Mayhew. His son Jeremy was aboard as crew. When they hauled their trawl net about two miles off the green buoy outside Menemsha, they found a shark in it. The shark was about five feet in length and Captain Mayhew thought it was a porbeagle, a species long prized for both its meat and its fins. While there is not a big market in the US for porbeagle, it is commonly exported to Europe and East Asia, where it fetches among the highest price at market of any shark species. As a result, porbeagles caught as bycatch are often worth the hassle of landing.

When the *Unicorn* returned to Menemsha Harbor later that morning, Karsten Larsen, another local fisherman, immediately identified the shark as a white shark. Since 1997 the white shark had been prohibited from retention in federal waters, and Captain Mayhew, worried about running afoul of the law, immediately put the fish in a pen filled with seawater. Using the harbormaster's phone, Mayhew called me—but I was in Delaware with Wes Pratt, Nancy Kohler, and a student named Abbey Spargo, working on a sandbar shark postrelease mortality study. I asked my friend Jesse Fuller, whose brother, Bobby, had worked for me, to drive over to Menemsha, take some pictures of the shark, and email them to me. When I saw the images, there was no doubt in my mind that the fish was a white shark. I estimated the shark was just a few months old.

In 2002, a white shark near the Vineyard was a rare sight, and the unusual catch generated a lot of curiosity and excitement. People came down to the harbor to see the shark before it was released back into Vineyard Sound near the Elizabeth Islands. I was interviewed by the *Vineyard Gazette*.

"It's an extremely rare species," I told the reporter. "We don't see a lot of these fish. We know they swim in New England, but it's unusual to see one. We know there are great white sharks off South Australia, South Africa, and the coast of California. Those are the hot spots for great white sharks, where attacks are most frequent. In the Atlantic, the shark is enigmatic. We don't have access to the animal in predictable ways."

In the sixteen years of the Oak Bluffs tournament, nobody had ever landed a white shark, and people wanted to know what the shark landed by the *Unicorn* meant. I theorized that the white shark may have been attracted to the area by a dead whale. "We know in Southern California they feed on marine mammals like seals and sea lions," I told the reporter. "In the Atlantic they don't seem to work intensely on marine mammals. We think it's very different. . . . We know that juvenile great white sharks feed on bluefish and other schooling fish. When they get to be big enough, they eat larger animals. They don't want to eat popcorn when they can afford something bigger. We think it's dead whales."

But of course the big question lingered on everyone's minds: Should the beachgoing public be concerned about shark attacks? I thought back to hearing Jack Casey handling the media. I wanted to give accurate information but also avoid creating any unwarranted hysteria. I explained that a whale had washed up on South Beach earlier in the week and that the shark might have been attracted to the easy meal. "My concern is that when you put a dead whale on the beach," I cautioned, "you increase the probability that there will be shark and human interaction."

My message, which would later become my mantra, was that common sense dictates not to go swimming around a white shark's food source.

A Knock at the Door—2003

It was a Monday night in August, a little before midnight, when the police pulled up to my house. The officer at the door had a handwritten note instructing me to call the Barnstable police on Cape Cod as soon as possible. There was a report that a white shark had washed up on Long Beach, and public safety officials wanted to get ahead of the news. They knew, as I did, that it didn't take much for a shark report to turn to panic, and the last thing a place like Cape Cod, Martha's Vineyard, or Nantucket needed at the height of summer was shark hysteria.

Be careful when wishing to be the real Matt Hooper.

I called the Barnstable police station and listened as the officer described the shark. I asked a few pointed questions. I told him that I believed the five-foot, eighty-pound shark was most likely a brown shark or a dogfish. He apologized for bothering me, and I assured him that it was no trouble at all. He'd done the right thing. "It's best to get ahead of these things," I said.

A photograph the next day confirmed that it was a sandbar shark, locally called a brown shark, that posed no risk to beachgoers. Even so, the rumors began to circulate. I had come to appreciate in my sixteen summers on Martha's Vineyard that this was the time of year when shark sightings were relatively common. As the regional biologist and the state's shark expert, I tracked down each report and researched it in an effort to allay fears. It's why I had started studying sharks in the first place, and it's what drew me to white sharks specifically. I wanted to set the record straight. Yes, *Jaws* was filmed on Martha's Vineyard, and yes, white sharks showed up offshore, but they were not the killing machines the movie and the local media made them out to be.

When a white shark was spotted ten miles off Chatham, I was interviewed on television with a line of questioning focused on the risk the shark posed to beachgoers. It doesn't pose a risk, I said. A local reporter agreed, writing in the *Cape Cod Times*: "While not impossible, the likelihood of a shark attack on Cape Cod is improbable. The last fatal shark attack in Massachusetts was in Buzzard Bay in 1936. Since then, there have been no confirmed bites."

Other shark experts agreed. George Burgess, the director of the Florida Program for Shark Research at the Florida Museum of Natural History, said regarding the risk white sharks posed to the public on Cape Cod, "The chances of encountering one up there are markedly lower than in other parts of the country."

Chatham's harbormaster, who was responsible in large part for public safety on the town's beaches, took a realistic approach. "It is the ocean," he said. "People are going to see sharks, they are going to see turtles, they are going to see seals, and they are going to see whales. It's their marine environment, and they need to understand that."

The Shark in the Salt Pond—2004

By 2004, I was "living in the key of G." I wasn't, of course, working directly with white sharks, but I'd largely made peace with the fact that I wasn't going to study that apex predator in any sort of intensive way. My self-imposed five-year time limit for the job on Martha's Vineyard had long since passed, as I'd fallen in love with both the job and the island. I'd returned to school at the urging of Boston University professor Phil Lobel, who had convinced me I was well on my way to earning a PhD from the Boston University Marine Program (BUMP) based on the work I was already doing as part of my job. The shark program I'd started was humming along, and the Monster Shark Tournament had grown to a point where the comprehensive data I was collecting rivaled any tournament with which I'd been involved in the past. As the agency's regional biologist, I was also reviewing coastal alteration projects for the state and regularly collaborating with local anglers and other researchers on local fish species. Like Jack, my relationship with fishermen, both recreational and commercial, was my greatest asset, although it meant I received calls at all hours of the day and night. If there was a fish story to be had, especially if it involved sharks, it would surely get to me.

In late 2003, I was also still very much focused on the basking shark work, but funding was tight. Nick was shaking all the broadcasting trees, like Discovery Channel and National Geographic, with no success. We were able to keep the tagging going for the next year with a small grant from the Massachusetts Environmental Trust, which paid for six tags, the spotter pilot, and vessel time. In the proposal to MET, I partnered with the agency's then-whale biologist, at the time, Ed Lyman, to make the argument that the basking shark could be used as a proxy to study the critically endangered North Atlantic right whale. For a variety of reasons, not least of which was the fact that they are an endangered species, right whales were difficult if not impossible to tag. Scientists had been looking for other ways to learn about right whale movements in the western Atlantic.

While much about basking sharks remained a mystery, it was now known that both basking sharks and right whales occupied the same biological niche. It was at least plausible, we thought, that the movements of basking sharks in the western Atlantic were similar to the movements of right whales. As a result, we were awarded the grant to buy tags to track basking sharks, which would at least keep the research going in 2004. But to achieve a significant sample size, I knew we needed more funding.

In April 2004, that funding would be delivered in a call I received from my friend Erin Summers. Both Erin and I had overlapped at BUMP, where she earned her master's working on Atlantic white-sided dolphins in 2002. Anyone who has been to graduate school will say that it is pretty common to develop tight bonds with your fellow graduate students, and this was certainly the case at BUMP. The structure of the program at BUMP was academically intensive because each class was taught in a three- to four-week block. This meant that you spent every day with your classmates, working together and hanging out. This was the case with Erin and me as well as the rest of our contemporaries.

I was the oldest student in the program, and I think Erin came to think of me as a sort of big brother. Because of my age and my experience, I did my best to mentor the other students, including Erin, who was really bright, ambitious, and well on her way to being a good scientist. I also benefited from hanging out with Erin and the younger graduate students because I was older, rusty, and hadn't been in an academic setting for more than a decade. To say it was awkward is an understatement. But Erin was able to set me at ease and draw me out of my shell.

After she graduated, Erin headed south to work as a research technician with the NOAA Cetacean and Sea Turtle Team out of the Beaufort Laboratory in North Carolina, but now she was back in New England as a marine mammal research specialist at the University of New England, where she was working on a project to track and predict right whale habitat use and distribution. She was working with oceanographer Stephan Zeeman, who had just secured a pile of funding from NASA to study right whales, but tagging that species was not going to be possible. We discussed the similarities between right whales and basking sharks and both agreed that tagging the latter seemed to make a lot of sense. She went back to Steve with the idea, and he loved it.

By late summer of 2004, all the pieces were falling into place and Erin was planning to come down in late September to help tag some basking sharks. I had lined up Captain Billy Chaprales again, although this time we would be using his commercial harpoon fishing boat *Ezyduzit*. I was looking forward to it—not only to revitalizing the basking shark project but also to spending time with Erin. However, just before Erin's planned arrival, I received a call that would change my life.

On a late September day, Tuck Hayashi was chasing false albacore amid a tangle of islets in the Elizabeth Islands with his girlfriend, Joanne Manning, and his friend Drew Colella when they came upon the largest fish any of them had ever seen. The Elizabeth Islands trail southwest more than fifteen miles from Cape Cod's southernmost point at Woods Hole, where the Woods Hole Oceanographic Institution and a variety of other labs are located. To the northwest of the islands is Buzzards Bay and the mainland, and to the southeast, across Vineyard Sound, is Martha's Vineyard. Tuck kept his sixteen-foot, teal-colored skiff at Woods Hole. That morning, they had motored out under the Eel Pond drawbridge, between the Woods Hole Oceanographic Institution's wharf and the Martha's Vineyard ferry docks, across Great Harbor, and into the channel separating Cape Cod's southern tip from the most easterly of the Elizabeth Islands, called Nonamesset Island. Tuck, who had fished and roamed these waters by boat since his youth, paralleled the north shore of Nonamesset heading west into Hadley Harbor looking for fish.

Hadley Harbor is tucked up neatly between Nonamesset and Uncatena Islands, with Bull Island standing proud at its head and guarding the northeast approach to Naushon Island's inner harbor. Like most of the Elizabeth Islands, Naushon Island is a private island belonging to the Forbes family and, at the time, was in the news because senator and presidential candidate John Kerry vacationed there. Tuck didn't think much of the wealthy residents as he passed to the east of Bull Island and entered a maze of lagoons and channels that snake between convoluted necks of land, rock-strewn shorelines, and pockets of sand. Tuck more identified with the pirates who had lurked among the islands, taking advantage of commercial ship traffic transiting through Buzzards Bay—famed scallywags like Thomas Pound, Thomas Hawkins, and William Kidd, whose name now graced the Woods Hole watering hole.

False albacore are much revered by anglers for their speed and fighting ability. Similar in looks and design to their larger cousins, the tuna,

The star of the show in all its glory: the white shark. As a young child, I was fascinated by the beauty and magnificence of the white shark. Only in my wildest dreams did I ever imagine I'd be able to study this creature of my obsession for a living, let alone in my own backyard of the northeast United States.

Geof Pay

Peter Benchley, bottom, author of "Jaws," examines great white shark caught off Block Island Friday. George Benz, a fisheries biologist at the University of Connecticut looks at the shark in disbelief.

Above: My friend and first mentor, Jack Casey (*second from right*), with ten juvenile white sharks sampled off New Jersey in 1964. This discovery of a shark nursery just off the coast was highly significant for shark research in the region. The photograph itself brings a smile to my face, as I recall developing that image for Jack many years later in a darkroom while working for him at the Narragansett Lab in the mid-1980s. In the resulting 1985 paper, my name appeared for the first time in the scientific literature, although it was only in the acknowledgments.

Left: "This kind of wanton slaughter is madness," said *Jaws* author Peter Benchley (*kneeling*) of the white shark harpooned by two Noank fishermen in 1983. My friend and colleague George Benz looks on. This event was so significant that this photo appeared on the front page of the *Hartford Courant*.

Here I am suspended to photograph the sixteen-foot Noank white shark at Mystic Aquarium in 1983. I had photographed a lot of sharks that summer, but this one felt different, as it was my first white shark specimen. We raced against time to study it before it decomposed.

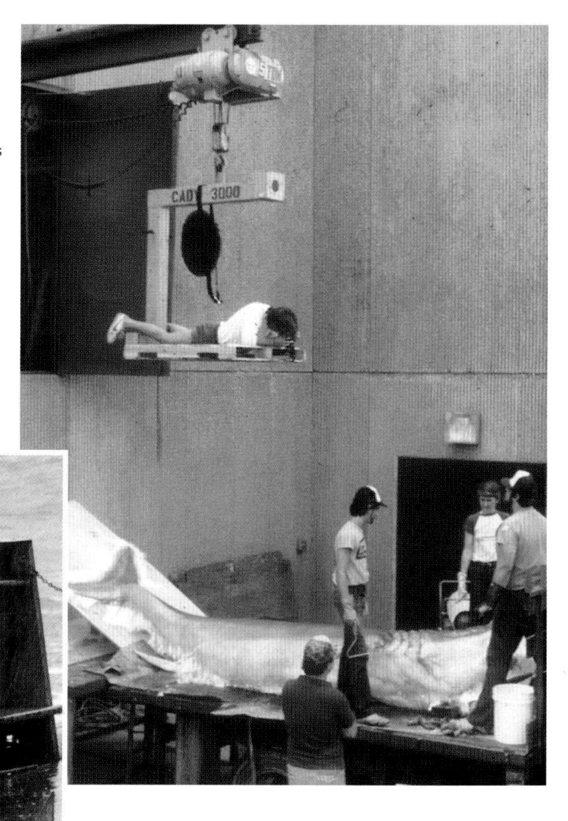

Left: Here's me removing the jaws from a shark during my first research cruise on the F/V *Wieczno* in 1984. A trip aboard the *Wieczno* was like boot camp and a rite of passage for new biologists working with the Apex Predators Program.

Shark researcher and colleague Chuck Stillwell lies next to a sixteen-foot white shark that was harpooned by fishermen off the New York coast in 1983. Like other white sharks harpooned that summer, samples like this specimen provided some of the first insights into the biology of the white shark in the northwestern Atlantic.

The F/V *Wieczno* takes a wave while hauling longline gear during a shark research cruise. Rough weather, language barriers, tight quarters, and lots of sharks all made for a memorable, formative experience.

The shark research team with a large tiger shark aboard the F/V *Wieczno* in 1984. The goal of this trip, to catch and study sharks, was a huge success. *Left to right*: me, Chuck Stillwell, Jack Casey, Nancy Kohler, George Benz, and Wes Pratt.

Cage diving off the coast of Rhode Island on the F/V *Snappa* in July 1990. My annual cage-diving trip with this crew and others became one of the highlights of my early career, because I was working with live sharks instead of dead ones on the dock. *Back row (left to right)*: Nancy Kohler, Brian Skerry, me, Brian McCarthy, Lance Stewart. *Front row (left to right)*: Wes Pratt, Jerry Prezioso, Captain Charlie Donilon.

Right: Me helping Jack Casey tag a shortfin mako on the deck of the F/V *Wieczno* on my first research cruise in 1984.

My lifelong colleagues Lisa Natanson and Nancy Kohler with a great hammerhead shark on the deck of the F/V *Wieczno* in 1986. Like me, Lisa knew from the time she was a child that she wanted to work with sharks. Nancy would eventually take over the Apex Predators Program from Jack Casey upon his retirement.

Left: A Greenland shark under the Arctic ice in 1999. During my trip to the Canadian Arctic, I had the opportunity to dive below six feet of ice in 28°F water with these massive, ancient sharks. It was one of the first times we leveraged funding to make a film while also conducting shark research.

Below: Me and George Benz measure a Greenland shark on the Arctic ice before tagging in 1999. This was the first time any animals had been tracked using acoustic tags under ice.

A 1,221-pound world-record mako shark landed during the Oak Bluffs Monster Shark Tournament in 2001. Shark researchers have been collecting vital data at shark tournaments for many decades.

Above: A white shark extends its upper jaw to bite the carcass of a seal as I film it for identification and research purposes. The seal population's return to Cape Cod was an ecological success story that brought with it another unexpected conservation achievement: the restoration of an apex predator to the ecosystem.

Left: A white shark charges after a gray seal in the shallow waters off the coast of Cape Cod.

Up close and personal with a white shark as it devours a seal carcass.

Captain Billy Chaprales tags one of the first white sharks off Cape Cod from the pulpit of the F/V *Ezyduzit* in 2009. Billy was a pioneer of tagging from the pulpit with a modified harpoon, and his contributions to my bluefin tuna, basking shark, and white shark studies were monumental.

A white shark carries an acoustic tag and a camera tag (pink) attached by me off the coast of Cape Cod. The larger camera tag will record for two days, then detach and float to the surface for retrieval. Tagging sharks with these cutting-edge, high-tech tags provides us with movement data and behavioral observations.

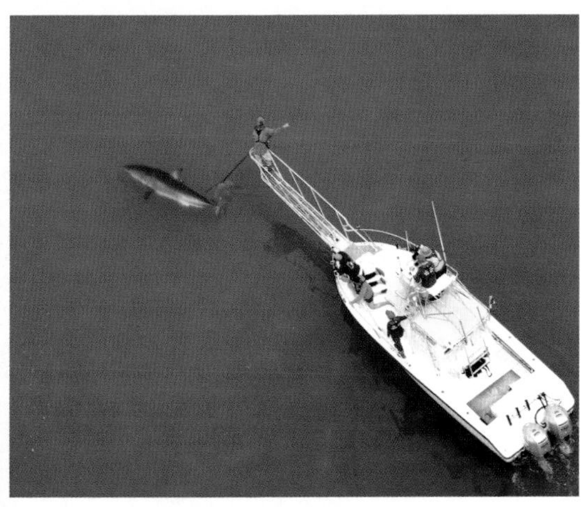

Me tagging a white shark off the pulpit of the F/V *Aleutian Dream*. Our tagging technique evolved from throwing the tagging harpoon from a large fishing boat to placing the tag with a tagging pole from a smaller, faster vessel much closer to the water.

An eighteen-foot white shark we'd later dub "Curly" thrashes around as it becomes entangled in the shark cage with me and Nick Caloyianis inside in 2010. A terrifying but memorable experience.

Above: With her new satellite tag attached, Curly circles the cage with me inside. Not only was this my first underwater experience with an Atlantic white shark, but Curly was also the first mature female I tagged.

Right: A white shark is hauled onto the OCEARCH research vessel lift for tagging.

I prepare to dissect a fourteen-foot white shark that stranded on Cape Cod in 2015.

Below: I'm dissecting a white shark in Scituate, Massachusetts, with Lisa Natanson in 2018. Dissections of stranded sharks, sharks landed during shark tournaments, or sharks caught as bycatch by commercial fisheries give us fantastic opportunities to study their anatomy, natural history, and physiology.

The West Gutter, where white shark "Gretel" was trapped for thirteen days in 2004. Note the net used to keep her from reentering the shallow estuary. I took this photo from a Coast Guard helicopter (*see shadow on the bottom right*) while we were looking for the shark.

Right: Ed Lyman and I, in an act of desperation, after twelve days of exhausting all other options, attempt to herd Gretel from the salt pond using a high-powered hose from a local cranberry farm.

Left: Me briefing my boss on the phone while Gretel swims in the shallows of the West Gutter estuary. Here, you can really see how shallow the waters were in which poor Gretel found herself.

Above: When numerous beachgoers at Nauset Beach saw the dorsal fin of a white shark behind Walter Szulc's kayak, and then when this photo went viral, I wondered if this would be the incident that would finally make people wake up to the reality of white sharks on the Cape. The next day, people were back in the water—most seemed unfazed by the event.

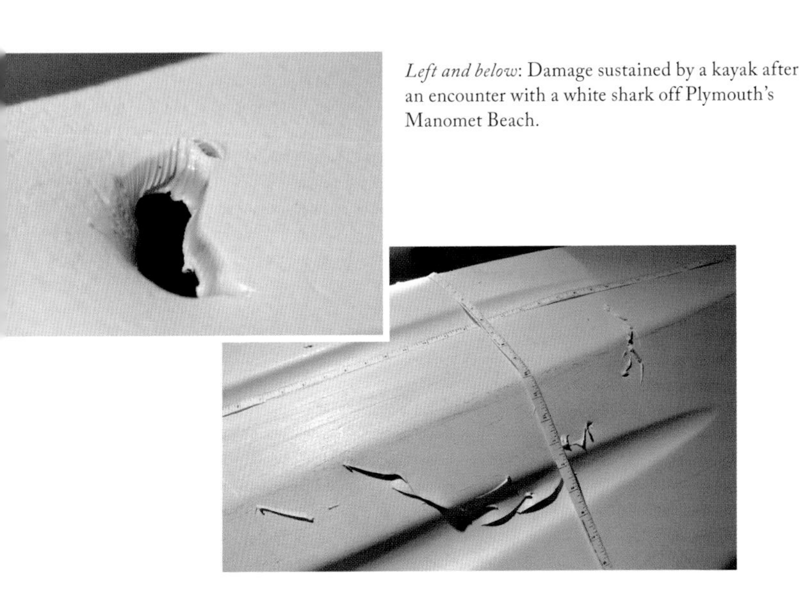

Left and below: Damage sustained by a kayak after an encounter with a white shark off Plymouth's Manomet Beach.

John Chisholm prepares to release a sand tiger shark after tagging it off Massachusetts. We first worked together when Gretel swam into that salt pond, and we've been close friends and colleagues ever since.

Left: My colleague Megan Winton analyzes white shark identification photos for our population study. Megan is an Atlantic White Shark Conservancy researcher who is making huge contributions to the future of shark studies.

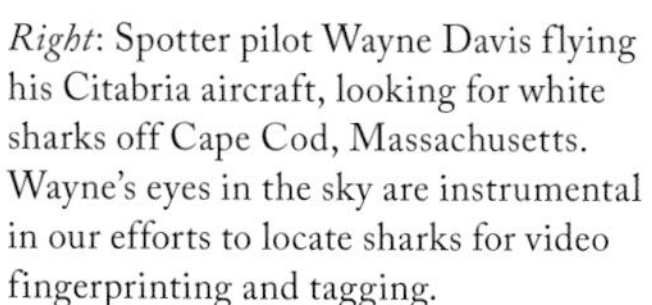

Right: Spotter pilot Wayne Davis flying his Citabria aircraft, looking for white sharks off Cape Cod, Massachusetts. Wayne's eyes in the sky are instrumental in our efforts to locate sharks for video fingerprinting and tagging.

White shark "Turbo" checking out the camera off the coast of Cape Cod. Here, we can see distinct scars and other physical features used in our population study to ID and catalog individual sharks as they migrate to the Cape each year.

A white shark hunts for seals in shallow water, extremely close to a Cape Cod beach, drawing a crowd of onlookers. This charismatic creature has the rare ability to simultaneously fascinate and strike fear into the hearts of humans across the globe.

Suspended in the deep, beautiful blue of its ocean: a perfect white shark specimen.

they are sometimes called *little tunny*, but they do not belong to the genus *Thunnus* and are not true tuna. Fusiform in shape, with powerful tails and stiff bodies capable of great bursts of speed, they also share many characteristics with white sharks, including the ability to retain heat and the lack of a swim bladder. I was particularly interested in these similarities among species that evolved separately because they provide scientists with an opportunity to study the relationship between genetics and physical traits. For anglers like Tuck, however, they simply provided some of the most exciting fishing available. Although false albacore are, like tunas and white sharks, a highly migratory pelagic species, they often opt for warm inshore waters where baitfish, their primary prey, are plentiful—which is why Tuck was probing the inner sanctuaries of the Elizabeth Islands on that gorgeous late September day.

The nervous shaking of the water ahead belied a school of fish. Birds circled overhead, confirming the presence of the fish, which were moving deeper into the watery labyrinth. Tuck followed them around Goats Neck, but the water's surface had once again gone still and glassy. He dropped the boat into neutral, scanning ahead. There were some birds working the far side of a small bridge that crossed from Nonamesset and Veckatimest Islands to Naushon Island. Tuck accelerated and turned the skiff, sliding under the bridge into a small salt pond known locally as West Gutter.

West Gutter is not a true pond in the sense of an impounded body of water. Directly south of the bridge, West Gutter opens to the east at high tide over a shallow sandy spit that connects to Monohansett Island and hems in the salt pond along one side. On the other side is an arm of Naushon Island that hooks south then southwest to a point. Here West Gutter opens onto shallow Lackeys Bay through a narrow channel between the point and the tip of Monohansett Island. From the bridge at the north end of the salt pond to the outlet is about a quarter of a mile, and at no point is the width greater than 250 feet.

As Tuck motored under the bridge and into West Gutter, he didn't see the false albacore, but something else caught his eye off to his left.

"Is that a fin?" he wondered aloud.

It was. And it was huge.

He estimated that the fin stuck out of the water about two feet. He turned the boat toward the animal. He was thinking basking shark, but then he saw a second fin. Two sharks separated by seven or eight feet?

As he closed the distance, there was a thrashing in the shallows, and, as the fish rolled and turned for deeper water, Tuck realized it was not two sharks. It was one very large shark—about as long as his boat.

Tuck followed the fins until they dropped below the surface when the shark found deeper water toward the middle of the salt pond. The deepest part of West Gutter is only fifteen feet deep, but much of it is far shallower—far too shallow for the large shark. They waited, engine idling, searching the water. Soon the shark surfaced again and was swimming straight toward the skiff in a scene reminiscent of *Jaws*. It seemed to swim deliberately toward the port side of Tuck's small boat, diving moments before reaching it and then quickly coming back to the surface on the opposite side. They all got a good look at it as its shadow passed under them. Tuck and Drew debated whether it was a white shark or a basking shark. Joanne simply marveled at its girth. "It was wider and fatter than a shark should be," she later recalled. With hindsight, she wondered why she wasn't frightened—why she didn't insist on getting to shore at once.

They stayed and watched the shark until late in the afternoon. Tuck's skiff had no running lights, so, with darkness approaching, they reluctantly headed back to Woods Hole. That night, Tuck called a couple of fishermen friends to tell them about the shark. Through a friend of a friend, a sportfishing captain named John C., usually just "J.C.," Burke got wind of the shark-in-the-salt-pond story, and he called me while I was watching the Red Sox, who were just wrapping up a series with the Baltimore Orioles. I was immediately skeptical. Fishermen are storytellers, and this sounded like a heck of a story. Nonetheless, the Elizabeth Islands were part of my region, and clearly there was some kind of large animal over there. Whether it was a shark, a sunfish, or a dolphin, I didn't know, but I decided I'd go check it out. The worst that can happen, I figured, is I end up taking the day off and doing some fishing with J.C.

Early the next morning, I took my boat, a twenty-one-foot center console Parker, across to Falmouth Harbor, where I picked up J.C. Before leaving the Island, I'd dropped by my office at the Lobster Hatchery and grabbed one of the PSAT tags and a tagging pole slated for use on the basking shark project . . . just in case. I was still pretty sure I'd be using the fishing rod more than the tagging pole.

Tuck was also up early, and he was planning on fishing as well. With him were Adam Lazarus, who was an amateur filmmaker hoping to film the shark, and Drew, who was still convinced the shark was a white shark. Tuck was still undecided on what species it was, but if it was a white shark, Tuck was going to try to hook it. While landing a white shark had been illegal in federal waters three miles offshore since 1997, the law in state waters was murky at best. Tuck told Adam he wanted to land the shark as a trophy, like Frank Mundus, and then give the meat to the homeless. In reality, Tuck knew there was no way he could land a shark that size with his boat and the fishing tackle they had on board, but that wasn't going to stop him from dreaming.

Upon entering the salt pond, they immediately spotted the shark still circling. Adam, who had admitted to being afraid of sharks the night before, when Tuck called him, wanted off the boat immediately. Tuck, on the other hand, was envisioning a man-versus-the-ocean moment—like a scene out of *The Old Man and the Sea.* He acknowledged that he wasn't sure if hooking a white shark was legal, but he simply couldn't help himself. He had to try. They dropped Adam off on the shore, and he started filming as Tuck readied his rod.

Tuck cast a little in front of the shark and hooked it on its dorsal fin as it swam into the big treble hook. He was adamant he hadn't snagged the shark; rather, it swam into his hook. "It was like hooking a car and having it drive away," Tuck later remembers. The rod was set up for bluefish with twenty-pound test, and there was no chance he was going to land the shark, but still . . .

As Adam filmed from shore, he sensed this was a profound moment for Tuck. "It was a culmination of years of fishing," Adam said. "It was kind of a sacred moment to see him there like that." Or as Greg Dubrule put it in regard to the Noank shark in 1983, "It's like being one of Christ's disciples. Many are called, but few are chosen." When the shark turned again in the shallows, it threw the hook. Tuck was "a little bit bummed out and a little bit relieved."

J.C. and I arrived in the salt pond not long after Tuck had broken off. Almost immediately we too saw the fin. I knew right away from the shape and color of her dorsal fin that it was a white shark. There were very few animals of this size in the ocean and, more specifically, in New England. I had seen my share of ocean sunfish, dolphins and porpoises, small whales, and basking sharks to know it was none of them. This was definitely a great white shark. My heart started to race, and I was grinning like a kid. I'd seen a fair number of dead white sharks, and I'd been cage diving with a couple of live ones, but this was something entirely different. This huge shark in the small salt pond, swimming the perimeter and probing the shoreline, was almost like seeing a shark at an aquarium. It was obvious to me that the shark was looking for a way out, but my initial impression was that it was otherwise healthy. It likely swam in on a high tide, and then couldn't get back out. Being in this animal's presence filled me with childlike wonder. It was a dream come true. I waved over to Tuck, and thanked him for the heads-up. Tuck, unaware of who I was, said, "You should have been here five minutes ago! I had him on the line." I somehow maintained a warm smile as my mind weighed the murky legal terrain surrounding what he'd just shouted at me.

I knew I needed to compartmentalize and get into work mode. I caught my breath and pushed my childlike enthusiasm aside. I would have plenty of time to revel in this experience when I remembered it later, but I had to get to work. I was thinking the shark might find a way out of the salt pond, and I needed to get the tag on it before that happened. I was well aware that a white shark had never been tagged with a satellite tag in the Atlantic before. This was the opportunity of a lifetime, not only for me personally but also for white shark science in general. A lot was riding on my actions in the next few moments, and I was already feeling the pressure. I needed to first notify my bosses about the big white shark swimming in a small coastal salt pond. By that point, I'd been around sharks long enough to know that this was going to cause a massive media reaction, and I didn't want my bosses turning on the news to find out what was going on before they heard it from me. I filled in the director, Paul Diodati, and we briefly discussed the lack of state regulations related to white sharks and the possible need for him to implement emergency actions. Before hanging up, I asked him if he'd be okay with my plan to tag the fish. The tags, after all, were grant-funded for basking sharks. Paul gave his blessing, thanked me for the heads-up, and sent the news up the ladder to the commissioner and secretary.

J.C. took the helm while I mounted the tag on a six-foot sling spear designed to spear fish underwater. I then watched the shark closely to determine the optimal time to tag. Given the length of the spear, we needed to get close. I noticed that the shark was routinely probing the shallow edges of the pond, with occasional forays into the center, where it was too deep to tag. It was on one of those passes into the shallows that I noticed that the shark didn't have male reproductive organs, called claspers—it was a female. We needed to intercept her when she tested the shoreline, so, as the shark emerged from the center of the salt pond, I instructed J.C. to follow it into the shallows. He maneuvered the boat in close, and I leaned over the bow, drew back the sling, and jabbed the dart into the base of the shark's dorsal fin. The shark, which we estimated to be about thirteen feet long and weighing somewhere in the neighborhood of 1,700 pounds, barely reacted. As soon as I was certain the tag was secure, I breathed a sigh of relief.

It was, without a doubt, the high point of my career.

Tuck hadn't exactly kept the shark a secret, and when he headed back to Woods Hole after having hooked it, he was as eager as ever to tell the story. As J.C. and I monitored the shark and took photos, we began to notice other boats approaching the area. At one point, a nine-foot Zodiac cruised into the salt pond with a man at the tiller and two children aboard. When the Zodiac passed behind the shark's fin, the children's awestruck faces were completely eclipsed by it. I began to think that the shark's situation was more serious than I'd originally suspected. This shark seemed to be stuck, and it was going to take some time, tide, and good luck to get her back out to deep water.

More boats with eager onlookers showed up, and I was amazed at the level of interest and how fast word had traveled. By late afternoon there were as many as twelve boats in the Gutter, having entered with their shallow drafts over the grass flats at the southern outlet that the shark was too leery to cross. I was worried about public safety, but I also thought that with so many boats already in the pond, maybe we could drive the shark out. I asked them to line up with me to try to herd it out of the southern outlet. I hoped that all the commotion and engine sound would motivate it to leave, but the shark simply dove below the boats in the center of the pond. It began to dawn on me that this was going to be a multiday job.

The next several hours were a blur. At 4:00 p.m., I headed back to the Vineyard, packed a bag, and by 5:30 I was back in Falmouth Harbor doing interviews for several local media outlets. A few hours later, I had met up with Erin and we were configuring PSAT tags with my fellow graduate student Cate O'Keefe—they were still going to head out with Billy the next day to try to tag basking sharks even though I had to stay behind and deal with the white shark. Finally, at around midnight, I headed over to my friend Chuck Martinson's house, across from Falmouth Harbor. I finally had a chance to lie down, realizing that I probably wasn't going to make it home for a few days.

The shark in the salt pond on Naushon Island blew up in the press, as I had anticipated. It rapidly escalated from a local story to a national story and then to international news, and I found myself at the center of it all. Having become the region's de facto shark expert, I'd gotten accustomed to doing interviews with local press and even the larger Boston media outlets, but this was different. It was a feeding frenzy. By the second day, it was clear to everyone that the shark was trapped, and it was going to need assistance to get free. Later that morning, Division of Marine Fisheries director Paul Diodati put together a meeting at the NMFS Woods Hole Laboratory with the commissioner of the State Department of Fish and Game, Dave Peters; my colleagues Nancy Kohler and Lisa Natanson from the Apex Predators Program; and their boss, Frank Almeida. We discussed our limited options and quickly concluded that the first step had to be to secure the area. As news spread, there had been a steady stream of boats with eager onlookers filling the Gutter, so Diodati consulted with the state's Environmental Police, the agency charged with enforcing the law when it comes to Massachusetts's natural resources. There was concern both for public safety and for the safety of the shark, and the Environmental Police quickly agreed to cordon off the area. That didn't stop the boat traffic from getting as close as they could and trying to get a glimpse of the shark. In the meantime, Division of Marine Fisheries deputy director Dan McKiernan and I assembled a team to assist us, including John Chisholm. John worked for the Division's Sportfish Program but had the primary responsibility of creating and distributing the annual Massachusetts Saltwater Fishing Guide. I didn't know John well, but I did know that John was a shark buff with an intense interest in white sharks. He had been personally documenting the occurrence of white sharks in the Atlantic for almost twenty-five years, and he's amassed a substantial historical database. With a white shark trapped in the salt pond, John was the perfect colleague for this adventure.

That afternoon we had our first of many formal press conferences and the media was relentless. For most of their questions, like "How did the shark get in there?" and "When will it leave?" there were simply no answers. I was realizing that this was not going to be easy, and I had to muster all my explanatory skills, with some levelheadedness and my public relations savoir-faire to keep the media on point and away from hyperbole and excessive *Jaws* references. As I negotiated the media frenzy, I thought back often to Jack fielding media calls in the Narragansett Lab. One thing had not changed: the media loved a shark story, and their penchant for exaggeration remained unabashed. The difference, however, was the way in which the public responded.

Public perception was more mixed than it once was, and it was leaning ever toward a "save the shark" mentality instead of a "the only good shark is a dead shark" attitude. Ironically, the Monster Shark Tournament that past summer, which was airing on ESPN, helped fuel the former. Given the wider network viewership and the editing that emphasized the "monster" part, along with ample manufactured drama and gratuitous blood, the four-part series would be an impetus for activists and animal rights advocates who would eventually bring about the end of the Martha's Vineyard tournament. I had always emphasized the legality of the event and the scientific value of the tournament and the fact that, of the thousands of sharks hooked, only a fraction were brought to the dock. Far more were tagged and would contribute to science through the tagging program, while the ones that were killed provided valuable insight into the biology of the animals through dissection before being butchered for food. Nonetheless, I had my haters, who criticized me not only for being part of the tournament but also for "giving cover and legitimacy to a blood sport." It didn't help that other researchers, eager for attention, jumped on the bandwagon, criticizing shark tournaments—and, at times, me.

As the days started to drag on and the shark remained trapped, the overwhelming sentiment from the public was to save the shark, and I was doing my darnedest. People were absolutely fascinated by the animal, and a few young guys with boats and an entrepreneurial spirit cashed in on that fascination by offering paid tours to Naushon Island. The Environmental Police had a fairly large area cordoned off, but people were willing to pay a premium for even a long-distance glimpse of the fin through binoculars. Seeing the glint of binoculars pointed toward us wasn't helping with my stress level. For their part, the regularly private residents of the island had had enough of the spectacle on their doorstep and were eager for the state to do something. This, in turn, put more pressure on me from my superiors, who were in turn getting beaten up by their superiors.

Back in the salt pond, John and I were stumped. We first thought the shark would leave on the next high tide, but it didn't. The eelgrass bed at the southern outlet to Lackeys Bay is peppered with large boulders and just too shallow even at high tide for the shark to cross it of its own accord. To motivate any animal, even your dog, you need to get its attention. To do that, we needed to appeal to one or more of its senses. The boat test on day one had already told me that spooking it with sound wasn't going to work. Over the following days, we tried to obscure the shark's vision by setting up a seemingly impenetrable cloud of lime, but the shark ignored it. We then decided to try a device called Shark Shield, which creates a powerful electrical field designed to overwhelm the shark's electrical receptors. As a shark approaches it, the shark should feel the strong electrical field and hopefully turn away, much like we would turn away from a very bright light. We placed the Shark Shield on a line between two boats, directly in the path of the oncoming shark. If the shark turned away, it would head for the southern inlet. As the shark neared the device, it suddenly exploded in a burst of spray, abruptly turning away. We were thrilled.

"This is going to work," I said to John. But on the second approach, the shark swam right past the Shield with absolutely no reaction. Perhaps it was no longer charged, someone suggested, but upon placing his hand in the water, he received a shockingly powerful jolt. It was working all right . . . just not on the white shark.

I had worked so hard since arriving on Martha's Vineyard in 1987 to become the local shark guy, but there I was with the shark of a lifetime stuck in a salt pond and my every move being scrutinized by the media. I was spending all my time trying to get this shark freed, and day after day I felt like I wasn't getting any closer. We tried baiting the shark using a seal carcass at one point, but the shark by then appeared too stressed to feed. My frustration started to turn to worry. What if the shark died? So much of my work was out of the public eye, but with this situation, I was center stage with a bright spotlight tracking my every move.

I met Erin at the Kidd that night for a beer. I don't think she'd ever seen me so preoccupied and glum. The two of us usually bantered like siblings. We joked easily with one another because we both had thick skins. She was hoping I'd be able to join them the next day on Billy's boat to try to tag basking sharks, but I told her I couldn't go with them because of the shark. I remember being quiet that night—at a loss for words. I was exhausted and stressed. I barely touched my beer.

"I guess your job has been reduced to babysitting a shark," she jabbed.

I didn't laugh or have a witty comeback. She was right. I simply exhaled deeply. Erin later said she'd never seen me like that—and she knew me better than most.

We were now more than a week into the ordeal, and I was beginning to think about nuclear options. One was to tail-rope the shark and drag it out of the Gutter with a larger boat. The other would be to tranquilize it, then tie a rope around its tail and tow it to deep water. I knew both options could potentially kill the shark and would most definitely create an unwanted spectacle. I was regularly updating my bosses at the Division of Marine Fisheries, including Dan McKiernan, who was equally anxious to resolve the situation. The agency had a federal grant called Fishermen Helping Scientists, which, Dan thought, could provide funds to hire some additional help. He called Mark Simonitsch, a Chatham-based weir fisherman, and started to explain the situation, but Mark was already well aware of what was going on because of the media coverage. He readily agreed to help and suggested that Ernie Eldredge would also be an asset. Ernie was another weir fisherman who was especially skilled at utilizing the ancient fishing technique whereby heavy nets are constructed along a shoreline to move and ultimately trap fish. The next day Mark and Ernie met up with us to see the situation firsthand. When they arrived, Ernie coolly observed the shark with the eye of someone who's spent a lifetime thinking like a fish. "This looks like a no-brainer here," he said matter-of-factly. "This looks easy."

My roller coaster of emotions was again on an upswing.

The next day we all rendezvoused with Mark and Ernie at the salt pond. The general idea was to methodically reduce the swimming space available to the shark by moving nets from north to south, thereby forcing the shark to the shallow southern inlet. The fishermen brought all their gear, and Ernie carefully picked his spot to string the heavy weir netting across the pond. The first set would effectively reduce the size of the pond by almost half, but his timing had to be perfect so as not to trap the shark in the northern half. He waited for the shark to swim south, and they set the net. With the barrier in place, it was the moment of truth when the shark turned and came back toward the net. We all held our breath. As she approached the net, she appeared unaware of it, and she hit it. But she didn't struggle. Instead she spun around and headed back away from it toward the outlet. This was going to work, I thought. This was going to do it!

Ernie and Mark continued leapfrogging nets, reducing the area in which the shark could circle. I was amazed to watch the effort—set, haul, reset. Set, haul, reset. With each set, I was feeling more and more optimistic. Finally, this was going to end. The shark would swim away with the satellite tag, and I could get back to my job. I feel like the shark's survival was always my top priority, but I was also pretty excited at the prospect of deploying the first satellite tag on a white shark in the Atlantic. The data could certainly result in a paper and, just as Frank, Jack, and Wes had done with the white shark they tagged back in 1979, I could have an opportunity to contribute meaningfully to the knowledge about the species.

As Ernie and Mark moved the shark down the Gutter, I started to notice that the shark was turning at about the same place in the shallows at the outlet. The ever-encroaching nets were limiting her swimming space, but they were no longer pushing her farther south. She just wouldn't go out between the boulders and over the shallow eelgrass bed. Instead her circling in the smaller and smaller space became more and more frantic. I watched in desperation. I was at my wits' end.

One of the Environmental Police officers who was helping was also a cranberry farmer and suggested a new strategy. He'd seen at one point how the shark had turned away from his propeller wash when he'd throttled his boat up near it. He wondered if we might be able to direct the shark using a high-pressure cranberry farm pump. Desperate for a new approach at that point, we all agreed that it was worth a shot, so we loaded a pump and hose onto a boat and returned to where the shark was hemmed in at the outlet by the last weir net Ernie and Mark had set.

We experimented blasting the surface of the water on one side of the shark. It turned. When we blasted it on the other side, it turned again. We kept at it, alternating the jet of water from one side of the shark to the other. "There we go!" I yelled. "Keep it going!" Finally, the shark had had enough, and rocketed over the eelgrass shallows, as I let out a whoop with hands extended triumphantly over my head.

We had done it.

Adios, Gretel, I hope our paths cross again. I muttered to myself. I'd started to call her Gretel, by the way, even though the scientist in me generally knew better than to name the animal experimented upon or studied.

Unfortunately, like every other small success we'd had with Gretel, this one was short-lived. Instead of continuing to her left and southeast around the tip of Monohansett, toward Vineyard Sound and deep water, Gretel circled right into Lackeys Bay, which, while larger than West Gutter, was much shallower. A feeling of utter dread and despair overcame me. It was entirely possible, I realized almost immediately, that Gretel could drown when the tide drained the broad, shallow bay. Like many species of sharks, the white shark must swim forward to breathe. If the bay became too shallow, she could strand.

"Shit," I muttered, as I sat on the boat's gunwale and hung my head in defeat.

The first day of October was a Friday, and it had been eight days since I'd tagged Gretel. In the back of my mind, I was starting to get an uneasy feeling. PSAT tags are very sophisticated electronic devices. Most are applied to animals that have been captured and handled, but not all animals survive this process. So, the tag manufacturer built in a feature that tells the tag to pop off prematurely if the animal dies. The tag does this based on depth data collected every ten seconds by the tag. If the shark's depth doesn't vary enough over time, the tag assumes it's dead and pops off. The amount of depth fluctuation and length of time can be preprogrammed by the researcher, and the tag on Gretel was set for basking shark work in deep water. If the shark's depth didn't fluctuate by five feet for three days, the tag would release. In the West Gutter, the shark had moved routinely through fifteen feet of water, so this hadn't been a concern. But now, in the much shallower water body, I knew the clock was ticking.

The relentless schedule, media frenzy, political pressure, and gravity of the situation were taking their toll on me. I was utterly exhausted and emotionally spent. And we were back at square one. Over the weekend, we periodically checked on the status of the shark. She was moving throughout Lackeys Bay but appeared unable to exit on her own. We assembled the team to brainstorm our options, but nothing seemed feasible. The shark had more room to move in Lackeys, making the effective use of the weir nets close to impossible. But the weir fishermen still felt that some kind of barrier was needed and came up with a curtain composed of a mainline with short lead-weighted drop lines every couple of feet. By extending the curtain between two boats, they thought it could be used to corral the shark toward the inlet.

As the new week dawned, the media interviews were getting harder and harder. The ordeal had gone on longer than any of us had expected, and it was becoming increasingly difficult to face the reporters and their questions. They—like me, like my bosses, like everyone working long days to resolve this situation—wanted to know what was going to happen. I was, quite frankly, running out of things to tell them. We were doing everything we could, short of the nuclear options like grabbing the shark by the tail and just dragging it out into the ocean. One evening I was doing a radio interview with WBZ and the reporter kept asking if we'd named the shark. I knew that my bosses at the Division of Marine Fisheries were dead set against naming the shark, but the interviewer was persistent, and I was exhausted. I eventually just thought to myself, *screw it*, and told him we'd named the shark Gretel. "It's a play on GREat white," I explained, "and she's lost, like Hansel and Gretel."

As I knew they would be, my bosses weren't happy, so I lied and told them the interviewer had made up the name. I fully understood why the agency didn't want to name the shark. For one, I was a biologist for the state, and I oversaw the state's shark research program. With perhaps marine mammals like whales being the only exception, it was basically unheard-of to name study animals in fisheries science. But it was more than just policy. It was the fact that we didn't know what the fate of the animal would be. People tended to embrace creatures that are named. If we named it and it died or hurt somebody, there would be worse repercussions than if those things happened with an anonymous fish. There would be blame, and the agency would feel the brunt of it.

On Monday, October 4, a team of more than ten people in four boats descended on Lackeys Bay. We were prepared to make every possible effort to free the shark, even if that meant employing one of the nuclear options. In our toolbox was the fishermen's lead curtain, more water pumps provided by the Falmouth Fire Department, and a specially designed loop for tail-roping should we decide there was no other choice. Upon arriving at the bay, we found the shark well up inside the farthest reaches—pretty much a worst-case scenario. We tried to coax her out and then we tested the iron curtain, but she swam right through it. We then turned to the water pump again, but it was hard to move the shark in a specific direction given the broadness of the bay. With powerful pumps and hoses mounted on two boats, we were finally able to get behind the shark on both sides and blast the water, keeping her moving where we wanted her to go.

Ernie observed that it was kind of like cattle herding, as the boats swept into position behind Gretel and pummeled the surface with high-pressure jets of water. The shark was responding, but experience with this shark had taught me not to count my chickens. We prodded and coaxed and cajoled the animal from the far reaches of the bay into deeper water near the opening until finally, after almost two weeks of turning and circling and cautiously probing, the shark had had enough and rocketed herself over the remaining eelgrass beds and into the open water of Vineyard Sound.

"Go! Go! Go!" we yelled as the shark we'd named Gretel headed to deeper water. The shark had been trapped for a long time—nearly two weeks, during which I had slept little and worried much. I was exhausted but elated. This was undoubtedly the high point of my career and likely a point, I realized, that would be tough to top. *I must be the luckiest guy around*, I thought. As a kid I'd been enamored with an animal made famous to my generation by a movie shot on the island on which I now live. I'd let my dream of studying white sharks subside, like so many children's dreams, but that was okay because I had just about the coolest job I could imagine. But now, seventeen years after I'd moved to the island and become the state's shark expert, the very animal that had started it all for me showed up right in my backyard. I had been responsible for its well-being, and we'd saved it. Although it was mentally and physically exhausting, I soon realized that I'd been given an opportunity few white shark researchers had ever been given—two weeks to observe the animal up close.

And I'd tagged it!

I barely had time to savor my euphoria before I was again plunged into anguish. Less than an hour after she swam free, the tag at the base of the shark's dorsal fin popped off. The tag assumed that the shark was dead because it had remained in the shallows for too long. I argued with the tag company that it hadn't even been in the shallows for three complete days, but it really didn't matter. To me, the study only really started when the shark left the pond, when I was collecting data about natural behavior. Had the tag popped off an hour earlier, I could have replaced it with a new one. Had it come off an hour later, I would have had at least an hour of data. Now I had nothing.

Seeing how dejected I was, Dan McKiernan said, "You know what? You never know what's going to happen. In a few years, you may be tagging these things left and right."

Highly unlikely, I thought, dejectedly.

But I was dead wrong.

CHAPTER 8

IT BEGINS

As the holiday weekend approaches, we just want the public to realize sharks are in the area and to exercise caution and good judgment.

—Lisa Capone, Massachusetts Executive Office of Energy and Environmental Affairs, Labor Day weekend, 2009

AUGUST 25, 1986—43°52'09.0"N, 69°05'29.8"W (GULF OF MAINE)

The twelve-foot female white shark passes Monhegan and Metinic Islands, swimming in a northeasterly direction through the Gulf of Maine. It's summer, and the days are long. The water has warmed from the thirties in the winter to nearly 60°F, but the shark's internal temperature can be as high as 70°F. It's the reason the shark can both venture here in the first place and, more importantly, hunt successfully.

As she swims, water moves into her open mouth, then flows over her five gills and out her long gill slits, allowing her to breathe. Sharks lack lungs, but like human lungs, the shark's thin gill filaments provide for gas exchange, whereby oxygen is extracted from the water and carbon dioxide is released. Because water possesses far less oxygen than air per unit, the shark must breathe ten to thirty times more water to get the same amount of oxygen that a human would get from the air. As a large, active, and fast-swimming shark, she relies on having more gill surface area and a larger heart to collect and then pump oxygen-rich blood throughout her body. The downside to this so-called *obligate ram ventilation* is that she must keep swimming. Otherwise, unlike sharks that can breathe by using the muscles of their mouth to draw water in and over their gills (a process known as *buccal pumping*), the white shark will drown if she stops swimming.

At this point in her life, the shark is, for the first time, venturing into these higher latitudes. She spent her first several summers within the relative safety of the New York Bight, where she fed largely on ample fishes and squid in the middle of the water column or on the bottom. As the water temperature cooled in the autumn, she headed south, moving on the continental shelf to the Carolinas, where she spent the winter, feeding largely on other species making the same migration.

But changes in her body over time have prompted changes in her behavior. She is larger now, and thick body walls acts like insulation surrounding her internal furnace, allowing her to move farther north. She now has ventured northeast along the coast of Long Island, past Block Island Sound, east along the southern shores of Martha's Vineyard and Nantucket Islands, and then north along the Outer Cape and finally into the Gulf of Maine. It is here where, like her ancestors, she is learning to hunt seals along the rocky coastlines.

As she passes inshore of Little Green Island and enters Penobscot Bay between Owl's Head and Vinalhaven, something attracts her attention. She's unfamiliar with exactly what it is. A scent? A vibration? The waters of Penobscot Bay are warmer than the Gulf of Maine but colder than the New York Bight. It's noisy with the thrumming of diesel engines and the bottom is littered with lobster traps, their vertical lines soaring up through the water column to colorful buoys above. The shark passes the entrance to Rockland Harbor, with its mile-long breakwater sheltering it from storms.

The tides here are much larger than along the New Jersey coast, and there are few sandy beaches. The tidal range, which can exceed ten feet, twice a day exposes rocks and ledges against the backdrop of an evergreen-dark shoreline. The tide is rising now, and the shark swims closer to shore, exploring the drop-offs as turbulent water fills the crevices across a low ledge near the entrance to Rockport Harbor. Rockweed undulates. Gulls circle above. As the ledge gives way to the tide and submerges, there is a disruption in the water. Harbor seals bathing in the afternoon sun slide into the water, unaware of the shark's presence.

Because of the location of her eyes, the shark can see almost 300 degrees as the seals enter the water, but the image in her mind's eye is devoid of color. Instead, the seals appear as dark, contrasting shapes moving quickly against a bright sky. For most of her life, the shark's diet has consisted primarily of fish, but as she's grown, the energy expended to chase and catch fishes returns less and less of a caloric reward. In response, her teeth have become broader and distinctly serrated, better evolved for the type of hunter she will soon become.

She turns toward the nearest seal . . .

*Bruce Bean and Rod MacKinnon launched their sea kayaks from Cow-*yard Landing in North Chatham on the fifteenth of August 2009. Despite being a Saturday morning in mid-August on Cape Cod, it was still pleasantly quiet at 8 A.M. as they pushed off from the beach and pointed their boats south. Both Bruce and Rod were paddling eighteen-foot fiberglass sea kayaks, and it was clear to anyone observing them from shore that these guys were not weekend warriors. They were experienced kayakers paddling performance boats, and they wasted no time. Soon they were moving fast to the south, the sound of their paddles creating a cadence into which they settled as Chatham's well-heeled coastline of large homes and green lawns slid by to their right.

Chatham Harbor is shielded here from the brunt of the Atlantic by a low-lying swath of beach, sand dunes, and grasses known as North Island. Inside the harbor, and directly to the left of the kayaks, was Tern Island, with its extensive tidal flats and exceptional birding. The undeveloped island, which was a Mass Audubon sanctuary, came to a point on its southern end just across from the bustling Chatham Fish Pier, where Bruce noticed a family watching seals follow a fishing boat to the dock. Bruce and Rod continued southeast toward Chatham Inlet, between the tip of North Island and the mainland.

Once through the inlet, they were exposed to the open ocean that spilled away to the east. The shoals here were a constantly evolving labyrinth of sandbars and swales. As the tide dropped, the shallowest of the bars slowly emerged in ragged lines of white surf. The low-backed ribbons of sand standing proud above the tide attracted seals that hauled out on the wet sand and bathed in the summer sun. To the west was the iconic Chatham Light. It was first built in 1808 to protect sailors from these dangerous shoals and currents. The lighthouse's forty-eight-foot white turret was bathed in the harsh morning light, its black iron superstructure cutting a crisp silhouette in the unbroken August sky above a beach aptly named Lighthouse Beach.

Bruce and Rod had paddled here many times, but the trip that day was special. Their plan was to follow the sandy beach all the way south to Monomoy Point and a place called Pollock Rip, where the tidal current races as much as two and a half knots. It was a stunning spectacle of nature's power, and it was also a great place for experienced kayakers to play.

Monomoy itself is a spit of sand and shoals with an on-again, off-again relationship with mainland Cape Cod. It dangles south-southwest from the Cape's elbow for a little over eight miles. In the middle of the twentieth century, one could walk from Chatham to Monomoy Point and gaze across Nantucket Sound toward Great Point Lighthouse, some ten miles distant. A 1958 storm breached the peninsula's fragile sandy neck and created Monomoy Island, which was subsequently breached during a 1978 storm, creating North and South Monomoy Islands. In the 1990s, storms and currents coaxed South Beach, arcing out from Lighthouse Beach southward toward Monomoy Point, where it eventually merged into South Monomoy, once again making overland travel possible . . . at least for the time being.

Like many people who know Monomoy well, both Bruce and Rod considered it more than a stunningly beautiful beach. It was a special place. In a region where so much of the coastline is densely populated, Monomoy is a refreshing respite, like stepping back in time. It's a unique place on many levels, which is one reason it was designated a National Wildlife Refuge in 1944 and then a Wilderness (with a capital *W*) in 1970 under the Wilderness Act. Per the Wilderness Act, Monomoy is federally recognized as an area that "generally appears to have been affected primarily by the forces of nature with the imprint of man's work substantially unnoticeable." While a decommissioned forty-foot lighthouse and an assemblage of supporting structures still stands on South Monomoy, much of the rest of Monomoy's human history is erased, leaving an environment that looks much as it did to the region's earliest inhabitants, the Monomoyicks.

Monomoy is the only nationally designated Wilderness on the southern

New England coastline, a region most people don't associate with the word *wilderness*. Not only does the designation offer long-term protection for remaining wild landscapes, but it also offers a sort of figurative antidote to civilization, even for those who have no intention of ever actually going there. On a philosophical level, having Wilderness means there is *wildness* in a world where our relationship with nature is often so severely mitigated. It's one reason people living in New York City became passionate about wolf reintroduction in Wyoming. It's not that they wanted to see a wolf necessarily—certainly not up close. But the knowledge that something wild still exists, something that could be argued is higher than us on the food chain, is important.

Some would argue elemental.

Monomoy may be the best glimpse of what the Outer Cape looked like before the first European settlers, and it's tempting to look out over the windswept dunes bristled with American beach grass, beach pea, and dusty miller and imagine this is exactly what Samuel de Champlain saw when he named it Cape Mallebarre in 1604. It's tempting to also assume that the ecosystem, the assemblage of flora and fauna, is the same as it was before Europeans settled. After all, rare and endangered species, like black-capped roseate terns and the piping plovers with their bell-like whistle, that were once plentiful in New England can today be observed at Monomoy. But this ecosystem is forever altered. As big as the Monomoy National Wildlife Refuge is—it's almost twelve miles square—it's but a fragment of a much larger ecosystem that is no longer intact. It is by any measure a shadow of the productiveness and bounty that existed when the Monomoyicks hunted and fished here and returning it to what it once was is likely impossible because of that. Instead, it exists as a pocket of critical habitat—a critical stopover and refuge for endangered migratory shorebirds. In 1990 it became part of the Western Hemisphere Shorebird Reserve Network, a complex of protected habitat that helps usher birds along a north–south axis on their annual migrations.

It's not only birds, though. For kayakers like Bruce and Rod, the most obvious animals to have returned to Monomoy are the seals. First it was the harbor seals that arrive in the fall and depart in May for the Gulf of Maine. Then it was the gray seals, which were slower to return but which are now present year-round in larger and larger numbers.

"I must have passed many thousands of seals that day," Rod told Bruce regarding a solo trip he'd made that spring from Chatham Light to the tip of Monomoy. Rod had stopped paddling in order to pull out his camera. "As I approached the tip of Monomoy," he continued, as he fiddled with the camera's settings, "the beach was confluent with seals, and I had to go all the way around the bulbous tip—a lot of beachfront—to the sound side, where there were no seals, to stop for a brief rest." The two kayaks lulled in the swell as Rod photographed the dozen or so seals hauled out on an adjacent sandbar.

Many of the seals returning to Massachusetts were doing so by migrating south from Canada. At first they would simply haul out at various locations as they spent time in the region, but soon, thanks to both state and federal protection, they started reestablishing breeding colonies. It was something that no living New Englander had ever seen, and opinions were mixed. Wildlife managers tended to see the return of seals in large numbers as proof that management strategies were working, which, they assumed, would lead to a healthier ecosystem overall. Tourists were willing to travel and even pay to see the seals. Fishermen were not so sure.

By the late 1980s, the harbor seal population in New England had nearly doubled, and in 1990, the first gray seal pup was observed at Monomoy. The gray seals that Rod was photographing were a much larger animal than harbor seals. Male gray seals can potentially grow to eight feet in length and weigh as much as eight hundred pounds. By comparison, the largest harbor seal may sometimes exceed five feet in length and seldom push more than 250 pounds. A smaller juvenile gray seal is readily differentiated from similar-sized harbor seals by their long snouts, which have earned them the moniker "horse head" seals. While they were slower to return, the gray seal population was now increasing dramatically. In the nineteen years since the first gray seal pup was documented at Monomoy, the annual surveys had recorded fewer than ten pups in any one year, but in 2009, a total of sixty-eight were documented.

It was about 8:45 A.M. when Rod paddled his kayak over to Bruce to hand him the camera. In addition to the seals hauled out on the sandbar, there were several in the water swimming near their kayaks. Bruce always felt more relaxed when seals were around, because he figured if the seals were acting relaxed, there was little chance of a big predator lurking nearby. Bruce was not afraid of white sharks, but he was aware that they were out there. They'd both seen half-eaten seals in the past, and frankly, they thought it was pretty cool. Rod had done some work with Lisa Natanson, and knowing she was a shark expert, he'd asked her if he, as a paddler, should be concerned about white sharks. She told him that white sharks will not intentionally attack a kayaker, but, she said, "a kayaker could get bopped." Rod had relayed the exchange to Bruce the next time they were heading out on the water. "By 'bopped,'" Rod explained, "she meant they are very curious and could check you out to see just what you are. It turns out sharks feel things with their mouth." Rod told Bruce that Lisa had also said not to urinate in the water when white sharks were around. "Mammal urine will really get them activated." They both got a good laugh out of that.

"What did she say to do if we actually see one?" Bruce asked.

"She said to look as large as possible," Rod replied.

Rod nestled his kayak in beside Bruce's boat in a maneuver boaters call "rafting." He passed the camera to Bruce so he could shoot some pictures of Rod. Although the day was generally calm, Rod was aware of a larger wave pushing toward them. The wave crested as it hit the shallower water, and it caused Bruce's boat to side-surf into Rod's. Being unprepared for the jolt, Rod's kayak, a performance design that is notoriously tippy at rest, flipped. Because they'd been rafted, Rod was not holding his paddle when he went over, and therefore couldn't perform a classic Eskimo roll. Instead he made several "thrashy" attempts to do a tricky hand roll. Finally, he found his paddle floating beside him and rolled upright. He was still wiping salt water from his eyes when both men heard a loud splash, followed quickly by a second. It was a seal, and it seemed to be trying to jump from the water. A few seconds later they understood why as they both saw the dorsal fin of a shark slice through the water. Rod was amazed at the sound the fin made, not unlike his boat at speed. The fin itself was about a foot out of the water and Rod clearly saw a large dark mass beneath it. The seal, which they could see was likely a large male, disappeared beneath the water again and then, after several beats of confused silence, surfaced right beside their kayaks in a plume of bloody water. As quickly as it appeared, it was gone again. As was the shark.

Remembering Lisa's advice, Bruce and Rod rafted up again and watched the water around them uneasily. Rod's kayak was nearly impossible to reenter in the water if the paddler had to bail out because he or she could not roll upright, and he was thinking about what might have happened if he had not found his paddle. It was likely he would have been treading water amid the blood. Eventually the adrenaline started to ebb, and Bruce and Rod decided there was no reason not to continue to Pollock Rip as they had originally planned. They resumed paddling, their eyes scanning the surface for fins.

"You know how we always assumed we were safe in the water around seals because we thought they must be watching," Bruce eventually said to Rod.

"That's clearly wrong," Rod replied.

Bruce remembered seeing my request for anyone who had a possible white shark encounter to contact me, and so later that day, Bruce placed the call. After listening to Bruce's account, I was excited. I was convinced the report was credible, which made it the first white shark sighting of the summer. "We think it's a great white shark," I told a reporter at the *Boston Globe*. "There's not many species of sharks in New England that would attack a seal." I was quick to point out, however, that this was not a cause for concern. Like when the dead juvenile white shark washed up on a Nantucket beach the summer before. "It's a natural event that's been going on for a few years," I said. In 2007, two seal carcasses examined at Monomoy were confirmed to have been attacked by a white shark. "It's certainly not a new phenomenon," I told the reporter, but that didn't stop him from closing his article with a reference to *Jaws*:

> A giant, implacable great white shark was featured as the villain in the 1975 hit movie *Jaws*, making an indelible mark on popular culture and heightening many people's fear of the animals.

With Labor Day weekend 2009 just around the corner, citing *Jaws* in an article about a white shark close encounter was perhaps just too rich to pass up, but as the state's shark biologist, I was careful to again tamp down hyperbole.

"White sharks do occur in New England off our waters; however, they aren't very abundant." *How not-very-abundant?* the reporter wanted to know. "We'd be hard-pressed, for example, to go out and try to find one," I said.

"Getting Sharky"

By the mid-2000s, the *Jaws* effect still held sway over some beachgoers, but computer-generated imagery had long ago surpassed in realism the larger-than-life models Spielberg used to film the 1975 movie. To millennials, the opening scene was, largely by omission of the actual shark, still every bit as gripping, but the later shots of the "monster" were increasingly seen as just that—a monster that, while based on a real animal, did not actually exist. It was fake, and an ever-growing demographic thought sharks were "cool" and nothing like the movie's antagonist. They were more likely to talk about the white shark being maligned by Hollywood than express fear based on a movie.

In January 2009, a South African naturalist named Mike Rutzen called white sharks "the most misunderstood animal on the planet" in a BBC documentary featuring him swimming with white sharks without the protection of a cage. In some camps, white sharks were even portrayed as an animal more akin to a golden retriever than a wild apex predator. For its part, the media doggedly continued to hype the *Jaws* narrative whenever it saw an opportunity, especially on Cape Cod, given its proximity to where *Jaws* was filmed, but for much of a younger generation born after the film came out, the image of Quint (Robert Shaw) sliding over the transom into the shark's manic jaws was not the first image that came to mind when they heard news about a white shark. Public opinion was shifting.

Then, in the spring of 2009, a new iconic image of the white shark appeared when the BBC released the feature-length documentary *Earth*. Based on the 2006 BBC series *Planet Earth*, the film featured dramatic footage of a white shark breaching in pursuit of a seal. The scene was filmed with a state-of-the-art, high-speed, high-resolution camera capable of shooting a thousand frames per second, and the shark attack was subsequently slowed down forty times to create one of the most compelling pieces of footage of a white shark ever produced. It was spectacular to see the power of the animal launching from the depths to engulf a seal in its fully protruded jaws. Those large white triangular teeth cut against a cavernous black throat, massive pectoral fins outstretched, caudal fin slashing air as the shark turned in midflight and reentered the water with the seal in its jaws. It was truly awesome. "When you slow it down to that speed," said Alastair Fothergill, the film's director, "it becomes a ballet really." But it also was so much more. "It says look at this beautiful animal and admire it," said Fothergill. "Yes, it's about to eat Bambi, but that isn't the point."

For some people, however, that was exactly the point. The way the shark ambushed its prey from beneath without warning was terrifying, far more terrifying than any scene in *Jaws*, and the quality of the filming blew away any CGI shark (or monster for that matter). For those swimming in waters inhabited by white sharks, there was new fodder for nightmares despite knowing that the odds of an attack were very low. And this was happening at the same time that people on Cape Cod were beginning to come to terms with the fact that a white shark in shallow waters off the Cape was perhaps no longer the anomaly it had been for more than a generation.

The seal attack observed by Bruce and Rod on the fifteenth of August was followed by the confirmation that a seal carcass examined by us on the twenty-eighth was the result of a white shark. Then, in the week leading up to the Labor Day holiday, a slew of "large shark" sightings were reported. As I had learned, reports of white sharks in the media often lead to more white sharks reported. Just as was the case with the New Jersey shark attacks in 1960, when the media put sharks on the front page, everyone and their brother suddenly started seeing sharks in every shadow in the water. But in the week before Labor Day 2009, there was enough evidence that I was pretty certain things were getting sharky on Cape Cod.

It was the Thursday before Labor Day, and the Marine Fisheries Advisory Commission was meeting for its regular monthly meeting at the Radisson Hotel in Plymouth. The purpose of the commission, which was primarily made up of recreational and commercial fishermen, was to discuss upcoming regulations and advise the Division of Marine Fisheries whether they should move forward on proposed regulations. I was there at the behest of my boss, Paul Diodati, who was still the director of Marine Fisheries. Paul, who was also at the meeting, had asked me to summarize some proposed fishing regulations regarding sharks, but I had yet to speak. As I listened to the commission discuss another topic, my phone vibrated in my pocket. I fished it out and flipped it open. It was Captain Billy Chaprales. I took the call, but kept my voice low, so as not to be disruptive.

"Hey, Billy. I'm in an Advisory Commission meeting. Can I—"

"George saw white sharks!" Billy was excited, and he must have been yelling into the phone. I covered the speaker, but could clearly still hear Billy say, "He saw white sharks!"

I got up and left the room. After the door closed behind me, I asked, "What are you talking about, Billy?"

"Call George right now," Billy said.

"Okay, I will." I hung up the phone and momentarily considered going back into the meeting and calling George later. But white sharks? Plural? I dialed George. "Hey, George. It's Greg."

"Billy called you?" George asked. George was a commercial airline pilot turned spotter pilot who had worked with Billy both spotting for bluefin tuna and for the basking shark work.

"Yeah. He did," I said. "What's up?"

"I was flying the Outer Cape," George said. "I had a photographer with me. She wanted to just photograph the beaches, but I'm pretty sure I saw two white sharks."

"Where?" I asked. "Were they in the shallow water?"

"Yeah!" George said. He was clearly excited. "It was down off Monomoy, and they looked like they were patrolling right up against the beach."

"Holy crap!" I felt my heart race.

"If you wanna come down to the airport, I'll take you out to see them," George offered.

I checked my watch and glanced back into the room, where the meeting was still taking place. I caught Paul's disapproving eye, and I motioned for him to come out. "That sounds great," I said to George. "Let me just tell Paul, and I'll jump in the truck."

"What's up?" Paul asked, coming out of the room with an annoyed expression on his face.

"I just got a call that there could be a couple of white sharks swimming off the Outer Cape. I think it's credible, but I'd like to go confirm it," I said.

Paul nodded. "Hang on one sec, though." Paul opened the door and got Mary Griffin's attention. Mary was the new commissioner of the Massachusetts Department of Fish and Game, and she joined us in the hallway.

"What's going on?" Mary asked.

Paul looked to me.

"I think we may have a credible sighting of a couple of white sharks swimming off Monomoy," I said. "I want to go have a look."

"White *sharks*? Plural?" Mary asked.

"That's what I'm hearing, and I trust the guy who called me. He's willing to take me up in his plane right now to try to locate them again."

Mary looked from me to Paul and then back to me. "A couple of great whites swimming off a Cape Cod beach right before Labor Day? Yeah," she said. "I think it's a good idea."

I drove down to the Falmouth Airpark, where George lived in a house adjacent to the runway, his 1955 red and white Piper Super Cub parked in the garage. We took off and headed for Monomoy, where I hoped to confirm the sightings and perhaps even the species if we spotted the sharks. I was excited by the prospect of multiple white sharks showing up in my backyard, but I tried to remain objective. A single white shark feeding on seals close to shore was a possibility—really a logical conclusion I'd been talking about for a few years now. Multiple white sharks feeding on a whale carcass, as had been observed many times now in New England waters, was also a reasonable possibility. But several large white sharks hunting near each other in shallow water close to shore over a period of several weeks? That would be a game changer, both for beach managers and for the science.

As the plane banked off the eastern shoreline of Monomoy, I saw the first of them—a distinct gray shape like a shadow in the water, and it was shallow enough to clearly see the bottom. The fish was swimming slowly, and judging by its body movement, it was undoubtedly a shark. A big shark. Could it be a white shark? The plane circled so I could get a better look. *That sure looks like a white shark*, I thought to myself. As we made another pass, I became certain.

"That's a white shark, damn it!" I yelled, followed by a euphoric hoot. "That's a great white shark!"

George continued down toward Monomoy Point and then circled back low over the shallow water breaking onto beaches thick with seals.

I spotted four additional large sharks. This was remarkable. I tried to call Paul from the plane, but the noise was too much to communicate anything clearly. When the plane landed at Falmouth and taxied back to George's house, I thanked George and immediately called Paul.

"Yes," I told him. "Five. And I'm confident they were white sharks. I need to get on the water," I said. "We need to try to get tags into them."

I called Billy, who said to meet him at his boat, the *Ezyduzit*, in Sandwich Marina first thing in the morning. Once I was off the phone with Billy, I called Paul back and explained the plan. Paul gave the trip his blessing.

"Um," I hesitated for a moment, holding the phone to my ear. "You know, I don't really have a lot of funding for this, just some tags."

"Don't worry about it," Paul said. "We'll sort it out later."

"Get the Damn Tag in the Fish"

On Friday morning, I walked off the first ferry from the Vineyard to Falmouth and headed to the town lot where I often kept a state vehicle during the summer months. It's about a forty-minute drive from Falmouth to Sandwich, and by the time I arrived, Billy was ready to roll. I was joined by my research assistant John Chisholm and UMass graduate student Jeff Kneebone, who was working with us on sand tiger sharks. We boarded and headed out of the marina, through the east end of the Cape Cod Canal, and steamed across Cape Cod Bay toward Provincetown. By the time we ran along the Outer Cape south to Chatham, it was close to 11 A.M. and George's red and white plane appeared overhead. The plan was for George to spot the sharks from the air and then direct the boat to them, just like we'd done for tunas and basking sharks.

George put us on the first shark almost immediately, and I was ecstatic to see it was indeed a white shark. Unfortunately, it was too deep to tag. As the day progressed, fog and cloud cover hindered the search, but in the afternoon, we found another white shark, but it also remained too deep. By midafternoon, heavy fog grounded George. Without eyes in the sky, we attempted to draw the attention of sharks with chum, which included bluefish and tuna remains. Although a white shark moved through the chum slick, it did not pass close enough to tag. As dusk approached, Billy pulled the plug and steamed to Saquatucket Harbor in Harwich for the night. Although no tags went out, I was excited nonetheless—there was no doubt now that there were white sharks in my backyard!

With the Labor Day weekend upon us, the state made a series of public statements. State officials confirmed, based on my report, the presence of several white sharks in the area. They reiterated that this was not a cause for concern. "As the holiday weekend approaches," Lisa Capone, from the Massachusetts Executive Office of Energy and Environmental Affairs, told the press, "we just want the public to realize sharks are in the area and to exercise caution and good judgment." The Office of Energy and Environmental Affairs is the parent organization for the Department of Fish and Game, under which the Division of Marine Fisheries operates. The higher-ups thought the news was big enough that they wanted to deal with it on a cabinet level. Capone emphasized that the threat posed by the sharks was small, and she assured people that sightings of large sharks, including white sharks, were somewhat common around Cape Cod.

Local officials took a similar tone. "We're not closing the beach," said the director of Chatham Parks and Recreation, Dan Tobin. "We would do that if we had any concerns. We're just encouraging people to use common sense."

As I was out searching for sharks by boat on Friday, a reporter writing about the sightings for the *Boston Globe* walked Lighthouse Beach interviewing beachgoers about the sharks. "No one interviewed said they were concerned by the threat of the sharks," he reported. From the philosophically tinged bravado of a forty-five-year-old firefighter ("If it happens, it happens") to the sage advice of a ten-year-old ("You're less likely to get killed by a shark than by a coconut falling on your head"), nobody seemed ready to pack up their beach chairs because of the official report of white sharks in the area. The grandmother of the ten-year-old opined, "There needs to be more control of the seals, but it would be better to have people doing the controlling. Sharks aren't very popular. They tend to scare too many people away, and that's not good for the economy."

On Saturday, I was again planning to head out aboard the *Ezyduzit*. As a result of his commercial bluefin tuna harpooning experience, as well as his scientific tagging work with me, Captain Billy was the best in New England when it came to deploying a tag into a fast-moving and agile fish—even so, he'd never tagged a white shark. In fact, I was the only one who had ever successfully tagged a white shark with a pop-up satellite tag in the northwest Atlantic, but hopefully that was about to change. I had six PSAT tags on the shelf; some of them were left over from a small grant I received from the Massachusetts Environmental Trust in 2006 to tag white sharks and some were for basking shark research. Once again I grabbed the earliest ferry to Falmouth and headed to Harwich. We left from Saquatucket Harbor and steamed around the southern tip of Monomoy.

It was about 8:30 A.M. when George's voice came over the *Ezyduzit*'s radio. "I've got a large shark off the southern tip of Monomoy Island," he said. The conditions could not have been better. Under a bluebird sky, the ocean was calm and Billy's son Niko maneuvered the boat into position behind the shark. Billy, a twelve-foot harpoon in hand, made his way out onto the boat's pulpit, a small aluminum platform at the end of a long aluminum pole, like a narrow walkway, that projected from the boat's bow. I leaned out over the starboard gunwale. The ten-foot shark was right there, swimming with long sweeps of its caudal fin. I had spent a lot of time close-up with Gretel, but she had not exhibited particularly normal behavior given the circumstances. This shark appeared to be doing exactly what I'd seen it doing from the air. It appeared unfazed by the boat and just continued to patrol the shallows. It was magnificent.

My eyes darted apprehensively from the shark's dark back to Billy's body leaning against the rail waiting for just the right moment. The shark was maybe four feet below the surface. In my mind, I was replaying Gretel's tag failure back in 2004. "Come on, Billy," I muttered to myself. "Get the damn tag in the back of that fish."

Billy waited. Years of harpooning and tagging fast-moving predatory fishes had taught him patience, and I marveled at how he possessed an almost sixth sense for when to strike. To correctly tag these fish, the pulpit of the vessel must be above the shark and the shark must be within four or five feet of the surface. Anticipating the shark's movements and figuring that into a calculus that included the boat's heading and speed, the sea state, and the shark's depth, Billy slowly inhaled the rich marine air into his nostrils and then struck with lightning precision. The tag seated into the shark's thick skin behind and to the side of the dorsal fin. It was perfect.

I couldn't believe it. We'd done it! We'd successfully tagged another white shark in the Atlantic with a PSAT tag. I was psyched by that fact alone, and then I started thinking about the data we'd get from the tag. Where would this shark go? What would it reveal to us? This was the beginning of something huge, especially seeing how far white shark research in the Atlantic lagged behind white shark research in other parts of the world.

Although George located additional sharks throughout the day, none were shallow enough to tag. However, later that afternoon, George put us on a shark about twelve feet in length, and Billy did it again. We had now tripled the number of white sharks successfully tagged with satellite tags in the northwest Atlantic, and we weren't done yet—after all, I'd seen five white sharks in the area just two days before. It was late in the day, though, so we decided it was time to head back to the dock.

As we steamed back to Harwich, I was already dreaming of January 15, when the first tags were scheduled to pop off, come to the surface, and send their data. But even if both tags somehow failed to work, we still achieved proof of concept. We'd shown we could do this, and there was little doubt in my mind that we were on the cusp of something big.

"From a scientific perspective, it's fantastic," I told a reporter from the *Cape Cod Times* about the successful tagging. "We're pretty excited to be putting together the pieces of the puzzle."

That night, as I took the ferry back to the Vineyard, I recalled how it was barely two weeks ago that I'd told a reporter, "We'd be hard-pressed to go out and try to find one."

That made me smile.

More Tagging, More Publicity, More Questions

The confirmation of several white sharks not far from the Cape's busy beaches on the biggest weekend of the summer elicited a somewhat more muted response from beach managers and town officials than it had for me. It wasn't that there was hysteria, but there was some confusion and concern. Beach managers, who had only days before said, "We're not closing the beach," now closed many of those same beaches. There were also rumblings about going out and "getting the sharks," an instinct that harked back to the days of shark tournaments intended to "rid the ocean of man-eaters." I recalled the scene on the Amity dock, and I reminded the public that white sharks were now a prohibited species under both state and federal law. "I don't think it would be prudent to catch one," I said. Although there were concerns about public safety, there was also a swell of people who made the drive to Chatham for the chance to maybe see one of these sharks. Beach parking lots, which were projected to be emptier owing to the shark reports and beach closings, were instead as full as ever.

On Sunday and Monday, Billy and his boat were unavailable, and I had interviews with CNN and ABC's *Good Morning America*, but I was anxious to get back on the water before Tuesday. So much of our work was weather-dependent, and it nearly killed me to not be on the water if there were white sharks about and the weather was good. I still had three tags, and I knew there were at least three more white sharks out there, so in Billy's absence, I called a friend who had a boat and had previously offered to help. The boat was not as big as the *Ezyduzit*, and it didn't have a pulpit—it was undoubtedly less than ideal, but I figured it was worth a shot.

With George in the air, we again quickly located a shark just off the beach. I made my way to the bow and prepared my own tagging pole, which I had used on basking sharks, with the dart and tag on the end. It was awkward, and I didn't have the visibility or the angle to make the kind of attempt I would have liked, but there I was . . . and there was the shark. I waited until I thought I had the best shot and then thrust the pole into the water. I missed the shark, and to make matters worse, the dart with the tag detached from the harpoon. I watched it sink, and although we could see it on the bottom, it was out of reach without someone going in the water. Jeff Kneebone was with us on the boat, and he offered to dive for the $4,200 tag. We talked it over and eventually decided that, with George overhead spotting, the level of risk was acceptable if no sharks were in the vicinity. Jeff went over the side but almost immediately had a change of heart. Overwhelmed by the combination of cool water temperature, murky New England visibility, and the memory of a twelve-foot white shark recently in the area, he propelled himself back into the boat faster than he'd gone over the side. Nobody faulted him or disagreed with his decision. I recorded the coordinates of where the tag sat on the bottom.

We planned to go out with Billy the next day, and I hoped we'd be able to recover the tag. I was frustrated and angry at myself. Losing that tag was a big deal beyond just its monetary value. I knew there were several other white sharks still in the area. That meant that even if we tagged all the sharks we could, and even if all the tags worked, we'd be losing out on 20 percent of the potential data due to the lost tag. When I told Billy what had happened, he was livid.

"What the hell were you thinking?" Billy shrieked. "Why would you attempt to do this without me?" He took a breath and shook his head like a disappointed father figure. "You guys don't know what the hell you're doing."

I didn't blame Billy for being so mad. The truth was that it was not the right boat to tag from. It was small and maneuverable, yes, but it had no pulpit, nor any kind of rail to support me. There was no tower to give the captain any kind of view of the shark, and it was clear to me, after the fact, that we spooked that shark too deep before I even took the shot. I was also a very inexperienced harpooner at the time—something that would change in the coming years.

Back aboard the *Ezyduzit*, I directed Nick Caloyianis, who was on another vessel, to the area of the lost tag. With George spotting, they put a diver over the side but were unable to locate the tag. After giving up on the tag, we once again found the sharks, and Billy was able to tag three more animals, ranging from eight to thirteen feet in length. It brought the total to five white sharks tagged in the past four days.

A few days later, the tag I'd lost washed ashore, and we were able to recover it and put the incident mostly behind us, except for a lingering tension with Billy that sat uncomfortably just below the surface.

The white shark news quickly became national news. Although some news outlets led with the science ("Great White Sharks Get First Atlantic Tags") others focused on public safety ("If you're sailing near Cape Cod this weekend, you might need a bigger boat"). Locally, my agency pushed the science, explaining that while great whites were a "high-profile" species, "nothing about their migratory patterns and behavior, at least in the Atlantic Ocean," was known. My perspective was a little more nuanced than the agency's spokeswoman's viewpoint, but that was because I was familiar with the previous work of Frank Carey, who had been the first to track a white shark in the northwest Atlantic in 1979, and Jack Casey and Wes Pratt, who had further defined the distribution of the white shark in 1985. It wasn't true that "nothing about their migratory patterns and behavior" was known, but what was known was, at best, spotty and opportunistic. If these sharks were coming to Cape Cod to hunt seals, and if this proved to be something that repeated itself in future years, the opportunities for study would be phenomenal.

In terms of the public safety angle, it had been nearly seventy years since a person was killed by a shark in Massachusetts. If these sharks that I had just tagged were simply doing what white sharks had probably done before the seal population had been culled from Massachusetts's coast (hunters in Massachusetts and Maine had, from the 1890s to the 1960s, been able to score a bounty of five dollars for a seal skin or nose), then that would have ramifications for those in charge of public safety. In 2008, a triathlete in California had been attacked and killed by a white shark, and more recently, at the height of the summer season in Australia, there had been a spate of shark attacks that had closed beaches and reignited debates over what to do about sharks inhabiting waters near popular beaches. I knew that some on the Cape were taking note. If the tagged sharks headed south as the water cooled and then returned north next season, that could be the first proof that Cape Cod had returned as a regular part of the species' migratory behavior in the Atlantic—meaning these sharks were no anomaly. Rather, I guessed, they would become the new normal.

There was ample evidence from other parts of the world that some white sharks predictably migrated between known locations, and it made sense to me that the increasing population of gray seals on the Cape, especially around Monomoy, would be more than enough to put Cape Cod on the map for migrating white sharks in the Atlantic. But I didn't want to get ahead of myself. The first thing I needed to do was to see where the five sharks we had tagged went after they left the Cape. The tags were programmed for two to release in January, two in March, and one in May. The waiting game began. If I fixated on it too much, it was maddening. So many things could go wrong—but then again, if even a couple of things went right, the payoff would be immense.

I did not sit idly by waiting, however. There was more than enough work to be done, not least of which was trying to figure out how I was going to tag more sharks next season if, as I suspected, more sharks appeared. I was confident we had figured out a reliable strategy to locate, approach, and tag white sharks off the Cape, but I was not nearly as confident about how we would fund the operation.

The first of the tags to report back pinged the satellite on schedule on the fifteenth of January 2010. It was fifty miles east of Jacksonville, Florida. When I shared this detail with the media, I couldn't help but anthropomorphize the shark. I deemed it a "snowbird"—the term used for people who travel south to Florida for the winter and then return to the North in the summer. The next tag popped up on February 4, about thirty miles away from where the first tag had appeared. A third tag reported in on the first of March, less than one hundred miles south of where the other two tags had emerged.

The tags showed that the sharks left the Cape and traveled south, hugging the coastline. The first shark spent most of October off the Mid-Atlantic and then moved south into the Carolinas in November, eventually reaching northern Florida in December. This activity, which was similar to the migration patterns of other fishes and whales that migrate along the east coast, surprised me. Based on reviewing a decade's worth of research in the Pacific using satellite and acoustic tags on white sharks, I had expected at least one, if not all, of the sharks to head offshore, but this trio remained largely on the continental shelf. While one shark did wander briefly off the shelf and dove to 1,500 feet, the animals generally seemed to prefer a relatively narrow temperature range between 59° and 67°F in water that was no deeper than 150 feet.

The data provided by the tags fascinated me, and while those data began to answer some questions, they also posed many new ones. It was thrilling and tantalizing and there was an almost incurable urge to draw conclusions from the data, but I knew I was dealing with a tiny sample size. My priority needed to be to tag more sharks, and that's exactly what I planned to do in 2010.

As the summer slipped into autumn and winter storms lashed the beaches of Monomoy, I reflected on what had transpired. During my life, I'd been at the right place at the right time more than once, but this coincidence was almost too much to believe. It was like a dream come true—and I half-expected to wake up at any moment in my childhood bed as just a kid, dreaming about being a shark scientist.

As a scientist, it's exciting when so little is known about the thing you want to study. There are scientists in California and South Africa and Australia who can tell you a million things about white sharks in those places, but we've learned that white sharks, while all belonging to the same species, have adapted to different parts of the world in fascinating ways. When we look specifically at white sharks in the Atlantic, there's almost nobody who can tell you much about them. Where do they go? Why do they go there? What do they do when they're there? What if we knew a place in the Atlantic where we could go and reliably find and study white sharks? It could open the floodgates. Would Cape Cod be that place? Would Cape Cod become a white shark hotspot?

As excited as I was at the prospect of being able to study white sharks in my own backyard, I was equally worried about how I would fund the work. As I've stated, tags cost $4,200 each, and I would also need to secure funding for Billy's boat and for the spotter plane. I think people have a sense that there must be plenty of funding for white shark research, but I quickly learned that was not the case. At the Division of Marine Fisheries, a lot of our budget is appropriated by legislators as is typical of most state agencies, and it became clear to me early on that sharks were not a priority. For almost all my career, I had to beat the bushes to fund the Massachusetts Shark Research Program. National Oceanic and Atmospheric Administration grants had supported a lot of my research, but the trouble was that white sharks, like seals, were already offered a high level of protection under both state and federal law. Unless you had some interesting twist, it was hard to get funding for white shark research because, once they had become protected, they essentially became a low-priority species. As a result, some of the grants like Saltonstall-Kennedy and Protected Species were therefore simply not applicable. You couldn't, after all, enhance protection of a species if it's already protected. On the east coast of the US, there were big problems with other species of sharks, like duskies and sandbars, and as I knew firsthand from my own research, it was much easier to get money to study them. The difference, of course, was that those animals were not nearly as high profile as an apex predator like a white shark.

Without traditional funding, I had to think outside the box right from the beginning with my white shark research. Thanks to Jack, I had learned some tricks along those lines, and those became very helpful, especially early on. I was able to leverage the relationships I'd formed with commercial fishermen and other nonscientists who were as passionate about the species as I was, and I was also lucky to have support from my bosses, who saw the value in the work and, on more than one occasion, approved a project as long as I found the funding to do it. Thinking back on it, the late summer of 2009 was one of the most exciting times in my career to date. It's when everything I'd learned both through my own experience and through the shared experiences of shark researchers like Jack and Wes started to come together. It was a new chapter, and I was excited to get writing.

CHAPTER 9

THE NEW HOTSPOT

I think that we are going to see more white sharks through the summer, and we can anticipate more for quite possibly many years to come.

—Greg Skomal, July 2011

There were so many great whites at the Cape that they were just about ordering Tanqueray and tonics around the pool at the Chatham Bars Inn.

—Brian McGrory, *Boston Globe* columnist

Just like the scientist in *Jaws*, [Greg Skomal] lives to study sharks.

***—Jaws Comes Home*, Shark Week 2011**

AUGUST 25, 1986—44°09'50.4"N, 69°03'27.1"W (PENOBSCOT BAY, MAINE)

She focuses on the closest seal. Her other senses will become more finely tuned as she gains more experience, but at this moment, it's only the sight of the oblong dark shape that elicits an instinctual, predatory response. There is no hesitation. No indecision or vacillation based on an odds calculus or risk assessment. She will learn to be a more judicious hunter, but right now, this is all about evolutionary predisposition as she whips around quickly and launches upward at an angle, tracking the silhouette. Closing the distance.

Since she's never observed one of her kind attacking a seal, she must learn to hunt primarily through experience. Both her successes and her failures will teach her, so long as her mistakes don't kill her. She'll become more cautious and less impulsive as she comes to better understand and utilize the full breadth of her sensory capabilities. She will become a better hunter.

Like an adolescent, the shark is not yet fully in control of her neuromuscular coordination, but she already possesses the speed and power to facilitate the hunting strategy for which she was built. Ram and bite. Her underslung jaw, armed with rows of blubber-tearing, serrated teeth and anchored by muscle, can generate an immense bite force. Her jaw protrudes as she closes on her target. Her eyes roll back an instant before impact.

Through the clear water, the seal sees the shark coming in the midday glare and turns sharply, propelling himself back toward the rocks with strong side-to-side flicks of his hind flippers. As he turns, he catches the shark's girth with the claws on his foreflippers. The seal is a large male weighing close to 250 pounds and approaching five feet in length. Now that he's seen the shark, he easily outmaneuvers the pursuit. The seal is no menhaden and flings himself out of the water and onto an exposed rock. The other seals, many of which are smaller and would have been better targets, scatter and make their way back to the safety of the rocks, too.

A gull that was eating an urchin looks on, distracted by the disruption. Nobody else has seen the drama. The summer day on the coast of Maine is undisturbed by the shark's appearance.

The shark turns and continues up the bay with long, methodical sweeps of her caudal fin. She takes water in through her nares, forcing it over her olfactory receptors, and while the oft-repeated "a white shark can sense a single drop of blood in a swimming pool" may be overstated, she definitely senses blood. As with so much of her environment, however, these signals are a result of human intervention—the chicken-processing plants in Belfast, Maine, that have for decades discarded their waste directly into the water causing a gelatinous blob of feathers, blood, oil, and viscera to undulate with the tide.

But it isn't just her sight and sense of smell—she also feels her way through the water, using specialized sensory organs called *lateral lines*. Essentially a hollow tube lined with sensory hair cells and running the length of her flank from head to tail on both sides, each lateral line allows her to detect disruptions in the flow of water. The tubes, a sort of extension of her inner ear, allow her to both hear and feel with her entire body, interpreting vibrations as much as three hundred yards away. Utilizing this sophisticated sense of "touch," she can determine the direction of potential prey—a flailing fish or cavorting seal—with deadly accuracy.

Sunset comes late to Maine in the summer, but the moonless night beneath a brush stroke of the Milky Way brings a placidity to Penobscot Bay. The shark glides through the dark stillness as wisps of bioluminescence create star trails through the dark tidal current behind her. This environment, like the shark's new dentition and changing behavior, is nothing like that with which she was familiar—the sandy beaches of the Jersey Shore, Cape Cod, or even southern Maine with their shoal water. Instead the rocky shoreline often drops off quickly, meaning the shark can get close in without the worry of beaching herself. She can hunt from deeper nearshore water, coming at her prey from beneath as opposed to charging it in the shallows. She will adapt. There is no other choice.

When we tagged the first five white sharks off Monomoy in September 2009, I wondered if Cape Cod would become a white shark hotspot. At the time, there were only a handful of other places around the world where shark scientists could go and be assured of finding white sharks to study. These hotspots, more accurately called *aggregation sites*, are where white sharks gather in large numbers to feed on high densities of seals and sea lions. In the eastern Pacific, there are California's Farallon Islands and Mexico's Guadalupe Island. In Australia and New Zealand, there are the Neptune Islands and the Chatham Islands, respectively. In South Africa, there's the south coast from Cape Town to Mossel Bay, including Gansbaai and Seal Island.

These white shark hotspots not only provide scientists a venue to study the species, but they also provide tourists a place to see white sharks up close. In 2010, South Africa was the dream destination for many white shark enthusiasts. The popularity of South Africa is, in large part, because of the remarkable breaching behavior that is the subject of the popular series of Shark Week documentaries titled *Air Jaws* (thirteen and counting since the release of *Air Jaws: Sharks of South Africa* in 2001). Although the water clarity in South Africa isn't on par with Guadalupe Island, one of the top shark tourism destinations in the world, the lure of witnessing a breach was enough to tilt the scales for many sharkophiles. As a bonus, a cage-diving trip to South Africa is both remarkably accessible and can easily be combined with a terrestrial safari, making it the perfect objective for anyone with a passion for wildlife viewing.

These were all reasons that Cynthia Wigren booked a trip to cage dive with white sharks in South Africa in May 2010. What neither she nor I knew at the time was that the trip would change both of our lives, ultimately providing me with the consistent funding I so desperately needed to continue my white shark research in Cape Cod.

Seal Island, False Bay, South Africa—May 2010

Cynthia Wigren was no stranger to traveling long distances to view wildlife. In Africa alone, by 2010 she'd already been to Tanzania, Uganda, and Rwanda. She'd been on safari, and she'd trekked through dense forest to observe gorillas. For Cynthia there was no better experience in the world than observing wildlife in its native habitat. Initially her love of wildlife led her to an undergraduate degree in wildlife management, but her career path took a turn. Eventually she worked remotely from her home on Massachusetts's North Shore for a Houston-based online trading company in the energy industry. Her day job may not have required a wet suit or binoculars, but it did provide the means to cultivate her love of wildlife through travel.

Five-acre Seal Island is just a short boat ride from Cape Town. In and of itself, the rocky island isn't much to look at—a low, guano-covered granite hump extending for about half a mile and never rising more than twenty feet above sea level. At its broadest point, the island is little more than 150 feet across. Despite its diminutive size, there are more than sixty thousand Cape fur seals that share the barren, rocky real estate with brown fur seals and several species of birds. With all that potential food available, it's not surprising that white sharks are drawn to the island between April and mid-September each year. In the 1990s, Chris Fallows was the first person to document white sharks breaching at Seal Island. In 2000 he led a team from Discovery Channel to capture the behavior on film, resulting in *Air Jaws: Sharks of South Africa*. Later, BBC came to Seal Island and filmed the stunning breaching behavior in high resolution and slow motion. In 2009 that footage became one of the showstopping moments in the feature-length documentary *Earth*. When she saw that footage, South Africa went to the top of Cynthia's bucket list.

Observing a large apex predator in its natural habitat can be a transformational experience. It's different than being at a zoo, where, even with the best care, the animal's behavior is altered. Seeing an apex predator going about its normal activities in the wild can reframe a person's perception of their own place in the world, and no activity inspires self-examination more than seeing an apex predator make a kill. That final interaction between predator and prey is as primal as any interaction in nature. It's the decisive moment between hunter and hunted that establishes the hierarchy in food chains, and it's what ultimately regulates ecosystems. Throughout history, humans have engaged in, and ultimately disrupted, the natural predator-prey bond to the point that witnessing the act of one animal killing another is, for most, reduced to observing one's cat kill a mouse or seeing a lion take down a gazelle in a wildlife documentary. Even in places where one can reasonably expect to see apex predators in the wild, witnessing a kill is an experience that most ecotourists won't see. At Seal Island, however, the predation of seals by large white sharks is as predictable as it is spectacular.

Cynthia booked their white shark experience with Apex Shark Expeditions, a white shark cage-diving company founded by Chris and Monique Fallows. The morning of their cage dive, Cynthia and her husband, Ben, woke early and boarded the *White Pointer II*, a thirty-five-foot catamaran purpose-built for observing white sharks. There were just twelve passengers, including Cynthia and Ben and the couple with whom they'd traveled from the States. It took about forty-five minutes to make the trip to the island, and Cynthia was excited . . . and a little nervous. Cynthia grew up on the New Jersey shore in Ocean Township, ten miles south of where Jack Casey had started his shark research at the Sandy Hook Lab in the early 1960s. As a kid, Cynthia loved the water and was in the ocean all the time, later becoming an avid scuba diver. Although she didn't outwardly fear going in the water, she'd be lying if she said she wasn't unnerved by *Jaws*—especially that opening sequence, where (spoiler alert) a skinny-dipper is attacked by a shark and killed. It wasn't that Cynthia had an irrational fear of sharks, but she did think of white sharks as something unto themselves—aggressive, dangerous, and unpredictable. It was a perception bolstered by things she had read, stories she'd heard, and footage she'd seen on YouTube and in documentaries.

As they motored out to Seal Island, Cynthia noted that the water was a muddled green color, not terribly unlike how it often looked off Cape Cod, where she spent as much time as she was able. In studying the predatory behavior of these sharks, researchers had learned that most successful kills around Seal Island utilize a very similar playbook. In some parts of the world, sharks need to patrol large areas looking for prey, but at Seal Island, the sharks need only orbit the south side of the island waiting for the seals to go to sea to feed and then return. Researchers have observed that most attacks at Seal Island occur in water that is more than sixty feet deep, deeper than the average visibility. They've also documented that the sharks tend to target young seals returning to the island alone or in small groups, animals that may be tired and less vigilant owing to their full bellies.

From close to the bottom, the sharks track their prey and then propel themselves in a rapid rush of strength and speed toward the surface. The vertical nature of the attack foreshortens the shark's bulk and presents the smallest profile to the seal as the jaws emerge from below. By the time the seal can see the shark—and in many cases, it probably never does—the seal's silhouette is starkly outlined by the sky, increasing the odds of an accurate strike. As the shark powers through the ocean's surface at up to twenty-five miles per hour, it crushes the seal in its jaws with nearly two tons of force per square inch.

It's one thing to see the footage of a breaching white shark on the screen, but Cynthia was completely unprepared for the breach when it happened in the wake of the *White Pointer II*. Although scientists, photographers, and filmmakers who spend the time to stay on-site have a very good chance of observing and documenting natural breaching behavior at Seal Island, Apex Shark Expeditions, along with other ecotour companies, often deploys a seal decoy to increase the chances of tourists seeing a breach during a half-day excursion. The shark that breached behind the *White Pointer II* hit the decoy, not a seal, but in that one moment, every image Cynthia had ever seen or imagined of a white shark, factual or fictitious, flashed before her eyes. Of all her experiences viewing wildlife, it was that moment, a moment that lasted no longer than a second or two, that shot through her like a jolt of electricity and instantly redefined the true meaning of the word *wild.*

Apex Shark Expeditions is an eco-friendly company, and while they strive to thrill their guests with a dramatic breach and an up-close look at a white shark, they also hammer home a message about the environment, conservation, and stewardship. The guide aboard Cynthia's boat said that, although the number of white sharks circumglobally is unknown, their numbers had decreased precipitously due to a variety of factors, including overfishing of their primary prey species and accidental catching in gill nets and in other fisheries. Cynthia learned that white sharks are listed as a vulnerable species under the Convention on International Trade in Endangered Species of Wild Fauna and Flora (CITES) and on the International Union for Conservation of Nature's (IUCN) Red List.

"It's believed there are," the guide said, "about the same number of white sharks left in the wild as there are tigers." The conservation status of white sharks is often overshadowed by the spectacle, and having just witnessed a breach, Cynthia understood in a new way how people may have a hard time reconciling that an animal that is so formidable can also be so vulnerable.

An hour later, Cynthia awaited her turn to enter the shark cage as the *White Pointer II* rolled in the swell. It was surreal to be in this place she'd seen on television, although she noted that one advantage to television is the lack of smell. Seal colonies are notoriously stinky places. As she fidgeted with her mask and snorkel, she couldn't get the image of the breach out of her head. Her heart raced at the thought of entering the water. The cage that looked so substantial earlier in the morning seemed somehow fragile as she lowered herself into it. The cold water momentarily took her breath away. She ducked below the surface.

The first white shark she saw was as graceful as it was quick. It was clearly aware of the cage and the humans in it—perhaps even curious, Cynthia thought. But there was no lunging or teeth on metal. The shark was the most awesome and capable predator she'd ever seen, but it didn't appear aggressive, dangerous, or unpredictable. Because of the water quality and lack of visibility, the sharks came and went, appearing and then disappearing, almost as if choreographed. It was beautiful and mesmerizing. She'd gone into the experience thinking it would be terrifying, but as she exited the cage, all she could think about was the immense gift of being in the presence of such an animal. Her heart still raced, but there was no longer fear—just passion and newfound appreciation for the white shark. It was a feeling she knew was going to last long after she returned to Massachusetts.

An Extreme Degree of Caution

As summer returned to Cape Cod in 2010, so too did the discussion about white sharks as a public safety issue. In late June, a seven-foot juvenile white shark was hooked, tagged with a conventional National Marine Fisheries Service tag, and released twenty miles southeast of Gloucester, Massachusetts. The event predictably launched a news cycle.

Here we go, I thought to myself.

The state was quick to issue a statement and reassure the public. "We don't believe it's a threat to public safety," Ian Bowles, Massachusetts Energy and Environmental Affairs Secretary, said. "People shouldn't be saying they want to avoid the beaches or anything like that."

But not everyone was convinced, especially after a June 29 story in the *Boston Globe* warning readers that "[a] growing population of seals off Cape Cod and rising water temperatures are combining to create conditions that could attract great white sharks to Cape beaches this summer." The seals were a known entity to Cape residents and tourists, but then, in the spring of 2010, we also had temperature data from the five white sharks tagged the previous September. We found that those white sharks spent most of their time in water ranging from 59° to 67°F, which is a really narrow temperature range for a fish. Those data were giving us a better sense of where we could expect white sharks to hang out and how long they might hang out. Jim Horna, Chatham's marine operations supervisor, noted that the water had warmed faster than usual in 2010, and that Lighthouse Beach was fluctuating between 58° and 64°F.

At Monomoy, the 2010 seal pup survey revealed 154 gray seal pups, a more than 220 percent increase over the previous year, which itself had been the first year in the past two decades that the number of gray seal pups reached double digits. Overall, the data showed the gray seal population in the northwest Atlantic had gone from fewer than 10,000 individuals in the early 1990s to over 350,000 in 2010. With the summer season just around the corner, the proximity of Monomoy to Chatham's beaches, combined with the warm water, had some officials anxious. The tension was especially high in Chatham.

The decision of whether to close a beach is up to the individual municipalities and beach managers. In Chatham, which had several guarded beaches where a more hands-on beach management approach was possible, that approach made more sense, but large parts of the Outer Cape's beach network are unguarded, making management decisions far more challenging. For example, of the forty miles of beach that exist within Cape Cod National Seashore, less than two miles are guarded. The national seashore's chief ranger said that no white sharks had ever been observed at the seashore's guarded beaches, but he also cautioned that "there needs to be an awareness that if those large seal populations are in the area, there possibly could be some inherent dangers."

The most common advice for beachgoers in 2010 was to use common sense—which translates, above all else, as don't swim near seals. Those wishing to venture into the water were also advised to avoid swimming alone and at dawn or dusk. In addition to the warnings and advisories, people were constantly reminded the risk is low and that the last human fatality attributed to a white shark in Massachusetts occurred more than seventy years ago. In advance of a hot and muggy Fourth of July holiday weekend, the United States Coast Guard stepped in and issued what it identified as its first-ever shark advisory. "I recommend an extreme degree of caution," said a Coast Guard spokesperson. They said the advisory was, in part, because of the expected record number of beachgoers and the sighting the previous week of a white shark offshore.

Even with all the talk of sharks, the advice on how to avoid them, and then the USCG advisory, beach managers and town officials remained largely undaunted. The general sense was that there is way more risk associated with the sheer number of people expected to flood Cape beaches than any possible shark encounters. "It's the perfect weekend," the assistant director of the Barnstable Recreation Department told a reporter. "It scares me—all the people more than the sharks." Because Chatham had been ground zero for last year's shark activity, Chatham's harbormaster, Stuart Smith, assured the public that beach managers would remain vigilant. Referring to the sharks that showed up in 2009, he said, "We don't know if that was an anomaly or whether that will be repeating itself."

Frankly, neither did I.

Shark Men

Cynthia was still riding her white shark high after returning from Africa in June. White sharks were top of mind for her, and she followed the news of what we were doing out on the Cape with immediate interest. The whole idea that white sharks were even part of the Cape Cod equation hadn't sunk in until she'd read the media reports the previous September. Even then, it was more a curiosity than anything else, which she admits seems strange to her now, since both white sharks and Cape Cod were a huge part of her life.

Cynthia's first trip to the Cape was in 2005, when a friend invited her. She was immediately hooked and soon bought a house. Through the same friend, she met her future husband, who had himself grown up spending summers on the Cape. They were married on Nauset Beach in 2007 and continued to spend as much time on the Cape as they could. Cynthia, after her white shark experience in South Africa, was having an aha moment, realizing that she'd traveled halfway around the world to see white sharks, while here they were off her beloved Cape Cod.

Because of her newfound passion for the species, Cynthia was excited when she heard that National Geographic was launching a documentary series about white sharks called *Expedition Great White*, which would later be rebranded as *Shark Men*. The series featured OCEARCH founder Chris Fischer, whom National Geographic introduced as "a well-known personality in outdoor television and conservation circles," and Michael Domeier, "a well-known white shark expert from California." *Shark Men* was promoted as "an unprecedented great white shark expedition," where "[a] brazen team of seasoned fishermen and an innovative scientist join forces to unlock the mysteries of the world's largest predatory fish." The first episode was scheduled to air the first week of June 2010. To Cynthia, it sounded like something she would really enjoy.

She didn't.

"Forget crabs," the narrator says, as large steel hooks drop onto the deck. The scene cuts to a white shark swimming through the water. A man excitedly points at it. Someone exclaims off camera, "Oh! A white shark right underneath your feet! Right there! Nice one!" The narrator continues: "They're going after something"—he pauses for effect—"*bigger* . . ." Large baited hooks are thrown into the water, and a white shark comes up and bites the underside of a boat; the sound of teeth on hull is clearly audible. ". . . the great white shark." The camera cuts to another white shark in the water and someone says, "Oh! That's a bigger one!" to which someone else responds, "Go big or go home!"

The reference to crabs is intended by the network to put *Shark Men* on par with Discovery Channel's popular Alaska-based reality crab-fishing show *Deadliest Catch*. Ironically, Fischer's 126-foot vessel is a former Bering Sea crabber like those seen on *Deadliest Catch*. The boat has an unusual feature, though: a hydraulic lift with a seven-ton capacity installed by a previous owner in order to bring his forty-five-foot fishing boat aboard. Fischer had transformed the lift into a 500-square-foot platform that could be moved out over the starboard rail and lowered beneath the surface of the water. The plan, as the first episode of *Shark Men* explains, is to catch a white shark with a baited hook, tire it out, and then drag it onto the submerged platform. The platform would then be raised out of the water, giving scientists unparalleled access to the shark. The whole procedure resembles a Formula One pit crew during a pit stop with the clock ticking.

Domeier, who was keen to deploy advanced tagging technology that required bolting the tag to the shark's fin, saw the potential in Fischer's boat as an ideal platform for tagging the biggest white sharks he was studying. In addition to bolting the tag to the shark's fin, the crew of scientists could also collect specimens ranging from blood to parasites. The challenge was how to get the shark onto the platform, which is where a smaller boat, baited hooks, chains, buoyancy balls, and big-game anglers came into play. During 180 episodes of ESPN's popular reality fishing show *Offshore Adventures*, Fischer had established himself as an expert angler and an engaging on-screen personality. Now he was ready to up his game. "We've spent a lot of time on the water over the past ten years or so," Fischer says in episode one of *Shark Men*, "but to catch things that have never been caught and put them on the deck and let them go, alive, in the name of science—bring it."

Cynthia had a hard time getting through the first episode of *Shark Men*, which was set off the coast of Guadalupe Island, Mexico. She was familiar with the way I tagged free-swimming sharks on the Cape, and OCEARCH's approach seemed entirely too invasive to her. The footage of white sharks with large hooks in their mouths being held at the surface and exhausted with buoyancy balls and then dragged to the platform, where their massive weight was lifted out of the water, gills gaping, and body thrashing . . . it was just too much for her. Sure, the show was peppered with some conservation messaging, and Domeier's work seemed important, but overall the cumulative effect turned her off.

She didn't tune in to episode two.

Something Truly Deserving of Every Ounce of Our Fear

Perfect weather was forecast for the Fourth of July holiday on Cape Cod. Despite the media coverage of the white sharks, not to mention the direct warnings and advisories, the beaches were packed. Those who worried that tourism would suffer because of the sharks seemed to have worried for nothing. "No one's asking about it," said Wendy Northcross, the chief executive of the Cape Cod Chamber of Commerce. "If anything, it might be a curiosity that draws people here."

It appeared that people were committed to their holiday weekend plans, which was fine by me. I felt good about our messaging—yes, there were white sharks off Cape Cod, but if people used common sense, the risk remained low. I was eager to get on with our research, but I would also have been content if there were no shark sightings over the Fourth. Even just the one white shark caught offshore so far that season had created more work for me than I felt I had time for—speaking to the press, updating my bosses, offering advice. After the previous September, I wasn't sure what to expect, but I believed that however it played out, we were ready—ready not only to manage the risk but also to advance the science.

Apparently not everyone felt the same way, and I heard about it in a particularly scathing op-ed published in the *Boston Globe* on the last day of June 2010. In a piece titled "State's Little White Lie," *Globe* columnist Brian McGrory likened Ian Bowles, the state's Secretary of Environmental Affairs, and me to Leslie Nielsen's character in one of the Naked

Gun movies. In the film, Lieutenant Frank Drebin is keeping people from gawking at a crime scene, saying, "Nothing to see here. Keep it moving. Nothing to see." McGrory goes on to write, "Behind him, there are gunshots, explosions, bodies hitting the pavement. This is what came immediately to mind as Ian Bowles and Gregory Skomal told the people of Massachusetts this week that there's nothing to fear from the great white sharks that have taken up residence along our shores."

Later in the piece, McGrory played the *Jaws* card. "Amity being an awful lot like Edgartown," he wrote, "*Jaws* providing a prelude to what could be happening in real life." McGrory recalled that the mayor of Amity said there was no reason not to swim, but after several people were killed by the shark, that changed. In McGrory's assessment, the residents and tourists on Cape Cod's beaches had "dodged fate" in 2009.

"[H]ow long can our good fortune go on?" McGrory asked.

Of course, I bristled at McGrory's hyperbolic ramblings. I was becoming accustomed to the press's desperate need to sell papers during the summer, and his column advanced that particular cause while garnering him much-needed attention. It was relatively easy to dismiss it as little more than theater. However, if I was honest—*really* honest—I knew there was a nugget of truth in McGrory's rant. It was something that had been eating at me ever since I spent time with Gretel in the salt pond. *How long can our good fortune go on?* I didn't have an answer. I didn't agree with McGrory's assessment that white sharks in Cape waters were "something truly deserving of every ounce of our fear," but as a shark scientist, I knew that if the white sharks were back inshore in any significant numbers, inevitably the odds would catch up.

They're Back

By the end of July 2010, there was no doubt that white sharks were indeed patrolling the shallow water off Cape beaches, especially from Chatham out to Monomoy. The first sighting was by George Breen, who was flying his plane over Nauset Beach on July 11 when he saw a fifteen-foot white shark hunting seals less than one hundred feet off the beach. The media labeled the shark as "aggressive." Was this the confirmation for which people had been waiting? Was this proof that 2009 was not an anomaly? Captain Billy Chaprales thought so. "I've been a commercial fisher for over forty years," he said, "and I only saw three until last year."

On July 30, George was again up in his plane, this time with a reporter and a photographer who were covering the shark story. George spotted two sharks swimming parallel to South Beach a couple hundred yards apart. The larger of the two, George estimated, was fourteen feet; the smaller one was probably twelve feet. George continued along the coastline and spotted another fourteen-footer about one hundred yards off the beach. "They'll even go closer than that," he said over the intercom to his passengers as he circled so they could get a good look. "They'll hang out in the white water." George leveled off and set a course to return to Falmouth, but before they left Monomoy's beaches, he spotted three more sharks. One of them possibly was a shark he'd seen earlier, but he felt confident the other two were ones he had not observed earlier.

George radioed Chatham's harbormaster to tell him what he'd seen, and the harbormaster, based on George's report, made the decision to shut South Beach to swimming indefinitely. It was the first time that year that a beach had been shut to swimming. Harbor patrols made their way along the four-and-a-half-mile-long, unguarded South Beach and informed people of the swimming ban, but many people continued to enter the water anyway.

I'm not a sociologist, but it's fascinating to me how people respond to swimming bans in the face of credible sightings near beaches. I'm certainly no alarmist, and I was glad it was not my job to tell people what they should and shouldn't do—but what happened to common sense? If I was on South Beach with my family, and a patrol boat came by saying that a swimming ban was going into effect because of credible white shark sightings, I would certainly not swim. When George called me later, he told me that when they'd spotted the first sharks, they were not far from a large gathering on South Beach.

"Seaweed was arranged on the sand to spell out 'Happy 30th,'" George said. "It was a birthday party. They were playing Frisbee and enjoying the day, oblivious to the large shark that swam less than a football field away." Later, when the party was alerted to the fact that a swimming ban had gone into effect, many still chose to go in the water. "We're careful," said the birthday girl when questioned by a reporter later. "But I think I'm faster than a shark anyway." It was a joke, but still.

"Listen," said Katie McCully, a forty-four-year-old lifeguard at Nauset Beach, "we swim the length of our protected beach on a daily basis, and we feel safe."

"We live in a fear-oriented society," said a minister in his fifties, who was training for a triathlon and was undaunted by the shark sightings. "I try not to."

Like me, George didn't get it, either. He'd been flying in the area for more than three decades, and he'd only seen a handful of white sharks from the air before the summer of 2009. This summer, he'd made eight flights and seen white sharks on seven of them. "People say there hasn't been a shark attack since 1936 in Massachusetts," he said. "Well, I've been flying out here thirty years, and I've never seen sharks near the beach. If the sharks aren't there, they're not going to attack you. But now they are, so it's a whole different story."

I wanted to keep an open mind because I didn't have the data to show anything conclusive yet. Yes, there were more shark sightings. It could be an anomaly, but my gut told me that was not the case. I was increasingly convinced this was the new normal, which is why I was gearing up for more tagging. Gretel had been a big wake-up call, but there were many other indicators. When I dove into the data of credible white shark sightings between 2001 and 2009, there was a clear upward trend. Some countered, saying that the number of white sharks wasn't increasing; instead, the effort put into looking for them was increasing. In other words, more sightings didn't necessarily mean more sharks. But the data told me a different story. Much of the data were based on interactions with commercial fisheries, especially the bluefin tuna fishery and the groundfish fishery that targeted cod, haddock, and yellowtail flounder. From 2001 to 2009, as white shark sightings in the region increased, those fisheries were declining by as much as half due to regulatory changes. Those data suggest there was less effort and more sightings.

There was another possibility as well. The sharks might be exhibiting a dietary shift. Instead of feeding offshore in the Atlantic, the population was adapting to the growing number of seals and moving inshore to hunt. In other words, there were not more sharks; the sharks were simply concentrating in an area where people were seeing them more. This was an interesting hypothesis, and it made a lot of sense to me. After all, Frank Carey had theorized way back in 1979, when they'd observed those sharks foraging on the dead whale, that white sharks in the northwest Atlantic relied more on whale carcasses and offshore feeding than white sharks in other parts of the world where pinnipeds were in ample supply. When the pinnipeds return, as they'd done in the Farallon Islands, so too did the white sharks.

Whether there were categorically more white sharks or whether the white sharks were simply shifting their dietary habits from offshore feeding to inshore feeding was an interesting question, but each was not mutually exclusive. It was a question I wanted to pursue, but from a public safety standpoint, it didn't really matter why there were more white sharks showing up in shallow water off Cape beaches. If they were there, they were there. And they were there—making me increasingly confident that I was looking at the next chapter in the white shark story rather than just glancing at a footnote.

Tagging and Television

In July 2010, we were back on the water with Captain Billy aboard the *Ezyduzit* and, with George overhead in the plane, we tagged our first shark of the season. The shark was spotted near Chatham about one hundred yards off South Beach. It was estimated to be about 1,500 pounds and twelve feet in length. As the summer progressed, white sharks became a regular part of the New England news briefs in the *Boston Globe*. There were more sightings, more beach closures, and more tagging.

In addition to tagging three sharks with pop-up satellite tags, we also started using acoustic tags that year. While the pop-up satellite tags would help reveal large-scale migratory patterns when the tags popped up and the data were transmitted, the acoustic tags would register a shark's presence every time that shark swam near an acoustic receiver. In essence, this was the same technology pioneered by Frank Carey, but with some modern changes that allow for long-term tracking. Instead of physically following the shark, we set out an array of acoustic receivers that detected the high-frequency pings of the transmitting tags. In 2010 we could only afford four receivers, which we placed from the southern tip of Monomoy to Nauset Beach. The receivers archive data and wouldn't transmit in real time, so we needed to physically go and retrieve the data from each receiver. I was hoping these acoustic tags would present a clearer picture of the sharks' movements when they were near Cape beaches. Specifically, I felt this technology was ideally suited to pinpoint white shark seasonality, including arrival and departure dates, and to examine residency in Massachusetts waters, as well as site fidelity to specific areas. If I could find patterns in the white sharks' movements, it could help identify areas that were risky for swimmers and areas that were less so.

My friend Nick Caloyianis had reached out to me over the winter and proposed making a film on the Cape's white shark phenomenon, and I thought it was a great idea. Caloyianis partnered on the project with Nick Stringer, with whom I'd worked on the parasite film with George Benz, and the Discovery Channel's Shark Week. I saw the film as an opportunity not to only educate the public about the presence of white sharks off the Cape, but also to get the Discovery Channel to support the research—and we needed the funding. The film would eventually be called *Jaws Comes Home* and would air on Shark Week in 2011. Not surprisingly, the focus was going to be on the public safety angle, and thirty seconds into it, I'm on camera saying, "It's time for some new decisions. The towns are going to have to really think about what to do next."

The tension between science and public safety plays well on-screen; we didn't need to play up the drama at all as the summer of 2010 progressed. I'd been telling people for twenty-seven years not to worry about shark attacks, but for most of that period the data bore out my assertion. In 2010, however, I was suddenly being asked to advise towns on public safety issues.

"There's no mistaking that the potential, however small it is, exists," I said on camera during *Jaws Comes Home* regarding shark attacks. "Something can go wrong, and I don't want that to happen on my watch."

The statement made for good television, but as my wife and closest friends knew, it was increasingly what was keeping me up at night.

We tagged four sharks during the filming of *Jaws Comes Home* and, unlike the sharks I tagged in the fall of 2009, we gave each one a name. As I would come to learn, the television shows always insisted on the sharks being named—something about which both I and my bosses at the Division of Marine Fisheries remained somewhat uneasy. (All this despite my experience with Gretel . . . but she was special.) I was a biologist, after all, and these animals were first and foremost study animals. As a scientist I thought a lot about the issues associated with naming study animals, and it was something that still bothered me about whale research. Did naming imply or even promote bias? And what if that animal were to subsequently be involved in an attack? Would the agency be accused of humanizing these animals instead of protecting the public, which funded its work, from a deadly monster? Would I be accused of treating a wild animal, an apex predator, as something akin to a pet? But in the moment, just like back on the salt pond with Gretel, people wanted a name—and I must admit that it fit the narrative and made the story more relatable to the viewer.

When Billy tagged the first shark during the filming of *Jaws Comes Home*, I immediately asked him what he was going to name it at the off-screen urging of the film crew. For Shark Week, the sharks were the characters, and like most characters—both the good guys and the bad guys—they needed names.

"We'll name it Ruthless," he said as he stepped off the pulpit and looked into the camera. "After Ruth, my wife." Billy and I laughed, basking in the hard-won success of tagging a difficult shark. "She'll like that," Billy said. "She'll like that a lot. It was a hard one to tag. We got it, though. She was Ruthless." Later, during the filming of *Jaws Comes Home*, I named a shark we tagged with an acoustic tag after my son, Wilson.

Having worked with both Nicks over the years, they knew me well and did a great job of also capturing my unabated passion for the species. *Just like the scientist in* Jaws, the narrator said of me, *he lives to study sharks.* While there is some fantastic footage of tagging sharks aboard the *Ezyduzit*, the best part of the film comes when I learned there was a dead whale offshore. If there were white sharks feeding on the carcass, the production team and I wanted to get a cage in the water. Nobody had filmed white sharks underwater in the Atlantic since Stan Waterman filmed the segment for ABC's *American Sportsman* in 1979. It was something for which I'd been waiting my entire life, and I happened to have one of the best underwater cameramen in the world at my side.

At first light, we all met at Stage Harbor in Chatham and loaded a shark cage onto the Nantucket-based charter vessel *Starrfish*. It was no ordinary shark cage. Nick had brought along one of the cages built by Peter Gimbel for the movie *Blue Water, White Death* but never used. Being good friends with Peter, Nick had inherited all three cages built by Peter, including the one that was almost destroyed by a white shark. With the cage loaded and us aboard, we pushed off the dock and headed out to sea with Captain Jay Starr, John Chisholm, and the film crew. The last credible report had the whale floating about ten miles east of Chatham, so we steamed south along the western shoreline of Monomoy, around the southern tip, then northeast. George was unavailable, and so we had Wayne Davis overhead in his plane. Wayne quickly located the carcass. It was a thirty-five-foot female humpback whale, and when we arrived, there were blue sharks up to nine feet in length feeding on it. I scanned the water.

"Come on up," I called to the white sharks that I assumed were nearby. There were massive bite marks on the whale's body. We waited, and sure enough an eighteen-foot white shark emerged from the depths and started circling the carcass and the boat. "That's a big shark," I said, adding (I couldn't help myself), "We're going to need a bigger boat." Nick had a camera on a pole, which he used to film underwater. It was the first time since 1979 that a white shark was filmed underwater in the Atlantic, a major accomplishment on its own, but the most important thing for me was to tag this huge shark. John and I mounted the tag on the same twelve-foot tagging pole used during my failed attempt in 2009 and waited for the shark to pass close. The shark circled the boat, eventually coming close enough for me to drive the tag dart into the base of the dorsal fin. A perfect shot, if I don't say so myself. John and I were ecstatic. Unlike Ruthless and Wilson, we named this shark with a descriptive name, Curly, because the very tip of her first dorsal fin was slightly curled. The day had just gotten better, but the real pièce de résistance was yet to come.

I thought that it would be amazing to get underwater with this shark—but admittedly I had mixed emotions. In these situations, the logical side of my brain is often at odds with the emotional side. To dive with a white shark off New England was a lifelong dream for me and here was my golden opportunity. But then there was the cage itself. The cage Nick brought was untested, as it had never been used by Peter Gimbel, but its sister cage is featured in a famous scene from the film *Blue Water, White Death.* In the scene, a white shark becomes entangled in the shark cage and breaks it apart while photographer Peter Lake is trapped inside—he makes it out alive, but the cage is in shambles. Then there was the fundamental question about whether or not it was wise to jump into the water with a three-thousand-pound white shark that was actively feeding.

My internal debate lasted only a few moments, as Nick stated matter-of-factly, "Let's get in the water, Greg." We manhandled the cage over the transom and into the water, and we suited up. The cage was against the hull, so we could drop through a hatch on the top as opposed to needing to swim into the cage as I had done during my very first Atlantic cage-diving experience with Wes all those years ago. I went first and Nick followed me into the cage. The crew closed the top hatch and lowered the cage several feet under the water. Four large orange poly balls bobbed at the surface, keeping the cage from sinking to the ocean bottom 180 feet below us.

I got a blast of cold water as my wet suit flooded, and I hunkered down on the bottom of the cage, peering out into the green Atlantic, stewing with whale oil and bits of flesh. Nick settled down beside me. Above the water we could smell the decaying whale—below the water we could taste it. Visibility was only about ten to fifteen feet, and I could make out the dead whale glowing in the turbid green light, but the shark was nowhere to be seen, and that made me a bit anxious. After a few minutes that felt like hours, the shark emerged from the darkness and slowly circled the cage on the limits of visibility. She was slow and methodical, seemingly checking out her dinner guests while pushing herself almost effortlessly through the water. Soon the shark whom we'd named Curly returned to feed on the whale, gouging massive, garbage-can-sized chunks of whale blubber from its flank, leveraging her side-to-side body movement to rend them free. It was extraordinary. I'd been cage diving with white sharks before, but it was back in 1995 and on the other side of the world, and those white sharks were not scavenging a whale carcass. In 2010 I knew so much more about these animals, and they were no longer just an animal in which I was intensely interested—they'd suddenly become the center of my career.

After each feeding bout, Curly moved away from the whale and slowly circled the boat and the cage, almost like she was sending a message that this was her whale and she was not one to share. The blue sharks received the message loud and clear and scattered quickly with her every approach. At one point she appeared curious about the cage, which was no doubt generating an electrical field that drew her attention. Frankly,

I didn't want her attention, but the filmmaker beside me was ecstatic. As she approached the cage, a quick adjustment to the angle of her pectoral fins sent her to the surface, where she opened her mouth and seemed to attack one of the orange poly balls. I looked up as the shark's incredible four-foot girth slid over the top of the cage.

"Dear God!"

With her white underbelly blocking out the sun above, Curly began to thrust her tail without moving away from the cage. Horrified, I realized she was stuck in the lines holding the cage to the surface, and my emotions quickly shifted from awe to sheer terror. Images of Peter Lake's experience in *Blue Water, White Death* flashed through my brain. Irony would have it that I was trapped in a cage identical to the one still in shambles from Peter's experience and now on display in the Tennessee Aquarium. I didn't want the cage I was in to end up on display anywhere. Curly remained entangled among the four floats and thrashed on top of the cage, which is all that stood between the shark, Nick, me, and the bottom of the ocean 180 feet below. I could sense Curly's rising anxiety, which was mirrored by my own. Seconds felt like minutes. I looked up at Nick, seeking a solution to this predicament. He was surprisingly calm, although the melee was happening a few inches from his head. He was so still, in fact, that my immediate assumption was that he must be dead; he must have had a heart attack.

Now what?! my mind screamed. But then I realized that Nick was not dead; he was calm, cool, and totally focused on filming the whole damn thing. We continued to be tossed about at the mercy of the shark, the cage jerking us about like we were trapped in a washing machine. I was slipping into full-blown panic, and there was Nick, doing his job—always the consummate professional. Seeing how Nick reacted calmed me down a notch and I started thinking about our options. At that moment, the way I saw it we either plummeted to the bottom nearly two hundred feet below when Curly destroyed the poly balls tethering the cage to the surface, or she smashed the cage apart and, in a fit of rage, went after us.

As I contemplated these dire scenarios, Curly's weight bore down on the cage, causing one of the bars on the top to break and strike Nick on the head. He recoiled, momentarily dazed, but kept filming. It was like time was standing still, and my mind was racing. The cage had started to implode, the metal flexing under the duress of the continued thrashing of the eighteen-foot shark. Just as I was about to throw away all hope of making it out of the situation alive, the side door suddenly popped open, a result of the metal bending. I looked from the open door to the shark and then back to the open door. *Is this a sign?* I wondered. *But if I leave the cage, then what?* I glanced up at Curly, who continued to throw her weight into trying to break free. I reached over and closed the cage door, deciding I'd rather take my chances in the cage than free swimming with this massive white shark.

As suddenly as it began, it was over. Somehow, some way, the seemingly impossible happened. Curly managed to free herself from the lines and simply swam away. The rocking stopped. The cage held together. We were still alive. My heart continued to race for some time—I could feel the pulsing as the adrenaline that had coursed through my veins started to subside. I'd been in fight-or-flight mode.

I'd seen the world through the eyes of prey.

As the cage bobbed beside the boat, my rational mind returned. As it turned out, Curly had only swum off a bit, but she hadn't left the area completely. Nick reached up, opened the hatch on the top of the cage, and made a quick break for the boat under the watchful eye of the crew. I waited. Curly came back around toward the cage but then veered off. It was my turn. I broke the water's surface under the watchful eyes of Nick. "You're clear!" he yelled, and I clamored aboard like I was Mr. Frog again, leaping onto the safety of a rock.

Sitting on the boat's gunwale, I tried to process what had just happened, and as my lungs filled with the fresh ocean air, with my feet squarely planted on the boat's deck, it started to sink in that it had been one of the best days of my life. As the sun sank lower in the sky, the *Starrfish* turned and headed back to the Cape, leaving Curly to enjoy her whale carcass in relative peace. I continued to have a hard time believing what had just happened. *You don't get to cage dive with white sharks in the Atlantic*, I kept thinking to myself. *It's just something you don't get to do.* Yet we'd done it. We'd also tagged the only mature female white shark in our study to date. There were endless questions about where white sharks in the Atlantic mate and pup, and this tagging had the potential to begin unlocking some of those mysteries. Nick's head hurt, and mine was just spinning. The icing on the cake was that we'd gotten the whole thing on film.

⚓

Overall, the 2010 summer research season was more than I could have hoped for in many ways. I'd personally had the opportunity to get in the water with an eighteen-foot white shark, an experience I didn't expect to top anytime soon. We'd tagged six more white sharks, including the first mature female, and we were well on our way to showing with data that what happened in 2009 was not an anomaly. With the first acoustic tag deployed, we also now had a strategy in place for addressing the public safety issue by getting a better sense of when white sharks arrive, where they spend their time on the Cape, and when they leave.

Even with all the successes, however, there remained concerns. White sharks had been spotted within Chatham Harbor, and popular Chatham beaches inside the harbor were closed for the first time. Beach managers and town officials were concerned. The sharks did not seem to be hurting tourism, but would that change? There were clear data showing that the return of apex predators in other parts of the country increased tourism and tourist dollars. In Yellowstone National Park, for example, the reintroduction of wolves was linked to an increase in both. Could the same turn out to be true for white sharks off Massachusetts, especially since white sharks were a lot more difficult for the average tourist to observe than a wolf in a national park?

Then, of course, there was concern about sharks attacking humans. If there were more sharks along the beaches, and if some of those sharks were taking up residency during the summer and early fall, was an attack on a human inevitable? I reasoned that an attack would probably happen near seals and would be a case of mistaken identity, which is the consensus within the scientific community. But would that matter in terms of public perception? Would that matter in terms of the guilt I would feel because it happened on my watch? I tried to push thoughts like those from my mind. But what if, for example, the shark I had named after my son at the goading of a Shark Week film crew was implicated in an attack on a person? What if it killed someone?

The bottom line was that it all came down to data. To answer the questions regarding the biology and life history of the animal, as well as the questions regarding public safety, we needed a lot more data—and getting those data was going to be expensive. Alone in my truck, driving to and from Chatham, I thought a lot about where I was going to get the money we desperately needed—which meant that when Chris Fischer called me and told me he wanted to bring OCEARCH to Cape Cod, it was a call I couldn't ignore.

OCEARCH

When Chris Fischer called, I was on my way home to Marion, Massachusetts, where my family and I had moved in mid-2010. The small town was just off the Cape, making my commute a lot easier, as I didn't have to rely on a ferry boat to go anywhere, as I had while living on the Vineyard. It had been a long day out on the water, and I was ready for a beer, but I was also interested to hear what Fischer had to say. I didn't know him personally, although I'd seen him around at some fishing events, and I was aware of his reality television show on ESPN called *Offshore Adventures*. I was also aware of Fischer's new project, *Shark Men*, in which he chased big white sharks with Mike Domeier, who then tagged them using Fischer's very unusual boat. I didn't know Domeier well, but I thought of him as a serious scientist doing good work on white sharks. So, when Fischer led with the news that he and Domeier had had a falling-out, I was already skeptical.

Fischer told me that he wanted to move his boat to South Africa for a project, but then he wanted to come to New England and work with me on white sharks in the Atlantic. It would cost me nothing, Fischer promised. It was a tempting offer, but from what I'd heard, working with Fischer brought with it hype and drama. I wasn't naive. Making television shows, even ones about science, often leaves a lot to the production company, which is beholden to its boss (the broadcaster), whose only measure of success is number of viewers. Many in the television industry believed the only way to attract a lot of viewers to a science show was to include a lot of drama. I'd seen this firsthand, and I didn't know if I wanted to be associated with that kind of circus in Massachusetts, especially since the white shark phenomenon in New England was still relatively new. I listened to Fischer on the phone but, ultimately, I decided to keep OCEARCH at arm's length.

In February 2011, I ran into Fischer at the Eastern Fishing and Outdoor Exposition in Worcester, Massachusetts. Fischer was giving a presentation about his white shark work, and I attended the talk. The two of us had a chance to catch up afterward, and Fischer again pitched the idea of working together in New England.

"I'm not a scientist," Chris said, "but what I bring to the table is our expertise as fishermen to capture and handle these sharks so that local scientists like you can get their job done. And it won't cost you a cent."

As I drove home from the expo, I thought over my conversation with Fischer. I still wasn't interested in the drama, but I sure as heck needed the support. I worried about the seeming negativity surrounding Fischer and OCEARCH and allegations about the outfit's methodology. In fact, reports were just surfacing that a shark that OCEARCH had foul-hooked in the throat in the Farallon Islands in 2009 had not escaped unscathed as both OCEARCH and NOAA had reported. An individual who identified him- or herself only as "Shark Enthusiast" had leaked photographs of the shark a year after the incident. The photos showed an ugly wound on the shark OCEARCH had foul-hooked and then cut loose by reaching bolt cutters in through its gill slits. Shark Enthusiast included a short note with the photograph:

> My concern is that other scientists or media hounds will try to do this again, heck, who knows how many of these sharks have already been severely impaired as a result of this same researcher's made for TV science.

When the first episode of season two of *Shark Men* aired in the spring of 2011, the entire foul-hooking incident was shown, which further polarized viewers into pro- or anti-OCEARCH camps. In May, NOAA released a video of the shark that clearly showed the wound was a result of another shark and not the foul-hooking incident, but OCEARCH's critics had already done plenty of damage through shark conservation networks, online forums, and social media. Whether the drama helped or hurt ratings was anyone's guess at the time, but again, I questioned whether I wanted to bring that circus to town. Chris reached out to me via email a couple of times throughout the winter and spring of 2011, offering his vessel for that summer. He told me he was also pitching the Cape Cod work to History Channel for a show, which, of course, contributed to my unease.

A couple of months later, I was at a NOAA Highly Migratory Species Advisory Panel meeting in Washington, DC, when I was approached by Bob Hueter. He was also a member of the panel and had been since 1997. Bob was a bit older than I and was generally considered to be a well-respected shark scientist associated with the Mote Marine Laboratory in Sarasota, Florida—the successor to the laboratory Eugenie Clark had founded. Although I'd heard rumors through the shark research grapevine about Bob's alleged behavior and what people believed was his self-serving approach to science, we'd had an amicable relationship for many years, and I accepted his offer to go out to dinner after the meeting. What I hadn't expected was for Hueter to encourage me to take advantage of Fischer's offer to work together on white sharks off the Cape. He told me that he'd worked with Fischer off the coast of Florida, and he vouched for the entire OCEARCH operation. "It's a top-notch crew," Hueter said. "Incredibly professional, and it wouldn't cost you anything."

After dinner, I went back over my conversations with Fischer and Hueter. The falling-out with Mike Domeier still gave me pause, but admittedly I didn't know the details. Now here was Hueter, one of the long-standing names in shark research in the Atlantic, encouraging me to work with OCEARCH. It was still hard for me to untangle how much of the television show was about editing and pumping up viewership and how much was Fischer's own attitude, philosophy, and ego. At the end of the day, however, I had to agree that the pitch was appealing. The idea of being able to significantly ramp up our own research and not have to worry nearly as much about funding was very attractive. When I got back to my office in Massachusetts, I gave Fischer a call. Chris proposed two expeditions during the summer of 2012 that would be financed by the History Channel, but the History Channel support fell through, and, assuming we could secure the necessary permits, we tentatively agreed to work together in September 2012.

PART 3

INTERACTION

I can't say I'm particularly surprised, but that doesn't make it any easier when it happens. Although I am fascinated by this species, I am also deeply committed to producing information that will stop the tragic loss of human life.

—Greg Skomal, July 2012

CHAPTER 10

SEVERE BUT NOT FATAL

Anything that brings people to town is good for tourism, and the sharks are bringing people to town.

—Lisa Franz, executive director of the Chatham Chamber of Commerce, August 4, 2012

Some shark scientists will tell you that white sharks are overrated, but they're pretty freakin' amazing.

—Megan Winton

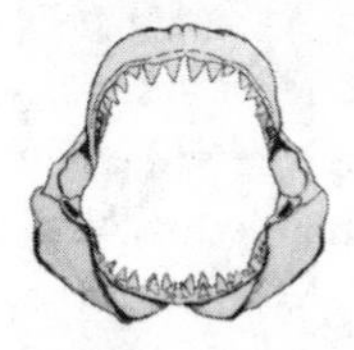

DECEMBER 8, 1990—31°54'30.85"N, 79°3'53.56"W (OFF THE COAST OF THE SOUTH CAROLINA–GEORGIA STATE LINE)

As autumn settles in the Northern Hemisphere, the sixteen-foot female white shark makes her way south through the Carolinas toward Georgia. She's mirroring the migration pattern of many temperate marine species, including other white sharks heading south, each on his or her own solitary journey. It's a ritual that is as old as the Gulf Stream and the seasonal shifts—as old as her species. Her journey was not a rushed journey, but it was deliberate. She had paused to probe several river mouths, killing and eating a dolphin at one. She had not, however, made her way east beyond the edge of the continental shelf, where a few of her kind ventured. Instead, she had largely sought out nutrient-rich water between the rush of the northerly trending Gulf Stream and the static coastline, favoring temperatures between 55° and 73°F. Most of her time was spent within sixty feet of the surface.

She has been able to make the long journey in part due to her body's efficiency. Like all white sharks, she possesses a unique digestive system, whereby food passes from her mouth into her esophagus and then on to her muscular stomach. Except for the liver, a shark's stomach is the largest organ in its abdominal cavity. The walls of the stomach have folds that allow the stomach to expand as she feeds. Once in the stomach, food, including anything from shells to mammal bones, is broken down by digestive enzymes and acids. As the food is digested, the stomach muscles contract and push a paste-like digesta into the spiral valve, a digestive organ unique to sharks and their close relatives. Although a relatively short organ compared to our intestines, the spiral valve's efficiency at absorbing nutrients is extraordinary. Additional enzymes and bile secreted by the liver and pancreas help break down the food, but it's also the enormous surface area created by layers upon layers of tissue that allows the shark to maximize the energy derived from the meal. Any waste is finally excreted through the cloaca, which for a shark acts as the common opening to the digestive, urinary, and reproductive systems.

Arriving in the Carolinas, the white shark is attuned to the migration of another, much larger animal—the North Atlantic right whale, which migrates to winter calving grounds off the southeastern United States. North Atlantic right whales, so named because they were the *right* whale for whalers to target, can grow to over fifty feet long and weigh as much as 140,000 pounds. Like the white shark, right whales can live to be seventy years old, but unlike female white sharks, who don't reach sexual maturity until as late as their thirties, female right whales may start producing young, when they're only around a decade old. After that, a healthy female right whale may give birth once every three to six years.

As the white shark has learned in the western North Atlantic, a dead right whale carcass here in the Southeast is a welcome meal. One reason that whalers targeted right whales was that up to 45 percent of a right whale's body mass is blubber. For the female white shark, with her small, tightly folded spiral valve and large liver, this means she can survive off one easy meal for an extended period, as opposed to exerting energy to catch and chase prey species such as dolphins. It's estimated that an adult white shark's liver can hold more than five hundred gallons of blubber, representing an impressive reserve.

As she ghosts through the Atlantic off the southeastern United States, she sees the distinctive stocky black body of a forty-foot right whale solidify in murky green water ahead of her. The whale, which needs to breathe air, is hovering near the surface. While a dead right whale makes an easy meal, a living right whale is not generally a viable prey item for even a large white shark, and so she has ignored the several adults she's seen. But this right whale is different, and she investigates. She senses the animal's heightened heart rate and erratic movements. She moves in and then turns quickly near the surface. Both her dorsal and caudal fins momentarily flash in the low angle of the winter sun. The whale suddenly rolls, and a plume of red erupts from her underside, further exciting the shark's senses as she circles. The tendrils of blood pull away from the whale with the prevailing current, revealing a newborn whale calf that is almost the same size as the white shark . . .

*In February 2012, we published a paper titled "Implications of Increas-*ing Pinniped Populations on the Diet and Abundance of White Sharks off the Coast of Massachusetts." In it I laid out a scenario whereby I expected the number of white shark sightings and seal interactions on Cape Cod to rise. To come to this conclusion, my coauthors and I took a hard look at seal–white shark interaction data from the west coast and found that places like the Farallon Islands could be a template for Massachusetts's future.

In the 1970s and early 1980s, the Farallon Islands, a known white shark hotspot, saw a rebound in the elephant seal population. Although the shark population in the area remained relatively static, the attack frequency increased as more seals appeared. By the late 1980s, both the seal population and the individual predation rates stabilized, but the number of white sharks attacking seals increased. "If the western scenario is applicable to the east," we wrote in the paper, "we are in the early stages of this relationship but can anticipate additional white sharks utilizing this food resource."

The paper didn't specifically address public safety, but public safety was increasingly on the minds of many Cape Cod stakeholders, especially beach managers, the Division of Marine Fisheries, and, of course, me. It was one thing to say that Cape Cod was experiencing the early stages of a phenomenon that likely existed in Massachusetts before seal bounties began in the nineteenth century. It was quite another thing, however, to say that "white sharks' natural predatory tactics provide ample potential for interactions with humans utilizing the ocean," which is exactly what another 2012 paper on which I was a coauthor stated. In this second paper ("Responding to the Risk of White Shark Attack: Updated Statistics, Prevention, Control Methods, and Recommendations"), we looked at the flip side of what happens when white sharks utilize a food source adjacent to popular beaches. The goal of the paper was to help stave off "sensationalistic, irrational, or ineffective responses by decision-makers" when responding to white shark attacks, and the first recommendation was to initiate a research program, which is exactly what we were doing heading into the summer of 2012. As a scientist, I wanted to fundamentally better understand what this new white shark phenomenon was all about—but as the state's shark biologist, I also wanted to make darn sure I didn't follow once again in Matt Hooper's footsteps—on the dock with a bloodthirsty, angry mob.

When the white sharks returned to the shallow waters of Cape Cod in June 2012, some people remained concerned about the "shark menace" and its potential effect on business and public safety. But even as the first beach closure of the season occurred in Chatham due to a fourteen-foot white shark chasing a seal thirty feet off the beach, residents and tourists appeared mostly undaunted. As the summer heated up, the beach parking lots were full, and downtown Chatham was as busy as ever. It appeared, in fact, that the sharks might have been becoming an economic boon.

"Great whites just bring people out," said Lisa Franz, executive director of the Chatham Chamber of Commerce. "People are interested, they want to know more about it, and they want to catch a glimpse of it if it's possible." She emphasized, however, that the shark's presence only affected a very small area. "We've got nine ocean beaches," she said. "Anything that brings people to town is good for tourism," said Franz, "and the sharks are bringing people to town."

The phenomenon fascinated me. On one hand, I got it. I'm as passionate about these animals as anyone, and I can understand why someone who doesn't get to work with sharks like I do might want to go buy a bunch of shark souvenirs, or maybe catch a glimpse of one. But I was also concerned that there was a very real risk associated with having these animals in shallow water off Cape beaches. Nonetheless, that didn't deter businesses from cashing in on the presence of the white sharks. One of the most popular items during the summer of 2012 was a T-shirt sold by the Chatham Clothing Bar featuring the jaws of a shark with the slogan "Every Week Is Shark Week in Chatham." It was the shop owner's daughter who had come up with the shark-themed line of souvenirs, and that shirt became the number one seller starting in 2009. She said the people who bought the T-shirts, as well as other shark-themed items, included everyone from locals to tourists to online sales. In 2012 the store announced it planned to air the Discovery Channel's Shark Week live in the store. "It's a lovely little formula," the owner said. "It's like the stars are lining up."

The summer of 2012 offered a couple of additional opportunities for local businesses to cash in on the feeding frenzy. In a twist of fate that seemed too good not to have been scripted, "JawsFest" was coming to Martha's Vineyard in August to celebrate the one hundredth anniversary of Universal Pictures. I knew it would bring *Jaws* fanatics to the island to celebrate the iconic film. As an invited speaker, I planned to attend JawsFest, where I hoped to hang out with folks associated with the film. Joe Alves, who designed the three mechanical sharks used in the film and known as Bruce, and Carl Gottlieb, who cowrote the screenplay with Peter Benchley, were slated to make appearances. Some of the actors were also scheduled to be there, like Susan Backlinie, who played the skinny-dipping victim in the first scene, and Jeffrey Kramer, who played Chief Brody's deputy, Hendricks. Also attending were Peter's wife, Wendy Benchley, and Robert Shaw's sister Joanna. Peter Benchley had died in 2006. Shaw died in 1978. I was excited, and I expected it to be an amazing experience. It also occurred to me, as well as Chatham's business owners, that many of the people in attendance would at least make a day trip to Chatham for a chance to see a real white shark. Although their chances of seeing a shark were slim, the opportunity to purchase some real-life shark swag from the newest white shark hotspot was a sure thing. After JawsFest was over, Discovery Channel's Shark Week was scheduled to begin, and it wasn't just any Shark Week—it was the twenty-fifth anniversary.

I was trying to temper the hype and my personal excitement because I needed to focus on the science. It was going to be a big year. For one, it was the first year I'd gone into the summer season with no doubt in my mind about white sharks returning to the Cape. What had happened in the previous three years was not an anomaly—Cape Cod was indisputably the newest white shark hotspot. I'd tagged seventeen white sharks between 2009 and 2011, and in 2012 I hoped to double that number. I also had plans to deploy the first real-time satellite transmitters on Atlantic white sharks through a collaboration with OCEARCH in September. As if that weren't enough, I had an ambitious plan to ramp up the acoustic array off the Cape that would allow us to accumulate data points on a much larger scale. I had also partnered with Discovery Channel, Big Wave Productions, and the Oceanographic Systems Lab at the Woods Hole Oceanographic Institution on the development of an autonomous underwater vehicle to track and film sharks. Dubbed SharkCam, the AUV would finally be ready to go live in July, and I'd coordinated with the science team, Billy, and the film crew to be sure we would capture it all for Shark Week in 2013. There was little doubt in my mind that 2012 was going to be the biggest year yet for white sharks and white shark research in Massachusetts.

Fin!

On Saturday afternoon, July 7, lifeguard Morgan McCarthy was guarding Nauset Beach when she suddenly went ramrod straight. With one hand she grabbed her binoculars and focused in on a kayaker who was paddling not far offshore. In the other hand she had her radio. "Fin!" she yelled into the radio. "Fin!"

Forty-one-year-old Walter Szulc had driven down from New Hampshire with his family for a relaxing weekend at the beach. Upon arriving at Nauset Beach, the family proceeded to lug all their beach paraphernalia, including a blue kayak, down to the sand. As they were setting up for the day, Walter's teenage daughter mentioned that she'd heard about shark sightings on the Cape recently. Walter laughed it off and dragged the blue kayak to the water's edge. He had never paddled a kayak before, and he was looking forward to trying it out.

Walter was about fifty yards off the beach when a nearby paddleboarder started gesticulating wildly, pointing with his paddle into the space behind Walter's kayak. Walter looked back, and the image he saw will be forever etched in his brain. It was like something out of a movie or at a Universal Studios theme park—the distinct dorsal fin rose out of the water directly behind him. Even more alarming than the fin was the bulk of the shark's body rapidly closing on the hull of his kayak. The shark's girth was massive. Walter paddled as if his life depended on it because, quite frankly, he thought it just might.

The fin was immediately confirmed by McCarthy's fellow guards, and they cleared the water without incident. Someone snapped a photo, which went viral almost immediately. The media didn't disappoint. The photograph of Walter, bare-chested and looking back over his left shoulder at the trailing fin, appeared in papers from Boston to Miami to Los Angeles with headlines such as "Kayaker Has Brush with Great White," "Man 'Paddled Very Fast' in His Own '*Jaws*' Encounter," "Shark Swims Too Close Off Cape Cod," and, of course, "I Think You Need a Bigger Boat."

Up until this point, a white shark–human interaction on the Cape had largely been an abstract concept for most. If someone had taken the time to look at the data or go up with George in his plane, he or she might have had a better sense of just how likely an interaction was becoming, but most people simply didn't get it—didn't get that these large predators were patrolling in very shallow water, very close to shore, looking for seals. The picture of the dorsal fin closing in on the kayak was anything but conceptual. It was a very tangible, concrete thing, and, upon seeing it, I immediately wondered if it would change the dynamic. *Is this the wake-up call? The slap in the face?* I'm not an alarmist, but I'd become concerned that some of the public were falling short of our best advice to use common sense. I wondered if this image would change people's perception of the risk posed by white sharks. It was one thing to see an aerial photograph of a large shark, even one close to beachgoers, but when the image is flattened into one dimension from above, it somehow isn't as real. To see that fin breaking the surface of the water just behind the kayak—now *that* was an image that was hard to dismiss. Better than any sign explaining the risk, that image clearly communicated that when you entered the waters off Cape Cod, you were entering the habitat of a large apex predator.

The next day, people were back in the water at Nauset Beach. Most seemed unfazed by the incident. Some said they trusted the system of lifeguards and spotter pilots; others said that the chances of being attacked by a shark were so low as to be laughable. More than one person made the common analogy that driving to the beach was far riskier than the risk posed by sharks. Later in the week, after John Chisholm interviewed eyewitnesses and reviewed the images of the shark's dorsal fin, we announced that the fin could also belong to a basking shark, not a white shark. I acknowledged that it can be tricky to tell the two shark species apart by the fin alone. It was a comment I would return to almost three weeks later when the fifth mission with SharkCam revealed a sixteen-foot female shark with a distinct notch in her dorsal fin. We named her Large Marge, and after studying the footage, John and I were certain that it was a white shark behind Walter Szulc's kayak.

The Petition

Cynthia Wigren first became aware of my plan to work with OCEARCH in September by way of an early July 2012 article in the *Cape Codder* claiming there was talk of *Shark Wranglers* coming to Chatham. By this point, Fischer and National Geographic had gone their separate ways, and in 2011, Fischer had taken OCEARCH to South Africa to film season one of a new series called *Shark Wranglers*, which he'd sold to the History Channel.

Through some of her white shark contacts in South Africa, Cynthia was aware of more controversy surrounding OCEARCH and the filming in South Africa of the first season of *Shark Wranglers*. A shark tagged by OCEARCH in South Africa named Maya had died after release, and soon after, OCEARCH was implicated in the death of a popular bodyboarder who was attacked by a white shark in an area near where OCEARCH was chumming in an effort to catch sharks several days before. Many white shark enthusiasts in South Africa, including big names like Chris Fallows, were appalled by OCEARCH's methods and the fact that the South Africa Environmental Affairs Department, which agreed with OCEARCH's position and ultimately determined the chumming and fatality were unrelated, allowed OCEARCH to continue operating despite these incidents. But some of the individuals opposed to their methods appealed to Cynthia to make sure that the same thing didn't happen in Massachusetts.

Using the website Change.org, Cynthia and her husband started a petition titled "Remove Shark Wranglers from DMF permit to capture and tag great whites." The petition read:

> On September 4, Ocearch *Shark Wranglers* will head out to capture and tag great white sharks off the coast of Massachusetts, under the Division of Marine Fisheries (DMF) permit issued by NOAA. DMF and NOAA have failed to answer the question, "what are the actionable conservation goals of the project here?"
>
> The white shark is already a protected species in the United States. Dr. Greg Skomal of the DMF has already been successfully tagging and tracking white sharks using minimally invasive methods working with local boat captains.
>
> A *Boston Herald* article mentions Ocearch and DMF interest in collecting bacteria samples to create an antibiotic for shark bites and conducting stress physiology tests. This data was just collected in South Africa. Furthermore, a simple Google search will show antibiotics to treat infection from shark bites already exist and NOAA's own guidelines on "how to maximize shark survivability" during catch and release clearly state: "reduce fight times, long fight times put stress on the fish." The Ocearch fishermen intentionally exhaust the shark through long fight times on the end of the line so that the shark is more manageable when it is hauled out of water.
>
> Many researchers are speaking out against the invasive and unnecessary methods used by Ocearch to collect data on great white sharks when other methods are available. In light of knowingly putting a protected species in harm's way, collecting duplicate or unnecessary data, having alternate research methods available, and not having an actionable/enforceable conservation goal in place, the DMF should remove Ocearch from its permit.

The petition was controversial, and OCEARCH maintained that their research methods were necessary to collect the type of data they were intent on collecting. But it received 750 signatures and soon got the attention of my boss at the Division of Marine Fisheries, Paul Diodati. I knew that Paul already was on the fence about OCEARCH, as he'd expressed concerns to me. Upon learning of the petition, Paul called me and said that he wasn't keen to have this controversy in state waters. I found myself in a tough spot. I agreed with what Paul was saying, but I also emphasized how working with OCEARCH could really help with the critical research the state was trying to accomplish but was unable to fund.

"What if OCEARCH operated in federal waters instead of state waters?" I suggested. Keeping OCEARCH at least three miles offshore seemed a reasonable compromise, and Paul agreed. When I notified Fischer, he was not particularly happy about it, but he also eventually agreed. I was relieved that we seemed to have successfully threaded the needle, and I was also relieved that this agreement pretty much guaranteed that I was going to get some important work accomplished that I otherwise would probably not have been able to fund. Cape Cod was emerging as the newest white shark hotspot, and I was the scientist in the thick of it. From the public to beach managers to politicians, it felt like everyone was looking to me for answers, but I simply didn't have enough data to even begin to answer those questions. But to get the data, I desperately needed funding, and OCEARCH seemed the most viable option on the horizon.

So I added Fischer to our NOAA tagging permit.

Cynthia left a message on my work voice mail letting me know she absolutely did not find the compromise reasonable. I called her back, and we had a lengthy conversation during which Cynthia made the case against OCEARCH, including the foul-hooking incident and the South African incidents. I countered that OCEARCH was cleared both by NOAA and by the South African government. I had this deep-down belief that Cynthia and I wanted the same thing—a better understanding of white sharks off Cape Cod—but we just disagreed about how to achieve that objective. Ultimately, we agreed to disagree.

Over the next day or so, I replayed the phone conversation I'd had with Cynthia in my head, and I kept coming back to the belief that it was important to try to get her on board before OCEARCH arrived. For one thing, I'd seen how capable she was at mobilizing opposition, and I didn't want the drama either from OCEARCH or from OCEARCH's opponents getting in the way of the science. But I was also still convinced that her passion for white sharks was not all that different from mine. Ultimately, we both wanted the same thing. So I called her back and asked if she would be willing to come into my office for a meeting.

A week later, Cynthia sat with me in my office. I told her that we really wanted to attempt working with OCEARCH for this one expedition. I assured her there would be no reality television show involved. "We want to see what they are capable of doing without the cameras and the drama," I explained. "This is about science and learning everything we can about these animals."

Cynthia said that she supported my work and she was definitely an advocate for more science, but she had one pressing question: "Why do you need these guys?"

I looked at her. "You know . . ." I paused and focused for a moment at the middle distance between us. Then I met her eyes and said, "You know what? We simply don't have a lot of funding for white shark research. I've basically stolen from my other projects to do the work I've done to date, and I can't keep doing that."

"So, it comes down to money?" she asked.

"Yeah," I said. "White shark research isn't a line item in DMF's budget that I can just keep drawing from. It's expensive work to acquire the tags, put the plane in the air, fuel the boat, pay the crew, and take the time to spend hours on the water chasing sharks." I sat back and exhaled. "Here's a group willing to donate their boat and crew and let us control the science. The amount of work we can get done with this model is potentially game changing."

Cynthia listened intently. What I was saying did make sense to her, but there had to be another way. She was an out-of-the-box thinker with more than her fair share of start-up experience under her belt. "What if someone started a nonprofit with the goal of funding your research?" she asked.

"That would be a fantastic," I said, "but the fact remains that we don't have the funding. There is no nonprofit, and these guys are willing to be here by the end of the summer and give us their platform for free." I leaned in. "I hear what you're saying, Cynthia, and I hear what OCEARCH's critics are saying, but you have to understand that I simply can't justify *not* giving this a shot."

Cynthia left the meeting visibly upset. This was obviously emotional for her. She clearly loved these animals, and she felt a weight on her shoulders from those in South Africa who had reached out to her. What if a shark was injured? What if a surfer or swimmer was attacked while OCEARCH was chumming nearby? What if either a shark or a person died? Before she'd even gotten home, she'd made up her mind that she was going to start that nonprofit.

It Seemed Like a Movie

For me, much of July 2012 was filled with tagging sharks and tracking them with SharkCam. Starting on July 9, we deployed the underwater drone on five missions. The first shark we tagged with a transponder, tracked, and filmed with SharkCam was a twelve-foot male shark we named Chex, off Monomoy. It was the first white shark ever tracked and filmed by an underwater robot, and I was pretty excited about it. Over the next couple of weeks, we tracked and filmed three more sharks off Monomoy—a sixteen-foot female we named Penny, a twelve-foot shark we named Braylon, and a fourteen-foot female we named Amy. Then, on July 27, we tagged and tracked Large Marge off Orleans and Eastham.

I knew right away that the SharkCam technology would be a big Shark Week hit, but I was most excited because it brought me a step closer to entering the shark's world. I knew that that was increasingly where I needed to be if I was going to give beach managers the best information on which to act. Satellite tags are great because they show us where the sharks are going on a broad scale as they migrate through the Atlantic. Acoustic tags are important because they paint a picture of where these sharks are going day-to-day and hour-to-hour as they take up seasonal residency along the Outer Cape. But tracking a shark with a camera allowed us to see exactly what the fish were doing. When Frank Carey, Jack, and Wes tracked the white shark off Long Island, it was groundbreaking work, but they had to fill in a lot of the blanks. They knew, for example, that the shark moved from point A to point B, but how was it swimming? They knew it dove to near the bottom, but they could only guess at what it was doing there. SharkCam was not a researcher's silver bullet, but it was a first step into the next iteration of the work I both wanted and needed to do: understanding the fine-scale movements of white sharks off Cape Cod.

On Monday, July 30, the sky was crisp blue above waves that pushed up onto a sandy beach festooned with colorful towels and beach umbrellas—a typical, idyllic summer day on Cape Cod. There were hundreds of people—families with small children, teenagers tossing Frisbees, young couples laid out under the sun. Chris Myers and his sixteen-year-old son, J.J., were visiting from Colorado—they come to this very beach every year on vacation. It's the beach where Chris, who grew up in Boston, learned to bodysurf from his father four decades ago. It's where he taught his children to bodysurf. He inhaled the salt air and felt the sand shift beneath his feet. He breathed in the timelessness of it all—the perfection of the summer beach vacation.

The beach is called Ballston, named for Sheldon W. Ball, who in 1889 purchased one thousand acres of oceanfront property and built a resort called Balls' Town Bungalows Colony. The resort's many structures were all but gone, but the name Ballston bears witness for those in the know. As Chris looked out over the beach, he wondered why more people weren't in the water. It was a hot day, after all. But he didn't ponder the question for long when he spotted a ragged line of white surf on a distant sandbar. He pointed to the break. "That's our spot," he said to J.J.

They entered the water a little after three in the afternoon and began the long swim out. After about ten minutes of swimming, they paused for a rest. Chris dove down to see if he could touch bottom. He couldn't. "What happens if a shark comes along while we're out here?" J.J. asked.

"Well," Chris said, "we'd be history." They both laughed.

They continued their swim, but a finger of doubt had curled its way into Chris's brain. What does happen if a shark comes along? He'd been going to this beach his whole life and he'd never seen a shark. He'd never even really thought about a shark. But . . .

"I think we better head back," he said to J.J.

As father and son treaded water, they looked back toward the beach. It was a long swim, and they hadn't even ridden a wave. That's when Chris felt something violently clamp down on his legs. It was like a vise.

He screamed. He knew, knew without hesitation, that it was a shark, and it was pulling him under. Trying to kick free, he realized only his left ankle was still in the shark's mouth. With his right leg, he kicked the shark repeatedly in its snout, along its head, in its teeth. He summoned all his strength and kicked again and again. The shark's massive body was like a refrigerator covered with skin. It took seven or eight adrenaline-fueled blows, but finally the shark released him. He felt unnaturally buoyant, the immovable weight that had moments before tethered him to another world was gone. J.J. was just five feet away, and father and son locked eyes. Then, as if on cue, the shark emerged in the space between them. Its dark back slid through the water, followed by a massive dorsal fin that, for a brief moment, caused them to lose eye contact.

Then it was gone.

Without a word uttered between them, they did what they instinctually knew they needed to do. They needed to get to shore. They swam as steadily as they could—J.J. yelled for help, but they were too far from the beach. They kept swimming. Chris felt no pain, but he was becoming lightheaded and was worried he'd pass out before reaching the beach. He could see people now clustered at the water's edge pointing at them. J.J. yelled again, and this time three men dove into the water to assist. Soon J.J. and his father were in water shallow enough to stand, but as Chris tried to get to his feet, he felt pain like he'd never known rocket through his body. He collapsed on the sand.

The 911 call came into the Truro Police Department at about 3:30 in the afternoon. A nurse from Boston's Beth Israel Deaconess Medical Center who happened to be there rendered aid on the beach. Both of Chris's legs were bloody, with cavernous lacerations and deep puncture wounds. One of his calves was splayed open, revealing muscle, and bleeding profusely into the hot sand. Paramedics arrived in less than ten minutes, and they transferred Chris to Cape Cod Hospital. Later that evening, they moved him to Massachusetts General Hospital in Boston for surgery.

In the aftermath of the attack, I took on a role I'd never anticipated being part of the job description of a marine biologist. Like my archetype

Matt Hooper, I found myself delving into the details of a shark attack, interviewing witnesses, looking at bite marks, and trying to make an official determination as to what happened. While I pursued this task with the objective, detail-oriented precision of a crime-scene investigator, in the back of my mind there was no question. There were reports of a fin circling the father-son duo. People talked about the number of seals in the water—some used the word *hordes*. There were four deep puncture wounds on each of the man's legs, necessitating nearly fifty stitches. To me this was obviously a white shark attack. The first on a human in Massachusetts in seventy-six years.

And it had happened on my watch.

"There's very few sharks that would administer such a bite," I said at a press conference on Tuesday. "I'm ninety percent sure. In all likelihood, these injuries can be attributed to a great white shark." I went on to explain that "the weight of the evidence, including eyewitness sighting of a fin, the presence of seals, and the extent of the injury, points to a white shark."

The beach reopened the day after the attack with temporary "Caution" signs posted. There was immediate talk of hiring more lifeguards for the town; Ballston was an unguarded beach. There were questions about how the presence of a lifeguard could affect the town's liability. It seemed that for the first time since 2009, beachgoers were actually altering their behavior. They were more aware. More cautious. On Wednesday morning, the same day that I made my findings official, Chris Myers and his son J.J. appeared on *Good Morning America*. "I heard him scream," J.J. told the host. J.J. said he turned around and "saw the back and the fin of the shark up out of the water. At that point it hit me—when it was happening. But at the same time, I thought none of it was real. It really seemed like a movie. None of it seemed real until I was on the beach." Chris Myers was lucky. He was able to get to the beach, where a trained medical professional rendered care until the paramedics arrived. The surgery was effective, and his prognosis was excellent. I knew it could have been a lot worse.

"I can't say I'm particularly surprised," I said, when asked if the shark attack caught me off guard. "But that doesn't make it any easier when it happens. Although I'm fascinated by this species, I'm also deeply committed to producing information that will stop the tragic loss of human life."

OCEARCH—2012

On August 25, just a few days before the OCEARCH expedition was set to begin, Cynthia sent another formal request to bar OCEARCH from operating off Cape Cod. She also posted an update to her Change.org petition:

AUG 25, 2012—ANOTHER REQUEST TO REMOVE OCEARCH FROM THE MASSACHUSETTS PERMIT WAS SUBMITTED TODAY BASED ON:

Ocearch record of foul-hooking white sharks

Duplicate study—post capture stress research

Potential to compromise the overall population

Success on Pacific coast using long-term, minimally invasive methods

Endangered-species request for Pacific white sharks

New source of funding

SUMMARY:

Long-term research using minimally invasive methods provides valuable information. Using the Ocearch vessel and techniques for a two-week period does not create enough additional value to overlook the risks to the white shark. Dr. Skomal's research work will not be significantly impacted by revising the permit. With the continued use of local boat captains and pilots, coupled with community supported funding through a newly established nonprofit, the DMF and NMFS can continue to manage, conserve, and protect white sharks without harming this valuable and vulnerable species.

The petition received more than eight hundred signatures. I also learned that, as the update alluded to, on August 16, she had filed paperwork to start a nonprofit called the Atlantic White Shark Conservancy. The stated purpose of AWSC was to support white shark research and conservation and to educate the public about the species through community outreach. Cynthia had impressed me and struck me as a go-getter with a deep passion for white sharks. However, even though she had told me about her plans to start a nonprofit, I was surprised she had already submitted the paperwork to the state. Even still, the work with OCEARCH was set to begin in days, and it was unclear to me and my bosses if the fledgling organization would be able to raise the necessary funds. We opted to go forward with the bird in hand, and I once again assured my bosses that I was confident we could work with OCEARCH. If I was completely honest, I was still nervous, but I was optimistic that I could ensure the science was done well, and that no unnecessary risks would be taken with either the animals or the public at large.

What I didn't fully understand at the time was that Chris Fischer and OCEARCH were not in a good place. The History Channel had just canceled *Shark Wranglers*, and Chris had subsequently told *Outside Magazine* that he invested his last $500,000 into what he thought could be his one last hoorah, saying "Fuck it. Let's go solve the mystery of *Jaws*."

The expedition was scheduled to be two weeks long, but more than a week in, the crew still hadn't caught a shark. Morale was running low. I was spending much of my time aboard OCEARCH's boat, the M/V *Ocean* (the following year Fischer would change the boat's name to R/V *Ocearch*), even though I could have easily headed to shore if need be. I had done everything I could to ensure the success of the expedition and to make sure that there was no unnecessary drama. I drafted bulletproof scientific objectives, called *science briefs*, and recruited top shark scientists like Jim Gelsleichter to study reproduction, Gavin Naylor for genetics, Nick Whitney for behavior tags, Simon Thorrold for feeding ecology, and my old friend George Benz for parasites. Because Bob Hueter was such an avid supporter of the expedition, as a courtesy I listed him as a co-investigator. I hoped we could realize the fullest potential of Fischer's *I-bring-the-boat-you-bring-the-science* mode, but what we really needed was a shark.

As one sharkless day turned into the next, I kept busy coordinating visiting scientists and entertaining Fischer's guests, which were mostly news media hungry for a shark story. During this time, I came to understand what I saw as Fischer's business model: generate media attention in any way possible, with or without a shark. His team created daily videos filled with hyperbole and hype for posting on social media, again without a shark. I was immersed in the circus, yet I was also captivated by Fischer's energy and drive.

Finally, nine days into the expedition, on September 13, word came over the radio that the smaller fishing boat deployed by M/V *Ocean* had a white shark on. The science team leapt into action. I was both nervous and excited. I'd been around more white sharks than many humans at this point, and I'd tagged more than thirty of them, but I'd never actually put my hands on a live white shark. Deep down, I felt the same way I felt when Jack invited me to help with the dissection of the Noank white shark, when I first entered a shark cage in Australia, and when I spotted Gretel in the salt pond. The difference is that this time, the buck really stopped with me. This shark would be tagged under my permit, and I was sampling for a team of top-notch scientists.

I learned later that Chris was driving the boat when his partner Brett McBride spotted the shark. I later saw in the video they shot that the two-foot circle hook was baited with a chunk of whale blubber and thrown over the side. The shark approached it.

"He's coming in behind your bait, Brett," Chris said. The shark's fin was out of the water as it approached. "He's picking up speed." With a snap of its scythe-like tail, the shark took the bait. "Is he eating it?" Chris asked excitedly.

"He doesn't have it all buried yet," McBride replied, but soon the hook slammed home into the corner of the shark's mouth. Brett attached a red float to the line, allowing it to slide down toward the hook. The line went taut, and the crew of the boat settled into a familiar routine, adding more floats to the line. It was a procedure right out of Frank Mundus's playbook—a technique replicated in *Jaws*. Each float, which is the equivalent of ninety pounds of buoyancy, created a lot of drag and kept the shark at the surface, tiring it out.

After seven days of waiting for the OCEARCH crew to work their magic, they were finally delivering a shark. At about 6:50 P.M., Chris called on the radio and told me that he and the crew on the smaller fishing vessel were towing the shark back to the mother ship, but it would likely take awhile. It was clear to me that he wasn't fishing close to us, but I had no idea from which direction the fishing team would approach. Regardless, we had plenty of time to prepare for the tagging. The goal was to keep the handling of this shark as short as possible. Although we had walked through the process many times, I was anxious, and the prolonged wait did not calm my nerves.

With darkness descending, I could barely see the small fishing vessel as it slowly approached from inshore just before 8:00 P.M. The shark had been on the hook for more than seventy minutes and dragged several miles, so I knew it was critically important to get it tagged and released as quickly as possible. The lift was dropped into the water and the smaller boat pulled the large fish alongside. In a signature move often highlighted in their television shows, Captain McBride jumped from the smaller vessel onto the partially submerged platform and guided the shark's towline so it could be dragged up onto the lift. As the shark was manhandled onto the platform, it struggled a bit, but it appeared exhausted and somewhat docile. The platform was raised, and the shark was high and dry, rolling about with its massive weight bearing down on its organs as it was exposed to gravity for the first time in its life.

Blood flowed from the shark's mouth where the massive hook was embedded, and the shark's belly was beginning to turn pink as tiny blood vessels under the skin started to rupture. I was in awe at the sight of this incredible animal, and I fully appreciated her mass as she writhed on the platform. For a moment I was frozen with reverence, but then suddenly wonderment turned to concern, as I saw a severely stressed animal that needed to be handled rapidly. I didn't want to kill this fish.

Brett inserted a hose into the shark's mouth to irrigate the gills and provide oxygen. The tagging team then jumped onto the platform and worked like a NASCAR pit crew, quickly and efficiently.

The shark was a female. She was fourteen feet, eight inches long, which meant that she might be mature. Based on her length, she was about two thousand pounds—every pound of which was sitting right in front of me. While Fischer and his crew tagged the shark, I took a blood sample from a large vein that runs through the tail to examine various components and assess her level of stress (which would turn out to be extremely high). It took us sixteen minutes to complete the tagging process, then the team jumped off the platform and the hydraulic lift was dropped back under the surface. Brett had the shark by the tail as he urged her back into the ocean. I was silently begging her to swim away. She was sluggish at first, but then, with an arch of her back and a sweep of her tail, she swam free of the platform and disappeared slowly into the depths below.

What must have seemed like an eternity for the shark was a flash to me, but as she swam away the whole experience began to sink in. I was elated as well as relieved—we had tagged and sampled a massive white shark, and she had swum away. I had stood next to her, touched her, drew her blood. For me, the experience was like none other. I'd never felt so insignificant as I did standing on the platform beside her. Although I'd known what to expect, I still couldn't quite process what it would actually be like to see an animal more than two thousand pounds just lying there, docile like a big old Saint Bernard. It was just remarkable.

With the whole team surrounded by the media and celebrating, Chris asked me to name her, and I chose Genie, after Eugenie Clark, the "Shark Lady." Genie is an icon in the shark world and an inspiration to me and thousands of others who aspired to be shark biologists. She spent a good part of her career trying to exonerate sharks from their bad reputation and I want to do the same. I thought Genie was a perfect name for this majestic animal.

Four days later, the team located another shark. This shark was even larger, about sixteen feet long and estimated at more than three thousand pounds. Chris named her Mary Lee, after his mother. He thought it was a fitting name, as she might be the last white shark OCEARCH tagged. As it turned out, nothing could be further from the truth. Mary Lee went viral and went on to become a media darling over the next three years—*People* called her "the Hamptons socialite of the season" and a "Hamptons celebrity," after her tag established that she liked to summer off the famous Long Island communities. By 2017 she had chalked up more than 100,000 miles. The photos of the team tagging her were all over the media. In some of the photos, Chris's white T-shirt with the "CAT Marine Power" logo on it is front and center. The ship is powered by Caterpillar marine engines, and the crew is frequently decked out in Caterpillar gear. Chris, always the savvy marketer, ran with Mary Lee's viral success and pitched Caterpillar, who agreed to a three-year, multimillion-dollar contract to sponsor OCEARCH's work.

After the 2012 experience with OCEARCH, and with Mary Lee's viral success, Chris asked if we were interested in an expedition to Jacksonville, Florida, since a lot of our PSAT tag data showed our sharks were going there in the winter. We agreed, assembled a team, wrote science briefs, and picked locations. The expedition took place in 2013 and, most notably, resulted in the tagging of a 14.4-foot white shark we named Lydia in March 2013. Lydia would go on to amaze us by becoming the first white shark documented to cross the Mid-Atlantic Ridge from east to west.

Chris told me that he wanted to come back to the Cape in August 2013, now with funding from Caterpillar, to help us tag in state waters. That put a lot of political pressure on the director and commissioner to get access, but given the success of two expeditions under our belts, we were excited about continuing the relationship with OCEARCH. However, I knew that I was going to eventually need to get Cynthia on board as well.

White Shark Discussions and the White Shark Working Group

As the summer of 2012 slipped into autumn on the Cape, I was feeling proud of what we'd accomplished, but I was also aware that not everyone saw our scientific successes in the same light. For me, the summer's work was a visceral success. I'd been there for each of the seventeen sharks we tagged, pouring my blood, sweat, and tears into the work. With a small team and limited funds, we had expanded the acoustic array to over twenty-eight receivers from west of Martha's Vineyard to Truro, adding thousands of data points to our research. We'd achieved something that nobody had ever done before with SharkCam, and we'd been the first to tag an Atlantic white shark with a real-time satellite tag with OCEARCH.

As proud as I was of our accomplishments, I understood why what was first and foremost in the minds of many was the first human attacked by a white shark in Massachusetts waters since 1936. In particular, politicians and beach managers all along the Outer Cape from Chatham to Provincetown were suddenly very interested in what the state could do about the risk of shark attacks. On October 24, the Cape and Islands Harbormasters Association called a "Regional White Shark Discussion" meeting, which was attended by officials and beach managers from towns along the Outer Cape and the National Park Service, as well as shark specialists and members of the public. In total, seventy-five people packed the Orleans Town Hall, and the meeting lasted two hours.

Those charged with managing public safety, mitigating risk, and assuming liability expressed their concern in no uncertain terms. The attack on Chris Myers was not fatal, but it was severe. Was it a harbinger of things to come? A wake-up call? Or was it an anomaly? Dan Tobin, director of the Chatham Parks and Recreation Department, said a lot of the concern came down to better communication. He said they all needed to do better. From my perspective, the meeting's purpose was to allow discussion and encourage the towns to be proactive and to work together. They needed to flush out and confront the issues regarding the risk posed by white sharks. I could talk about it until I was blue in the face—in many ways, that's exactly what I'd been doing—but the attack itself had changed the landscape.

"We're at very high seal levels," I told the audience. "I don't know if that will continue and whether it will result in more sharks being in the area, but that appears to be happening."

The towns decided to form the White Shark Working Group and, realizing that money was going to be critical, they applied for a Community Innovation Challenge grant. While they ultimately received only a fraction of the money they asked for in their proposal, the process of writing the proposal proved invaluable. It helped set priorities. Just about everyone was in general agreement that two things needed to happen, and they needed to happen fast. First, more public education was needed. If Chris Myers and his son had seen a sign at the beach parking lot, would they have acted differently? Would Walter Szulc have laughed off his daughter's question about sharks? Public education wasn't a silver bullet, but at least people could make a more informed decision. Second, there was widespread agreement that better and more widespread monitoring needed to be part of the solution. Money was needed to expand the network of acoustic receivers. Money was needed for more tags. While the state paid our salaries, there was no money earmarked for white shark research. We'd been able to do a lot with a couple of small grants and by leveraging television and film and partnering with OCEARCH, but that was nothing close to the kind of funding we needed to achieve the towns' goals.

As I got deeper into talks with the individual towns, I found myself drawn deeper into the discussion about public safety and liability. The acoustic receiver array we'd been establishing was great for the science, and it would benefit public safety down the road, but, as I had to continually remind people, they were not real-time receivers. Instead we downloaded the receivers' data opportunistically while we were out tagging, generally every two weeks. As such, a shark could quite literally be circling a receiver for hours on end within sight of a lifeguard tower, and unless someone saw a fin or shadow, nobody would know until the data were collected two weeks later. The other major challenge was that the acoustic array only detected tagged sharks, and I was definitely getting the sense that there were way more sharks visiting Cape Cod each summer than we could ever tag.

But people wanted a solution—something that would make it "safe to swim" again. Something that would return things to normal. I cringed a little bit each time I heard someone say they needed something that would make it safe to swim again. There was no solution out there that could ensure 100 percent safety. Instead, it was about understanding and then mitigating the risk. While there were systems that could get a real-time signal to a lifeguard when a shark was in proximity, they were expensive and not 100 percent reliable, especially on the Outer Cape, where the cellular signal on which the real-time receivers rely was often spotty or nonexistent. Once again, I reminded them, these systems would only notify a beach manager if a tagged shark was close enough to the receiver.

How many sharks are out there? someone always asked. While nobody had a handle on what the population size was, I was confident that there were now many more untagged than tagged white sharks swimming off the Cape in the summer.

New Funding and New Challenges—2013

In November 2012, the Atlantic White Shark Conservancy (AWSC) received its 501(c)(3) tax status, and by the start of the 2013 summer season they were up and running. It was a remarkable feat and says a lot about not only Cynthia Wigren's passion for white sharks, but also her ability to dig deep into her own experience working with start-ups. It was less than a year since she had sat in my office and asked me what I needed. "Funding," I'd said. "I need funding." And the next summer there was a dedicated nonprofit to help fund my white shark research.

The conservancy's goal for the summer of 2013 was to fund three tagging trips, including the spotter plane and use of the *Ezyduzit* and crew, at a price tag of around $3,500 per trip. They not only achieved that goal, but also funded the spotter plane at $1,800 for a fourth trip. In addition to funding the tagging trips, AWSC also spent about $4,000 to purchase ten acoustic tags. It was a great start.

Despite this new source of funding, however, 2013 was a challenging year for shark research in Massachusetts. Between 2009 and 2012, we had tagged thirty-four white sharks, with the momentum growing each year. In 2012 we tagged a record seventeen white sharks in a single season. By the end of July 2013, we'd tagged just four white sharks, which meant that we were really going to need to rely on our work with OCEARCH to keep up with our research goals.

The last thing I needed was any drama or controversy surrounding the OCEARCH 2013 expedition. As soon as I knew OCEARCH was coming back to the Cape in 2013 and would likely be working in state waters, I knew I needed to get Cynthia's support. I had invited her back to my office to meet with me and John Chisholm. We spoke at length about the science briefs I'd written to guide our work with OCEARCH, and I shared with her my firsthand experience working with Chris the year before. I pointed to the important data we were already receiving from the sharks we'd tagged through the OCEARCH partnership, and I assured her there was not going to be a reality television show angle. Cynthia listened, asked good questions, and ultimately agreed that collaborating with OCEARCH was the right move. A big part of the AWSC mission was public outreach and education, and Cynthia was savvy enough to see the reach OCEARCH had through social and other media. Cynthia trusted me and the science, and she saw there was a huge opportunity to have AWSC's message amplified through working with, as opposed to against, OCEARCH.

By spring of 2013, AWSC was working with OCEARCH in preparation of the upcoming expedition. OCEARCH was helping promote the Massachusetts white shark license plate campaign started by AWSC, and AWSC was running a contest to put two winners on the

OCEARCH boat for a day during the 2013 expedition. The organizations were supporting each other through their various social media channels, and the real winners were shark lovers and members of the general public who were able to learn through the combined education and outreach opportunities.

As the expedition kicked off on the last day in July, I was acutely aware that we'd put all of our eggs in the OCEARCH basket. I had committed the entire month of August to OCEARCH, essentially living on the boat, coordinating and rotating the science crew, and entertaining media and visitors. But when the anchor was pulled at the end of August, we had only tagged two white sharks. While we had accomplished a lot in terms of communication, education, and outreach, we were woefully behind on tagging. Fewer tags meant fewer data, and fewer data meant less answers for beach managers. With time running out, we used the conservancy funding to tag two more sharks with Billy on the *Ezyduzit* in two days. It was amazing that the conservancy had raised the money to get us out on the water, but it simply wasn't enough, especially as about a third of our receivers in the acoustic array were lost or damaged by storms, and some were apparently outright stolen. Those would need to be replaced as well.

It was easy to blame our disappointing season on the amount of time I had committed to working with OCEARCH, but what it really came down to was the same reason as ever: lack of funding. In 2011 and 2013, the Discovery Channel aired Cape Cod–based white shark films featuring our work as part of Shark Week. The making of those two films, *Jaws Comes Home* and *Return of Jaws*, paid for the bulk of our work on the water tagging white sharks from 2010 through 2012, allowing for as many as fifteen research trips per year. I worked with the Discovery Channel in the fall of 2013 off Guadalupe Island in Mexico filming white sharks with SharkCam (used in *Jaws Strikes Back*), but that project didn't directly fund white shark research on the Cape. Although white sharks were a priority for the state, they were not enough of a priority to make white shark research a line item commensurate with the costs of doing the research. I realized this was partly my own fault, as the Division of Marine Fisheries had gotten used to me creatively finding my own funding for all of the work.

I knew that future television opportunities would present themselves, and there would be other funding opportunities, but I made the decision to really lean into our relationship with AWSC and see what we could accomplish. I knew Cynthia was optimistic about future fundraising based on their success in the first year, and John Chisholm and I had talked a lot about focusing our research.

I was increasingly interested in a central question I got asked a lot: How many white sharks are there in Massachusetts waters? There was no way anyone could answer that question by counting every shark—but, with more robust data, the population could be estimated based on survey-based sampling and mark-recapture models. To accomplish that goal, John and I conceived a five-year population study and pitched the idea to Cynthia. The kind of money we were talking about was significantly more than AWSC had raised in 2013, but Cynthia felt there was momentum. As more people became aware of the nonprofit, the funding followed. To fund even the first year of the five-year population study would be a huge commitment for the fledgling nonprofit, but Cynthia believed they could do it.

The Five-Year Population Study

By the start of the 2014 season, we had tagged fifty-six individual white sharks in Massachusetts waters using a combination of pop-up satellite archival tags, acoustic tags, autonomous underwater vehicle transponders for SharkCam, and conventional NOAA Fisheries tags. When it came to shark research, tagging had been my focus since I first tagged Gretel in the salt pond, but things were about to change. We still wanted to get tags out, especially acoustic tags, but starting in 2014, I wanted us to identify as many individual sharks as possible and not be limited to just the sharks we tagged. Like many whales, white sharks can almost always be identified as individual animals based on unique markings and patterns, old injuries, and other features. In fact, we'd named some of the sharks we'd tagged already based on distinctive features, like Curly's dorsal fin. Our goal in 2014, and over the next four years, was to videotape as many sharks as possible and then catalog the footage in a massive database. I called it "video fingerprinting."

Aside from more frequent trips—AWSC planned to fund two per week from mid-June through October—our methods did not change appreciably despite our shift in focus. The spotter pilot found the shark and directed the boat to it, the boat's captain positioned the boat adjacent to the shark, then I would head to the pulpit. Instead of using a harpoon to tag the shark, I used a telescoping painter's pole with a GoPro camera on the end to film the shark. Once back in the lab, the videos were analyzed, and each individual shark was cataloged based on color patterns, scars, fin shapes, and other physical features. In addition, this allowed me to obtain the sex of each shark, which was very important, but virtually impossible to do from the high bow of the *Ezyduzit*. To determine the sex of a shark, I needed to visualize the claspers, an external appendage that extends underneath the shark from the pelvic fins and delivers sperm when inserted inside a female shark. Because of the location of the claspers, I needed to get the camera deep into the water, and even with the telescoping painter's pole fully extended, it was next to impossible to do from the pulpit of the *Ezyduzit*. For the camera work, I really needed a smaller vessel, and I knew that a faster boat would allow me to cover the entire Outer Cape over the course of a single day. The other advantage to a smaller boat was that I could tag the shark using a tagging pole instead of throwing the harpoon.

In addition to raising money to support the first year of the five-year population study ($50,000 in 2014), AWSC also provided critical logistical support. AWSC board member John King donated the use of his twenty-four-foot boat *Aleutian Dream*, as well as his time as captain. In addition, John outfitted the boat for white shark research at his own expense, adding both a tuna tower and a pulpit that was perfect for both the video and tagging work. John, a descendant of Chatham's founder, has deep roots on the Cape, but his background is more varied than most people I've met. His resume includes seven years as an Alaskan fisherman followed by an almost three-decade career in the pharmaceutical industry. Since retiring to the Cape, he and his wife have traveled extensively photographing wildlife but, starting in 2014, he was more often at the helm of his boat as Wayne Davis put him onto sharks for me to video fingerprint and tag.

Moving to the *Aleutian Dream* was a big change from working with Billy on the *Ezyduzit* and George in the plane above, and Billy was not particularly happy about it. I'd worked with Billy for about a decade tagging bluefin tuna, basking sharks, and then white sharks, and I considered him an excellent fisherman and a good friend. The decision to move to a new tagging platform wasn't easy, but it made the most sense from a logistical and cost perspective.

Even from the lower pulpit of the *Aleutian Dream*, there was a steep learning curve to getting the footage I needed of each shark. To do it correctly, I needed to account for light refraction and diffraction when adjusting the depth and angle of the pole and camera—it feels unnatural when you do it correctly. I always tell people to go deeper than they think they need to go with the camera. My experience with underwater photography really helped, but it wasn't just technique. There was also the fact that it took the whole team to get me in position to film a shark, and I felt pressure to not blow it. Every day when we would get back to the lab, I would painstakingly review the footage and then adjust my technique as necessary the next day. Today I have a sense for when I have enough footage of a specific shark, but early on, it was more difficult for me to judge. To get a good video fingerprint, we need footage of both sides of the shark, which comprises at least two sweeps of each flank. A lot depends on water visibility, depth of shark, behavior of shark, and angle of boat to shark. After doing this more than four thousand times now, I have a sense of when I have what we need, but again, there was a learning curve, and I learned a lot during the 2014 season.

The first year of the five-year study was a huge success, in no small part because of John Chisholm, who meticulously examined every video frame and identified sixty-eight individual white sharks. One of the advantages of documenting each shark by filming it was that we were able to give a far more accurate count as to the ratio of males to females. In 2014, forty-three of the sharks we documented were males and twenty-five were females. Of the sharks sighted, we tagged eighteen of them with acoustic tags. When we pulled all the data from the acoustic receivers at the end of the season, we learned that twenty-two white sharks pinged the acoustic array off Cape Cod from South Monomoy north to Truro. It was a banner year by any measure.

Our successes came at a cost, however, and the most obvious cost as we closed out 2014 was time. Both John and I knew we were going to need help. On average, I was getting roughly three to six minutes of footage of each shark I filmed, and every moment of that footage needed to be watched and cataloged in a growing database. Every individual shark needed to be confirmed as a new individual and not a shark already in the database. Moreover, all the data needed to be analyzed using new quantitative modeling techniques. I was already pushed to the limits, as was John, and neither of us really had the analytical skill set that was necessary. Over the years, I had learned that the best way to maximize productivity was to harness the energy, commitment, and efforts of graduate students. As an adjunct professor at the University of Massachusetts Dartmouth's School for Marine Science & Technology (SMAST), I routinely brought on students like Jeff Kneebone for various shark-related research projects. As far as data were concerned, I found the perfect student.

Her name was Megan Winton.

Megan Winton

Megan grew up in Florida but, as she readily concedes, Jacksonville is "redneck Florida," not the coral reefs and sandy beaches of the Keys. "It's basically southern Georgia," she says, but her grandparents lived fifty miles to the south, in Volusia County at New Smyrna Beach. Volusia County is famous for its beaches, but it also has the distinction of being the "world shark bite capital." On the one hand, sharks were just part of the landscape for Megan growing up playing in the surf at New Smyrna Beach, but on the other hand, they kind of scared her. She recalls going to the doctor with her mom when she was young. There was an oversize book about white sharks in the reception area, and after she dragged it into her lap, it flopped open to a page showing graphic pictures of shark attack victims. She was both mesmerized and horrified. As she turned the pages, she saw pictures of white sharks, some of which had been caught and were posed beside the anglers who caught them. They were massive animals. Megan was really into dinosaurs, and to her, the white shark looked a lot like a living dinosaur.

A couple of years later, she was bodysurfing at the beach with her sisters when a nearby angler landed a shark. Again, she was both horrified and fascinated. It dawned on them that they'd been swimming with sharks, which was simply insane to her and her sisters—but then there was the animal itself, and it was amazing. She became increasingly curious, and her mother fed that curiosity with a steady stream of books on sharks. The more she read, the more interested she became. She recorded Shark Week every year so she could watch it year-round. Her fascination with the animals slowly morphed into an intellectual pursuit, as she learned more about their biology. One Christmas in high school, she asked her mom for a compendium of scientific papers about white sharks. She was hooked.

Megan was still in high school when her interest in sharks turned into a passion she knew she absolutely wanted to pursue as her career. She was walking on the beach one day and came upon a couple of dead bull sharks. They'd been tail roped and apparently dragged to their deaths. As she looked at the dead sharks, she felt her stomach turn in revulsion. She wasn't against fishing, but these animals appeared to have been killed simply because they were sharks, and she knew this wasn't an isolated incident. While other fishes were actively protected, it was her perception that too few people cared about sharks. Whether it was shark finning or sharks dying as bycatch in other fisheries, it made her angry. It made her angry that these animals with which she was so enamored were treated with such disregard. It wasn't until 1993, within her own lifetime, that there were any protections for sharks in US Atlantic waters. Heck, the US government used to encourage people to fish for them as underutilized species. And don't even get her started on the continued media portrayal of sharks.

Through all of this, Megan realized there was a real conservation challenge that needed to be addressed. She knew she had the passion, and she'd begun to acquire some of the knowledge, but the more she sought to learn, the more surprised she was at what remained unknown. Much like me early in my career, Megan was shocked at how mysterious sharks really were—how much work still needed to be done. It was both mind-blowing and motivational.

In the fall of 2002, Megan headed off to Emory University to earn her bachelor of science degree. Emory is not known as a school for marine sciences, but Megan knew she wanted a strong background in biology. She then supplemented her classroom work with summers filled with marine science internships. After graduation, she moved to Panama City, Florida, where she started as an intern and then got hired on with the National Marine Fisheries Service's Shark Population Assessment Group. She loved her job. Like me, she also loved the dissections. She couldn't get enough of these animals and their biology. Throughout it all, she was surrounded by mentors who, like Jack and Wes had been for me, pushed her and inspired her. One of those people was a shark researcher named Enric Cortés, who said to her at one point, "A lot of stock assessment scientists are great with numbers but terrible with fish." From that moment on, Megan knew that she was going to focus her energy on being good with both numbers and fish.

Megan learned that David Ebert, who ran the Pacific Shark Research Center at California State University, Monterey Bay's Moss Landing Marine Labs, had a funded position and was looking for a graduate student. The work involved age, growth, and demography work on a deepwater skate species. It wasn't shark work, but population modeling was what she wanted to do, and this was a great opportunity. In 2008 she headed west and began her master's work at Moss Landing.

It was while living in California that Megan saw her first white shark. She had hoped that being in the so-called "red triangle" for her master's work would provide an opportunity to see a white shark in the wild, but that didn't happen. What did happen is that a friend of hers suggested they go to the Monterey Bay Aquarium on her birthday during the summer of 2009. She loved the aquarium, and they had a great time. As they were about to leave, one of the docents said to them, "You don't want to leave just yet." It was close to closing, but the professional aquarists were about to release a young white shark into the million-gallon Outer Bay exhibit.

Megan and her friend ran back to the tank, where the staff had all the bubbles going. They waited. The five-foot, three-inch white shark had been collected off Malibu and brought to the aquarium as "a way to change public attitudes and promote stronger protections for the ocean predator." When the bubbles slowly cleared, the young shark appeared, as everything else in the tank, even the tunas that were larger than the shark, sank toward the bottom. At eighty pounds, the white shark weighed less than Megan, but seeing it, even in captivity, was cathartic.

As part of her work at Moss Landing, Megan's co-advisor, Gregor Cailliet, sent her east to learn a new histological technique from Lisa Natanson. The new microscopic technique would hopefully help with her master's thesis work. While on the east coast, she both worked with Lisa in the Narragansett Lab and lived with her. One afternoon she drove up to New Bedford, Massachusetts, to meet up with a friend from her internship days, and her friend wanted her to meet a PhD candidate named Jeff Kneebone. "He only cares about sharks and fish and doesn't date anyone because nobody else gets it," she told Megan. "You two need to meet." They did, and they "got each other" and soon started dating. There were lots of things that Megan liked about Jeff, and while it wasn't really a motivating factor in their relationship, one thing was the fact that he was a PhD student with someone she wanted to meet.

That person turned out to be me.

Megan had always said she would never move for a boy—in fact, she'd ended several relationships for just that reason. But after she earned her master of science degree in marine science, she packed up and moved east in 2011. Megan felt it was important not to rush from one degree to the next, and this was especially the case with her PhD. She wanted to make sure that by the time she undertook her doctorate, the degree was serving her—a valuable lesson for all students. She didn't want to just take on whatever project was in line at a lab to which she got accepted. Once she moved east, she worked a couple of jobs, including a research assistant sampling bluefin tuna and a contractor for the National Marine Fisheries Service looking at winter flounder. Even though it wasn't shark work, she loved it, and she enjoyed living vicariously through Jeff and his work with me.

The first time I remember meeting Megan was at a party at UMass Dartmouth professor Diego Bernal's house, but Megan likes to remind me it was actually at an American Elasmobranch Society meeting in Providence, Rhode Island, in 2010. She was in graduate school and was sitting with Jeff and the rest of the Massachusetts crew at the banquet dinner. The party at Diego's house was for grad students like Megan and Jeff and professors such as myself. At some point in the evening, I was introduced to this grad student from California who had blue hair. She was getting her master's with Gregor Cailliet, whom I knew, and of course, she was staying with Lisa, who had been like a sibling to me since our days working together at the Narragansett Lab under Jack and Wes. So I remembered meeting her, but to be honest it was really her hair that stood out. I figured she might be more of a beatnik type or something. What I didn't know was that she was interested in Jeff, who was also at the party. Anyway, those parties could get pretty wild, so my wife and I left relatively early, and I didn't see Megan again for a number of years.

I got to know Megan a little bit through Jeff, and while her often-changing hair color always remained memorable, I ultimately came to know her as a talented researcher with a promising future. I also came to learn that she had a passion for white sharks. In fact, Jeff asked me if he could get her out on the boat with us on a research trip for her birthday, which I was happy to make happen. Unfortunately, it turned out to be one of the rare days that we didn't see a single white shark. From Megan's perspective, she was just thrilled to be on the boat while we were doing white shark research even if there weren't any white sharks. From my perspective, I saw a competent, easygoing, upbeat scientist who I thought had a lot of promise.

When Gavin Fay, whom Megan already knew and respected, took a job as a professor at SMAST, Megan reached out to him. Gavin was focused on developing interdisciplinary modeling approaches for the assessment and management of populations of marine mammals, reptiles, and fishes. Megan had proved herself as a biologist, but she was a biologist who remained very interested in data analysis and being Gavin Fay's PhD student seemed the perfect fit. I couldn't have been happier to see her come to SMAST, and John Chisholm and I made the decision to approach her about the five-year population study.

"Greg knew I was a nerd," she says, recalling the opportunity, "and they needed a nerd to help analyze and organize data."

CHAPTER 11

PREDATOR AND PREY

Great whites have become part of the fabric of Cape Cod in recent years. They're not quite as commonplace as tourists grabbing fried clams and french fries from a seafood hut—but for the most part, Cape denizens seem relatively unfazed by the shark's annual presence.

—Steve Annear, *Boston Globe*, July 11, 2018

Regardless of how much signage and information we provide, there still seems to be a concerning level of complacency.

—Nathan Sears, Orleans harbormaster, August 2018

DECEMBER 8, 1990—31°54'30.85"N, 79°3'53.56"W (OFF THE COAST OF THE SOUTH CAROLINA–GEORGIA STATE LINE)

Aroused by the blood and the thrashing, the sixteen-foot female white shark closes quickly on a whale calf. She extends her jaw, which is made of cartilage and is not connected to her skull. Instead the jaw is suspended by strong muscles and tendons that allow it to swing down and protrude forward, exposing rows of large triangular teeth with serrated edges and an opening large enough to feed on large prey items like marine mammals. Like a right whale calf. Because of the force needed to attack prey that is best rendered immobile on the first strike, the jaw is constructed of calcified cartilage that is stronger than the cartilage in other parts of the shark's body, such as its fins. This calcified cartilage is as strong as bone but much lighter, and it can exert tremendous force on whatever it bites, which in this case is the tail of the right whale calf just as it takes its first breath.

The attack is swift and decisive, and the white shark moves away as another plume of blood explodes from a severed artery. The hunting strategy is known as "bite and spit," so named by white shark expert John McCosker, wherein the white shark attacks its prey with a forcible initial strike and then waits for it to bleed out before feeding. It's been hypothesized that this strategy is the reason so many humans survive a shark attack, because somebody intervenes between the initial attack and the feeding. It may also be the reason that some large mammalian prey wash up on beaches after being carried into the shallows before the white shark moves in to feed. Regardless, the strategy is effective, and even the mother whale's usual heroic protective behavior is too little, too late to save her calf, especially since her fin is wrapped in rope from a Canadian crab pot she's dragged south from the Gulf of St. Lawrence after becoming entangled the previous summer.

The white shark is an apex predator at the top of her food web, which may suggest she is also selective in what she eats. While her hunting is now certainly more refined than her first thrashings at shoals of menhaden or lunges at gray seals, she is ultimately an opportunist. She takes what she can get, and she feeds when the feeding is good. She needs to consume, on average, as much as 3 percent of her body weight per day, which is actually not very much for a fish. Many fishes must consume at least twice that amount to meet their energetic needs, but the white shark has evolved into a much more efficient predator. Her feeding adaptations mean she can sustain herself on small and infrequent meals if she must, but being an opportunist, she won't turn down a larger, easy meal if the situation presents itself, as it has with the whale calf. Either way, her slow and efficient digestion results in an impressively low metabolic rate. The meal from this one whale calf may mean she will not need to feed again for weeks.

The associations between marine animals are complex and largely instinctual to the white shark. The disruptions to those associations, often the result of human intervention in the form of overfishing or some other environmental stressor, occur in a blink of evolutionary time. As she moves with the seasons from the Gulf of Maine to the warmer waters of the South, she proves time and again remarkably adaptable. She changes her behaviors and feeding strategies to match the environment she encounters even if it is different from that in which she is evolutionarily programmed to thrive. The immense shadow her girth casts as she patrols the ocean suggests she is too big to fail, but others were bigger and still fell prey to extinction.

When the Pilgrims landed near present-day Provincetown on the tip of Cape Cod, they didn't stay long. Instead they headed west, across the opening of Cape Cod Bay, and landed at Plymouth. It's almost twenty-three miles from the Race Point Lighthouse at Provincetown to the famous rock at Plymouth, a distance a white shark can easily cover in a few hours. It's not far "as the shark swims," but even as I was tagging a record number of sharks off the Cape in the first year of the five-year population study in 2014, people at Plymouth's beaches seemed to think white sharks were primarily a Cape Cod issue.

By September 2014, the number of gray seals in and around Plymouth had increased dramatically, so it wasn't surprising to me that white sharks would follow just as they had at Monomoy and then north along the Outer Cape. Nonetheless, when a Massachusetts State Police helicopter spotted a twelve- to fourteen-foot white shark swimming just outside the swimming buoys at Duxbury Beach in late August, a spokeswoman for the state's Department of Fish and Game said it was "an unusual place for a great white sighting." The beach, which is just across the bay from Plymouth, was cleared without incident and then reopened a couple of hours later after searches by aircraft and boat failed to locate the shark. Duxbury fire chief Kevin Nord said that, with one thousand beachgoers in the area, it was "too close for comfort."

For Ida Parker it was about holding on to summer when she decided to put her ten-foot kayak in the water at Plymouth's Manomet Beach a week after the white shark sighting at Duxbury Beach. It was a perfect evening in the way only a late summer evening can be—when you know the number of warm, sun-kissed evenings is waning and soon the ocean will turn a cold, slate gray, frothed by winter storms. The visibility that night was a good ten miles. Standing on Manomet Point, you could easily look northwest across Plymouth Harbor and see Gurnet Light standing proud at the head of the peninsula that extends out from Duxbury Beach—the peninsula along which they had searched for the white shark a week ago. Ida knew this water. She'd grown up here in a house on the bluff that had been in her family for eighty years and which overlooked Cape Cod Bay, with Wellfleet's Jeremy Point less than twenty-five miles to the east.

Ida, who was twenty-nine years old, asked her best friend, Kristin Orr, thirty-one, to join her for the paddle. As Ida made her way down the wooden stairs that descended the bluff from behind her house to the beach, she looked across the glassy, flat calm toward Stage Point. It had always been one of her favorite views in the world—it was her happy place. It was a little before five o'clock. The air temperature still hovered in the low eighties, but the humidity that could be so oppressive in the summer was almost unnoticeable. Ida dragged the two kayaks down the beach to where gentle waves lapped invitingly at the sand. She looked back up the stairs to see if Kristin was there yet. She was not, so Ida put a life jacket and paddle in Kristin's boat and, noting the tide was almost high, left it on the beach just above the high-water mark. The water was in the low seventies, and it felt wonderfully refreshing as Ida stepped in knee-deep, fastened her life jacket, and settled into her kayak's cockpit with her Olympus waterproof camera between her legs.

Earlier that day, and less than four hundred feet north along the beach from where Ida and Kristin launched their kayaks, Anna Kulmatiski, who was in her midthirties and whose family's house was just a few driveways up Manomet Avenue from Ida's house, had set out for a paddle in her nine-foot, five-inch, yellow sit-on-top kayak. Like Ida, Anna knew these waters well. She had spent many an extended weekend at the house on the bluff, and she loved to paddle out to near Stage Point and hang out with the seals. She enjoyed sitting quietly in her kayak, smelling the salt air and hearing the sound of the water. She didn't remember the seals from when she was a child, but they were sure there now. Sometimes she'd fish, but often she'd just wait for the seals to come close enough that she could hear them breathe. It was such a contrast from her city life in Somerville.

At a little after noon on September 3, Anna was watching the seals just to the south of Stage Point when she saw it. The fin broke the surface about fifteen feet away from her kayak, causing the seal she had been watching to quickly dive beneath the surface, soon to be followed by the splashes of many other seals sliding off the nearby rocks into the water. The fin was massive, and the dark back of the shark momentarily appeared like an island underneath it. Anna's mouth went bone dry. Her heart raced. She'd seen ocean sunfish before, and she knew that people often mistook them for sharks, but there was zero doubt in her mind about the animal she'd just seen. She told herself out loud that the shark was after the seal and wouldn't be interested in her. She started paddling, slowly at first, thinking the shark was still right there but not daring to look back. Soon she was paddling as fast as she could. When her kayak touched sand she was out of the cockpit and pulling her phone from the top of her swimsuit. She called 911, and the dispatcher connected her with the harbormaster.

At 12:46 P.M., after speaking with the harbormaster, Anna posted to the Facebook group "You Know You're from Manomet When. . . .": "Just saw huge shark eat a seal next to me in kayak! Stage point, Manomet!!!" Some of the first replies to her post surprised her given the primal fear she had just felt as she, for the first time in her life, considered what it might be like to be prey so close to a giant predator.

"I want to kayak there too!" someone commented on Anna's post.

"No!!" responded Linda Evans. "Jealous . . ." Then she followed up with, "I'm grabbing my kayak when I get home and heading out there . . ." Later she posted a picture of the view from the bow of her kayak with the caption, "I put in in Chatham last year after all the sightings. Didn't see anything. Hope tomorrow is a good day for shark finding!"

Other comments expressed concern and urged caution. Several people told Anna to call the harbormaster, which she'd already done. There were the requisite *Jaws* comments. "Where is Chief Brody when you Need Him!!!!!!!" and "Your [*sic*] going to need a Bigger Kayak!!!!!!" More than a few connected the dots between seals and white sharks. "Be aware!" one woman posted. "Stage Point is loaded with great white food supply." Another wrote, "It's about time!! Mother nature in action!"

Kristin arrived on the beach at around 5:15 P.M. and found her kayak waiting for her. She saw Ida out by where Ida's family's three boats were moored, just off the beach. The two fit, blond women were unrelated but often mistaken for sisters. They'd met at work a few years back and were now best friends, having traveled together to ski, white-water raft, and basically tackle any adventure they could dream up. As Kristin paddled out to join Ida, the hint of a breeze rippled the surface of the water, and Kristin thought it was one of those nights where the sky just melts into the sea. They started paddling north toward Stage Point, Ida in her yellow and white kayak and Kristin in her red and white one. Kristin's mom had packed chicken salad sandwiches for them, and they'd planned to eat out by Stage Point and then hopefully see the phosphorescence as they paddled back in the dark. Soon they passed where Anna had launched her kayak earlier in the day. They had heard about Anna's encounter in the same general area they were headed, but they found it hard to believe.

"I was like 'ha ha' you're so funny," Ida recalls later when thinking back on the experience. "I was like there's no possible way. I'm sure she mistook it, but there is no possible way it was a white shark."

Ida was aware of a growing number of shark sightings on the Outer Cape, but Plymouth wasn't Cape Cod. "I was just ignorantly thinking, *We're in Manomet*," she remembers. "*We're in Cape Cod Bay, we're so close to shore. It's not Chatham. It's not Wellfleet. We're not on the national seashore. We're not where white sharks are normally known to be*." She paused. "That was just pure ignorance." Despite the confirmed sighting at Duxbury Beach the week before, Ida and Kristin were not alone in their skepticism. Even Anna said she'd gotten the distinct impression that neither the 911 dispatcher nor the harbormaster believed her, but officials had nonetheless sent out a boat and looked for the shark. They found nothing. As Ida and Kristin approached the point, the notion that they were potentially in danger from a shark attack was about the furthest thing from their minds.

Soon they saw the seals they'd wanted to photograph. Growing up in Manomet, Ida remembered when it was a big deal to see a seal, but now they were always there, lounging on the rocks and bobbing up next to your kayak. Ida had paddled here more times than she could remember, but she never took it for granted, especially on such a stunning evening. She paused and let her kayak drift, as she listened to the sound of the hull cutting through the glassy water and felt the sun on her face. She turned to take a picture of Kristin, who was wearing sunglasses and a dark life jacket with an angled red stripe that matched the upper deck of her kayak. Kristin beamed in the warm glow of the late summer sun. Kristin took the camera from Ida and took some pictures of her. They continued paddling, chatting casually, taking pictures, and basking in the joy of their friendship and the summer evening.

Kristin's arms were tired, and she put her paddle down. Ida turned to get the camera back from Kristin just as the massive head of a shark came directly out of the water up to its first or second gill slit, directly under Kristin's kayak. It was just feet away from Ida's boat. Kristin's kayak was thrown upward and backward, and she fell into the water. At first she was unaware of exactly what had happened, but then she saw the shark biting her boat's hull before it disappeared as quickly as it had appeared.

"That was a shark!" Kristin yelled. She remembered its pointy nose.

"I know," Ida said. "I saw it!"

Kristin's kayak was right side up but swamped and, as a result, sitting low in the water. Even so, Kristin was able to scramble back into the cockpit, but Ida, whose boat had also capsized in the commotion and was now upside down, could only hold on to the bottom of the hull treading water. Disorientation and surprise erupted into fear. They were terrified, having a hard time reconciling what had just happened. They yelled to shore, but nobody heard them. Kristin's phone was in her drybag, and she got it out. She was shaking uncontrollably with fear. Their first instinct was to call Ida's house, which they could see from where they were in the water. They could see the two center console boats that belonged to Ida's family moored right off the beach beneath their home. Someone could easily come out to get them. It would be the quickest way to safety, they thought. The phone rang. Nobody answered.

They then tried Kristin's mother, but when she answered and heard them screaming about a shark, she thought they were joking because, to be fair, they did joke a lot. The call disconnected, and Kristin's mother called Ida's mom, who had heard about Anna's experience earlier. Ida's mom told her that she didn't think the girls were joking. Ida and Kristin looked nervously around them, not knowing if the shark was coming back or not. They decided to call 911.

"Help, help!" Kristin said to the dispatcher. It was a little before 6 P.M., and the fear was evident in her voice. "I was just on a boat with somebody and we're stuck in the water, and there's a shark."

"Okay, ma'am," the dispatcher calmly replied. "The harbormaster is on the way there. Where are you now?"

"How long are they going to be?" Kristin asked, her voice cracking.

"Where are you?" the dispatcher persisted. "Are you in the kayak, ma'am?"

"Yes, but he knocked me out of my kayak."

"Okay, so are you in the water now, or are you in the kayak?"

There was a pause, and then she said, "What?"

"Are you back in your kayak, or are you in the water?"

"It's sinking and my friend is holding on to hers—she's upside down in hers."

There was another long pause, and then the dispatcher asked, "Ma'am, where is the shark?"

"I don't know," she said. "He bumped me out of my boat."

"So, something knocked you out of your boat—did you see whether it was a shark or not?"

"Is that a seal?" Kristin asked Ida, nervously catching a shadow out of the corner of her eye. She scanned the water for any sign of the shark.

"It was a seal?" the dispatcher asked, thinking Kristin was speaking to him.

"No!" she said emphatically. "There's something in the water."

"Okay . . ."

"It was a fucking great white!"

"Okay. We have the harbormaster on the way. They are notified. We are on the way. Where exactly are you?"

"Right off of Stage Point in Manomet."

At 6:06, a local resident posted on Anna's Facebook post from earlier in the day, "Just heard on the news there is a rescue in progress off the point. Anyone have a scanner on? Anyone on the point to see what's going on right now?"

Within minutes, another responded, saying, "I live on stage point and a lot of rescue vehicles flew by." At 6:28, someone else confirmed a helicopter was circling overhead, with further reports that the helicopter flew halfway down Manomet Beach.

For Ida and Kristin, the wait was intolerable.

"We did know it was a shark," Ida said later. "Obviously that's a scary moment to know you're in the water with something that has that much force." Every time the kayaks bumped together, it reminded Ida of the sound the shark made when it hit Kristin's kayak. As a trauma therapist by training, Kristin was hyperaware. The water was deep and the bottom dark. The sun was getting lower on the horizon. Every sound and every movement sent shocks of adrenaline-laced fear coursing through their bodies. "It was nerve-racking," Ida recalls. To make matters worse, the kayaks drifted apart, and without being able to paddle, it was almost impossible for them to stay together. Kristin was thinking about how there was no warning prior to the attack on her kayak. No fin broke the surface. There was no circling. Just a rush from beneath that was forceful enough to upend a 275-pound kayak with a grown adult inside. Both women seriously contemplated death.

As they waited for help, one of Ida's neighbors who was on the beach had noticed something was up and paddled out in his kayak. He had also called 911, and as he approached, Ida and Kristin could hear him telling the dispatcher that there was no evidence of a shark. They started yelling at him that there was a shark, but he didn't seem to understand or believe them.

"It's going to come back!" Kristin yelled at the neighbor. "It's going to get you!"

"That's not going to happen," he said, trying to reassure them. "He's not going to get me, and I'm with you."

He was able to push their boats back together, and he helped Ida back in her boat, but Ida and Kristin were still hysterical.

Ironically, at about that same time, Anna was on her way to the Lobster Pound, a fish market on Manomet Point, to see if they had any shark meat—she thought it would be a fitting dinner given her experience earlier in the day. From the road, she saw the rescue occurring in almost exactly the same place she had seen the shark. She hoped someone had just been knocked off their kayak and it wasn't the shark. Ida's mom was also watching the scene unfold from outside her house atop the bluff. She had come straight home after Kristin's mom called her. From her vantage point, she could only see one kayak, as Ida's kayak was overturned and mostly submerged. She knew Kristin had made the call. She feared the worst for Ida.

The shark didn't return before harbormaster Chad Hunter, along with the fire department, arrived. Chad was at the helm of the harbormaster's boat, and the fire department was in a Zodiac. Much to the horror of Ida and Kristin, the fire department deployed a rescue swimmer to assist. Kristin lost it. She yelled at him to get out of the water because there was a shark, but as they would later learn, the 911 dispatcher had only told the harbormaster that there were kayakers in distress. He said nothing about a shark. Kristin, afraid the shark was still nearby, was shaking uncontrollably and sobbing. She didn't want to get out of her kayak to get into the harbormaster's boat. The first responders, who were still treating the rescue like a "distressed kayakers" incident and not a shark incident, were trying to calm her down. Ida, still terrified herself and desperate to get out of the water, said, "Just let them rescue you." Finally, the first responders got Ida and Kristin aboard the harbormaster's boat. The women hugged. They were crying hysterically as waves of fear and relief and disbelief washed over them. The rescuers tried to calm them. They tried to separate them to opposite sides of the boat and treat them for hypothermia. The neighbor said he was going to paddle back to the beach. "No!" Ida screamed. "There's a shark in the water!" She realized that nobody believed her.

They hauled Ida's kayak onto the harbormaster's boat first and then they began to recover Kristin's boat. As they slid the hull over the gunwale, a huge bite mark in the bottom was revealed. Everyone stared in disbelief. There was no denying what everyone could see, and a hush fell over the boat. The whole situation suddenly got very serious. The bite mark, which had pierced all the way through the boat's hull, was thirteen to fifteen inches across. John Chisholm later inspected the kayak and interviewed the women. He confirmed that both the size and the pattern suggested an exploratory bite by a white shark. He pointed to the striations around the tooth holes, explaining they were caused by the telltale serrated edges of a white shark's teeth. John estimated the shark to be twelve to fourteen feet in length.

"This is our notice that they are here," Chad said later. "It's definitely eye-opening."

Nowadays, Ida and Kristin agree. "It was definitely an eye-opening experience," Ida says. "It made me be a lot more mindful, and it was humbling to think I was like so freaking ignorant thinking there would never be a shark here." Ida had grown up loving sharks. She always tuned into Shark Week, and she was enamored with the white shark. "It's the shark—the king of sharks," she says. "They're just so magnificent, and I'd always wanted to experience the magnificence . . . just not the way we did." She says she was that person who probably would have said she wanted to go out in a boat and see one.

After the incident, which made national news, Ida was shocked by how many people contacted her and said, "Oh, God! I wish that had been me." She tried to impress on people how terrifying it was, but she continued to see people paddling out around seals and not taking the risk of white sharks seriously.

"It wasn't the shark's fault we were being stupid," she says. "Now I know better, and I'm actually glad it was Kristin and me and not a couple kids or a father and his child. We were very lucky. It could have been so much worse."

Megan's First Summer Aboard

The summer of 2015 was Megan Winton's first summer working with me and the rest of the team aboard the *Aleutian Dream*. She'd been out on the water the year before as an observer—it was the sharkless birthday gift from Jeff—but she was now part of the team. In 2015 she, like the rest of us, had responsibilities and with those responsibilities came a certain amount of anxiety. Being the newest member of the team, she didn't want to let anyone down. I knew the feeling well. I remembered how it was, working with Jack and Wes at the lab or aboard the *Wieczno*. I remembered how scared I was of screwing up. The trips aboard the *Aleutian Dream* weren't just day trips, and we were never far from shore like we were on the *Wieczno*, but working on the small boat can be just as stressful, sometimes more so. There is no place to get away on the twenty-four-foot center console boat, and it's incredibly easy to get in the way as the harpoon and poles with the cameras are passed back and forth. While we worked in all weather conditions aboard the *Wieczno*, on the *Aleutian Dream* there is nowhere to hide from it, whether it's sweltering heat or bone-numbing cold. If a squall comes up fast, you're getting wet either from driving rain or pelting spray. We were also working in shallow water around, in between, and inside of breaking waves—that scared me the most. To make it even more taxing, we were working with costly, high-tech gear in a harsh environment that pushes every piece of equipment to its limits. We tried to pick our days, but regardless of the conditions, the science had to get done. With a small team, nobody can afford to slack off.

We all also feel the financial pressure, especially working with AWSC funding. We know how expensive it is to put the boat on the water and the plane in the air. A lapse in focus can, at best, mean a missed opportunity or, at worst, an accident. Every time Wayne spots a shark from his plane and radios John at the helm of the *Aleutian Dream*, it kicks off a finely choreographed ballet that hopefully ends in us getting the footage we need and possibly tagging a shark.

Megan quickly demonstrated herself to be a valuable member of the team that first season. While she's small, she's fit, and although she's spunky with a keen sense of humor, she proved to be as tenaciously serious as a barracuda pursuing its prey when the team jumps into action. During her first season, she was shocked at how close John sometimes brought the boat to the beach in pursuit of a shark. John would be watching the shark and reading the water. I'd be in the pulpit, pole in hand. Megan would be readying the tag on the end of the tagging stick and looking aft, ready to give John a heads-up when a larger set of waves jacked up onto a sandbar and angled in hard toward the boat's transom, posing a very real risk to boat and crew. She did this while also doing the critically important job of constantly recording data on a tablet. It can get pretty intense.

The 2015 white shark season off Cape Cod wasn't Megan's first time doing field work with accomplished scientists, and she admitted it wasn't always easy. Over the years, she had observed her fair share of what she calls "expertism." Science, especially shark science, can too often become about ego, and some scientists work very hard to preserve the image of the unfailing expert and the power structure that goes along with it. In Megan's experience, these "experts" feel like they must have an answer for everything, and they don't like to admit when they're wrong. They're not particularly responsive to the opinions and ideas of students or technicians, and in turn, those students and technicians often learn to keep their heads down. This can especially be the case when the student is female and misogyny rears its ugly head.

The issue of misogyny and gender and racial inequality in shark research was addressed head-on in an August 2020 article in *Scientific American* titled "The Dark Side of Being a Female Shark Researcher." In the article, Catherine Macdonald wrote:

> Becoming a scientist should not require developing the grit to continually endure misogyny, discrimination, harassment, assault, or bullying. But from their earliest experiences, women scientists learn that if they complain, they'll be described as "difficult," or a "problem"; if they're heartbroken by how they're treated compared to their male peers, they'll be told they're "too emotional" and "need to grow a thicker skin"; and if they leave the field, the problem was them: they "weren't tough enough to hack it in shark science."

In the article, Macdonald cites example after example, ranging from professional conferences to remote fieldwork. I feel I've been fortunate in my career to work with female scientists like Lisa Natanson, Nancy Kohler, Erin Summers, and now Megan Winton. Some of these women—like Nancy, who became chief of the Apex Predators Program after Jack retired, and Lisa, who became a senior research scientist and who personally tagged over eleven thousand sharks as part of that program—reached the upper echelons in their field, but I know that is not the norm. "To call shark science a 'boys' club' is an understatement," Macdonald wrote. "Although more than sixty percent of graduate students in my field are women, the vast majority of senior scientists are men." As I get older and achieve that senior scientist moniker, I feel fortunate that I'm working alongside the next generation of shark researchers, many of whom are talented women like Megan and Erin.

Although I'd known Megan for a few years because of Jeff, I didn't know her well until that first summer she worked with us. There aren't many better ways to get to really know somebody than spending long days on a twenty-four-foot boat with them. I had known that Lisa was impressed by Megan's field skills and Enric was equally impressed by her analytical talent, and honestly that's all I needed to know, but it was a real pleasure getting to know Megan as a person, too. My approach with graduate students is to mentor and teach as best I can, but ultimately, I only know I've been successful in my duties when I start learning from them.

In the case of working with Megan during the first years of the five-year population study, I started learning from her almost immediately.

Although the videos of our work on the boat that you may see on Facebook or YouTube make it look exciting and even, at times, frenetic, the reality is that there is a lot of downtime when we're waiting for the spotter pilot to locate a shark or when we're heading to or from a location. During those times, we chat about all manner of things. Megan told me early on that there had been times when she thought she'd been born too late. She sometimes longed for the "Wild West" days of early shark research with the likes of Jack and Wes, the Cooperative Shark Tagging Program, and research cruises on the *Wieczno*. I totally related, and one of my goals became to make Megan feel that working with me made her part of all that. Even more important than that, however, I wanted Megan to realize she was coming of age as a scientist in an era of technology about which someone like Jack Casey could only dream. What she was going to accomplish with that technology in her career was something about which I could only dream.

The first white shark Megan saw in the wild was a fifteen-foot female we first spotted about one hundred yards off Orleans's South Beach. The shark completely took her breath away in the same way that I'd felt when I saw Gretel. Megan, like me, had seen so many images of white sharks during her life, from the book in the doctor's office when she was a kid to *National Geographic* and Shark Week, but seeing one in the wild was, to her, the equivalent of seeing a unicorn. As she hung over the rail of the *Aleutian Dream* watching the broad, slow sweeps of the animal's caudal fin, she marveled at the width of her back—her girth.

"I could just lay across her back," she remembers thinking. I was in the pulpit with the painter's pole filming the shark with the GoPro, but as I glanced back at Megan, I immediately recognized the look on her face. That was me and probably every other shark researcher who had studied white sharks and then finally saw one for the first time in the wild. It's impossible to describe. The shark's form diminished as she swam away from the boat, but Megan's enthusiasm did not. The broad back of the shark became a purple smudge and then she was gone altogether. As I stepped down from the pulpit, I smiled broadly at Megan. It was like the feeling I've gotten from giving one of my kids a present I can tell they really love and appreciate.

"Pretty amazing, right?" I asked Megan, who was giggling uncontrollably. I started laughing as well, and soon everyone on the boat was laughing. I think Megan was later embarrassed by how excited she was—how much of a fangirl she appeared—but I totally got it. We all did. She was euphoric because she'd just seen the most amazing animal she'd ever observed—the animal that represented so many of her hopes and dreams and goals. The animal that had inhabited her dreams and the one to which she wanted to dedicate her career. It's a poignant reminder to me that, while we work our tails off to gather the data and try to understand the species, there is a far more elemental connection we have to these animals.

We named the shark Freckles.

During Megan's first year with me, we identified 106 individual white sharks thanks to continued financial and logistical support from AWSC. Twenty-four white sharks were tagged, bringing the total number of tagged sharks to eighty individuals since 2009. Thirty-one individual sharks pinged the acoustic array from mid-June through October. It just kept getting better.

In early November, as the autumn temperatures began to drop, John King pulled the *Aleutian Dream* from the water for winter storage. The fieldwork was done for the season. Like me, Megan loved being in the field and working directly with sharks, but she was also that self-avowed nerd, and the development of the techniques that would allow us to efficiently analyze our rapidly expanding dataset was equally exciting to her. As the first of the winter storms buffeted the Outer Cape, Megan was in front of a screen coding and developing scripts that would be used to process the data collected over the course of the five-year population study.

As we progressed into the new year, Megan continued to work with the data, and while she'd seen it firsthand the previous summer, she couldn't help but be struck by how close these sharks were to popular beaches. The public safety issue began to loom large in her mind, as she knew it did in mine. Unlike me, when Megan got started in shark research, the mission was clear: shark populations around the globe weren't doing well, and sharks needed the help of scientists. She had thought that would be the motivating force behind her career, and in many ways, it still was, but what she hadn't anticipated was finding herself on the other side of a conservation success story. She didn't realize just how tricky that could be and how easy it could be for history to repeat itself if public perception changed, especially in the face of a severe or even fatal attack.

In addition to Megan joining the team, one other significant event happened in 2015 that made me more optimistic about our work moving forward. The state put in place a permitting requirement for any person or research group that wanted to attract, capture, or perform research on white sharks in the waters under the jurisdiction of the commonwealth. Becoming a white shark hotspot meant that we'd begun to attract a lot more attention for our work. The fact that people could rely on white sharks to be just off the beaches in the summer and fall had implications both in terms of ecotourism and science, and several Cape Cod municipalities and constituents had expressed concern that if the state didn't put regulations in place, something bad could happen.

If, for example, a tour company or a research boat decided to use chum to bring in white sharks, there was nothing stopping them from doing that right off a crowded beach, which could lead to dangerous shark-human interactions. While OCEARCH had been cleared of wrongdoing in the case of the fatality in South Africa near where OCEARCH had been chumming, the continued perception of their responsibility lingered like an ugly bruise. For my part, I felt it was particularly important not to bias the growing number of sharks included in our five-year population study. Any fishing, ecotour, or research activity that affected the sharks' normal behaviors could invalidate our study—and we'd worked too hard to allow that to happen.

"I Cannot Put Our Five-Year Population Study at Risk"

In late 2015, OCEARCH returned to the US after two years abroad and several lackluster expeditions, including missions to the Galapagos and Australia. Funding from Caterpillar was drying up and not likely to be renewed. Without a television deal or big sponsor, Fischer knew that putting big white sharks on the deck was the only way to drive social media, keep his business alive, and, perhaps most importantly for him, I was learning, stay in the limelight. Returning to Cape Cod could be just what

OCEARCH needed, especially as Mary Lee, one of the sharks they'd tagged off Cape Cod, was achieving social media stardom with nearly fifty thousand followers on Twitter.

In December, word got back to David Pierce, director of the Division of Marine Fisheries, that Fischer was considering returning to Massachusetts for an expedition. On December 11, David wrote to Fischer, in order that he not "be caught unaware of the agency's actions and intent as you make your future plans." Specifically, the director explained that the agency had "promulgated much-needed regulations to control activities around white sharks" and that "new regulations prohibit persons from attracting or capturing a white shark within the waters under the jurisdiction of the Commonwealth without having been issued a 'white shark special permit.'" He continued:

> The requirement for this new permit is designed with the specific goal of constraining certain activities designed to attract white sharks to persons, objects, or vessels to protect the sharks and safeguard public health. Examples of such activities include cage diving, shark chumming, baiting, and feeding, towing decoys, applying research devices on sharks, and attracting sharks to conduct these activities.

In short, OCEARCH's traditional methods wouldn't be allowed in state waters. "[W]ithout authorization to chum and attract white sharks," the director surmised, "we expect your success rate will be much reduced."

The director made it clear that this was not just about denying OCEARCH a permit to work in state waters. "[W]hile the multi-year, mark/recapture study is ongoing, we will not permit chumming for white sharks for any researchers given the potential to alter the natural behavior and distribution of white sharks in the area—an activity we believe will confound and compromise results of our ongoing research."

Despite the director's letter, in February 2016, OCEARCH formally requested permits to work in state waters in both 2016 and 2017. The director responded by saying that the agency remained concerned about any activity that includes attracting sharks. "As I mentioned in my previous letter," he wrote, "this agency is conducting a five-year mark-recapture study (which uses no attractants) to video document and tag white sharks in Massachusetts waters. This is our primary focus, and we will be continuing this work through 2018. The continued prohibition on chumming will protect the integrity and the sample design of the mark-recapture study and enhance public safety. I regret this may prevent your successful planning of expeditions to Massachusetts."

It wasn't just about my research, of course. There was also the growing concern regarding public safety that had been brought up by many of the towns. "Please understand," Director Pierce wrote to Fischer, "that the growing abundance of white sharks in our nearshore waters used by many for recreation has elevated the sensitivity of the situation. Municipalities and other public officials are apprehensive about future interactions with white sharks. While it would be scientifically useful and intriguing to get more long-term tagging results, we are balancing the needs of public safety with this decision."

Chris Fischer persisted, and in June, he and Bob Hueter met with the director to discuss permitting once again. In a follow-up letter to that meeting, where again the director denied OCEARCH the permit, the director was more specific as to how OCEARCH's work would affect the five-year population study. He pointed out that three of the four sharks OCEARCH had tagged off Cape Cod with me in 2012 and 2013 left state waters almost immediately after tagging. "Regardless of the manner in which great whites are attracted to your vessel (e.g., chumming or with bait), the outcome will be the same—capture and likely post-tagging behavior in violation of key model assumptions. I cannot put our five-year population study at risk."

I was thankful of David's commitment to protecting our work, and I knew it wasn't easy. In addition to holding the line, there was the potential PR fallout surrounding the denial of a permit to an organization like OCEARCH, which had such a large social media following. I understood why OCEARCH felt like returning to Cape Cod was in their best interest, but I and my team had been here since the beginning working with these sharks, and we'd built and funded this study using a tagging method that was very different from OCEARCH's. While some wanted to frame the whole issue as a turf battle, I simply wanted to get on with the science.

With my bosses dealing with the OCEARCH drama, I was able to turn my attention back to research as the 2016 field season got under way. One of the great challenges of working with white sharks is that both the sharks and the research work are hard for the general public to see. Other researchers working on other apex predators often had an easier time communicating their work to the public. For example, wolf researchers in Yellowstone could meet with park visitors at a pullout along the road, where the wolves they were studying could be viewed through a scope. We tried to provide as much access as we could to the media so that they could then communicate our work to the public, but it was not uncommon for something to be lost in translation. This is one reason I was pleased when, in May, the Atlantic White Shark Conservancy opened the newly renovated Chatham Shark Center.

"We really want this to be a great interactive space where people can learn about what's happening off the coast, and feel really connected to what's going on there," Cynthia told the *Boston Globe*. In addition to the Shark Center, the conservancy announced that it planned to launch an app in July called Sharktivity, which would allow users to get push notifications about shark sightings. The app would also allow users to become citizen scientists by submitting their own white shark observations, which would then be reviewed by John Chisholm.

While some continued to worry that highlighting the presence of sharks off Cape Cod's popular beaches would negatively affect tourism, I was all for getting the word out in any way we could. The days of burying our heads in the sand and saying, "What sharks? We don't have sharks over here," were over. It was time to move past that and be forthright and honest with the public about the presence of these animals. In addition to the work the conservancy was doing to get the word out, beach managers and public safety officials with whom I'd been meeting through the White Shark Working Group planned to begin utilizing purple warning flags with the silhouette of a white shark at many guarded beaches in 2016. Some towns also planned to erect large billboards at beach entrances that explicitly talked about the presence of white sharks and risks associated with their presence. "It's certainly not to sensationalize the situation," Nathan Sears, Orleans's natural resource manager, said. "You just really need to jam it down their throats."

What's in a Name?

On June 17, the first returning shark of the 2016 season pinged one of our acoustic receivers off Monomoy. It was a thirteen-foot male shark we had named Scratchy because of scratch marks along his body that were likely the result of interactions with seals. We had tagged white shark Scratchy on August 17, 2015. I'd become far more comfortable with the notion of naming sharks. In part this was because we were seeing some of the same sharks return, and with our growing database of video fingerprints, we were really focused on identifying these animals as individuals.

On June 24, we tagged our first shark of the season, and it was a particularly poignant event for me. Four days earlier, I'd been on the *Aleutian Dream* with Megan and John King looking for sharks to tag when Paul Cataldo called me. Paul was a friend, and his voice on the other end of the line was frantic, spitting out short sentences about our mutual friend Luke Gurney being lost at sea. Within an hour, I found out that a Coast Guard vessel had been dispatched only to discover that Luke had become entangled in pot line and was dragged overboard while setting gear off Nantucket. Luke had drowned. I was in shock. I contained my grief while on the boat for the rest of the day, but we didn't tag any sharks. I suppressed my emotions all the way back to the parking lot, where I dropped off Megan. I then, quite frankly, drove to a local boat ramp near my house, stared out at the merciless ocean, and wept like a child.

Luke Gurney had been one of my closest friends. He was a lifelong passionate recreational fisherman, whom I met in the mid-1990s when I lived on Martha's Vineyard. I hired him to work with me that first summer after meeting him, and our friendship quickly deepened. Even though he only worked for me one summer, Luke was very much part of my life. His talent at fishing meant that I regularly called on him for a lot of projects. Once, for example, he accompanied me to Virginia to tag bluefin tuna, and another time he helped me build tag floats for basking sharks. He also built part of my house on the island and later helped me refurbish my kitchen off-island.

Luke loved cooking and making sushi, which spawned almost weekly sushi parties at my friend Chuck Walker's house on Martha's Vineyard.

When I was working with George Benz on the National Geographic Explorer film *Body Snatchers* in 2000, I got Nick Stringer of Big Wave Productions to hire Luke and his wife to cook for us during the filming. Not unlike me, Luke's love of fishing was driven by a natural curiosity about fish and the ocean, so it was not unusual for him to call me to tell me a fish story, ask a question, or run a concocted hypothesis by me. Luke had become a very successful commercial conch (whelk) pot fisherman in 2009, and he adored the work—leaving the dock at 4 A.M. and landing his catch in late afternoon. There is no place I could imagine Luke being happier than on the water fishing.

Four days after learning of Luke's death, we were back on the water looking for sharks to tag. We had yet to tag a shark that season, so I was thrilled to find, video-fingerprint, and tag a thirteen-foot male white shark near Nauset Beach. It seemed fitting to name him Luke. I saw White Shark Luke several times that summer, and I've seen him every year since—he's pinged our acoustic array more than twenty-five thousand times as of this writing.

I told Luke's family and friends all about him, and as I did, I realized that what had started as a name to honor my friend morphed into more than that. White Shark Luke not only reminded me of my friend, but also was an individual I was getting to know better and better each year. With hindsight, I think that's when naming sharks took on more meaning for me. I was still somewhat concerned with assigning names to fish that could potentially hurt people, but I also saw the merit in doing so, and no longer was the agency pushing back as they'd done with Gretel. I'd noticed that people tended to embrace named animals, like pets, and I felt that was better than people being fearful of them. The conservancy also learned that naming sharks could help with fundraising—for a $2,500 donation to the conservancy, individuals could name a tagged shark. Names were also practical. They helped with keeping track of sharks day-to-day and in talking about them—it's easier to talk about "White Shark Luke" instead of "tag number 20905."

On June 29, for example, White Shark Jameson pinged a receiver off Monomoy, and people responded like the shark was an old friend. This shark had been stranded the previous July on a beach in Chatham, and I, along with several beachgoers, Chatham's harbormasters, and volunteers from the conservancy, had rescued him. Before I released him, I tagged him, as people cheered in the background. It was a profound moment for me—to see people cheering about saving a white shark rather than cheering about killing one. It marked a sea change in our collective relationship with this animal. The feel-good story made the news, and there was a lot of excitement when people learned that Jameson had returned. I think Jameson's fans would have been upset if we'd just said "Shark 23532." They felt like they knew him. It was personal.

There were other things we were learning about white sharks that made naming them feel natural. We had learned they could live more than seventy years, for example, and right or wrong, humans tend to have more respect for long-lived animals. Naming an animal you observe over a period of years or even decades seems somehow more appropriate than naming an animal you only see once. Personally, I was starting to see these sharks as not just part of a greater population, but also as individuals that returned year after year, like old friends and acquaintances. Naming them just seemed to make sense.

A Violation of Scientific Trust

By the end of August 2016, we'd tagged twelve sharks and video-fingerprinted many more for our database (by the end of the season we'd tagged twenty-one sharks and documented almost 150 individual white sharks). With the Sharktivity app up and running and the Chatham Shark Center seeing steady attendance numbers, I was really pleased with our work, and I was convinced we were making a difference. Unfortunately, 2016 still had some surprises in store, and they wouldn't all be positive.

While there was no hard-and-fast rule about interfering with another researcher's established study, in my experience, common courtesy and mutual respect generally sort out any potential issues. Unfortunately, courtesy and respect were gradually diminishing qualities between me and OCEARCH. Having been denied access to state waters, Chris Fischer radically changed his frequently touted mantra "to assist local scientists" and decided to move forward with Expedition Nantucket and his own scientific team, led by Florida scientist Bob Hueter, who was more than eager to fill my shoes. Unable to convince the Massachusetts Division of Marine Fisheries to grant them a permit because of fear of interfering with our ongoing five-year population study, OCEARCH sought and received its own permit from NOAA Fisheries' Highly Migratory Species Management Division to operate in federal waters at least three miles off the Massachusetts coast. When OCEARCH arrived in September, tensions quickly escalated. OCEARCH was operating in federal waters off Monomoy, and their work was certainly close enough that sharks that were part of our population study could be affected by an OCEARCH chum slick. I was worried. The media picked up the drama, using words like *feud* and *battle*, but for me it was not about turf—it was about the integrity of the population study and Megan's dissertation, and we were scared to death of introducing any bias into our own research.

While Chris Fischer and Bob Hueter maintained that the Division of Marine Fisheries' objections to OCEARCH's methodology were not "science-based," other independent researchers agreed with us. George Burgess, director of the Florida Program for Shark Research at the Florida Museum of Natural History, told the *Boston Globe* that he thought OCEARCH's actions were "a violation of scientific trust." Burgess agreed that chumming had the potential to either lure sharks away from our study area or attract them to it, since, at times, OCEARCH was operating within hundreds of yards of state waters off Chatham. "Any way you look at it, it's a modification," Burgess said. "The reality is, once an established researcher or team has a project underway, and it's known, standard operating procedure is to not intrude."

Unfortunately, my worst fears became reality when, on September 23, OCEARCH tagged a white shark they named Miss Costa, a nod to sponsor Costa sunglasses. After looking at images of the shark, John Chisholm came to me and told me that there was no doubt in his mind that Miss Costa was White Shark Island Girl, a shark we'd already identified and which was part of our study. Less than two weeks later, Miss Costa/Island Girl showed up off North Carolina having hightailed it from the Cape after OCEARCH released her. That was exactly what we were afraid might happen—a shark in our study would be attracted to the OCEARCH vessel via a chum slick, and then after the traumatic experience of being caught and lifted out of the water, the shark would dramatically alter its behavior. We'd seen the same pattern in sharks we'd tagged with OCEARCH in 2012 and 2013. Two of those sharks left the area and did not return for a year after being tagged, while another never returned. In contrast, our data showed that more than 80 percent of the sharks I'd tagged from the pulpit of the *Aleutian Dream* stuck around.

The entire incident left a bad taste in my mouth, which was only exacerbated by OCEARCH's continued penchant for media-attracting hype and hyperbole. For example, just before coming to federal waters off Cape Cod, OCEARCH claimed with great fanfare that it had found the first-known birthing site for white sharks on the North Atlantic coast. The headline read: "Researchers Discover Great White Birth Site off NY," and Fischer claimed that after twenty-six expeditions, this was the most significant discovery OCEARCH had made. "We think we found it," Fischer said on *CBS This Morning*. "Definitely the nursery and likely the birthing site."

I, as well as most other shark scientists, knew that Jack Casey was sampling and studying juvenile white sharks in the New York Bight back in the early 1960s. Jack and Wes published their findings in 1985, writing: "The occurrence of small and intermediate size white sharks in continental shelf waters of the Mid-Atlantic Bight suggests this area serves as a nursery area for juveniles." In that 1985 paper is a photograph of ten juvenile white sharks caught on a research trip in the New York Bight in 1964. The photograph itself brings a smile to my face, as I recall developing that image in the darkroom for Jack while working for him at the Narragansett Lab in the early 1980s. As a result, that 1985 paper was the first time my name appeared in the scientific literature, although it was only in the acknowledgments.

With all that had transpired in 2016, I was pleased to be able to close out our field season on October 3 by tagging a thirteen-foot male white shark I named White Shark Casey in honor of Jack. White Shark Casey was the one hundredth white shark we'd tagged since 2009, an important milestone in my career and a significant accomplishment for white shark research in Massachusetts.

The Population May Be Large

In March 2017, at a meeting of the regional White Shark Working Group, I released our most recent data. I told the audience that we had identified 147 white sharks in 2016 between June and October, and I told them that eighty-nine of those animals had not been documented by us previously. I compared those numbers to 2015, when we identified 142 white sharks, with 101 of those being new to us. In 2014, the first year of the five-year population study, we identified eighty sharks.

"We kind of hope that we're seeing a lot of the same repeat customers and starting to get a strong handle on the population," I told them. "But we're still seeing a good proportion of new individuals. That means the population may be large." I was sure to emphasize that the numbers I was reporting did not indicate a change in the risk level. "The best way to avoid any kind of interactions with these sharks is through public education, and that's been a big push from the Working Group, the towns in these areas, the conservancy, and us."

One of the other things that I reported at the meeting was that we saw more juvenile sharks in 2016. "Last summer," I said, being very careful with my wording, "we saw greater numbers of smaller sharks, including juveniles, and that tells us that the population is rebuilding." This was exciting to me and Megan, but I knew these data may be seen in another light by the public. I was beginning to believe that what the data might be telling us was that we were witnessing a conservation success story, and a very unusual one at that. If the population of white sharks in the western Atlantic was, in fact, rebounding, that was a momentous accomplishment. Restoring an apex predator to an ecosystem doesn't happen very often, and I, for one, wanted to celebrate that fact if we could prove it. I knew that others, however, might be concerned about a population rebound because more sharks would surely mean more risk. I got it, and to a certain degree, I agreed. Because we were just three years into our five-year population study, I didn't want to draw any major conclusions yet, but I was certainly aware that how we communicated our data was going to be very important moving forward.

With the 2017 research season fast approaching, I was feeling good about not only our research but also how the towns were working together on the public safety front. The feedback I received from beach managers through the regional White Shark Working Group was that the mitigation strategies were generally working. Primarily, these strategies focused on outreach and education. The goal has been to get the public "shark smart," and the informational signs, flag warning system, and community engagement seemed to be working. *Sharks eat seals, so don't swim near seals* was at the core of every communication. We emphasized that it was about using common sense and understanding that the sharks are probably there whether you see them or not.

In addition to common sense, technology was an ever-present topic when it came to discussing shark mitigation strategies, especially at the Working Group's last meeting before summer. Shark mitigation technology does not come cheap, however, and some of the towns were reluctant to spend their revenue on technologies that may or may not prevent a shark-human interaction. For example, a proposal to use tethered balloons outfitted with cameras to monitor beaches came in at $300,000 for a twelve-week trial program. As expensive as that was, it was a drop in the bucket compared to another proposal involving smart buoys that used sonar and shape-recognition software to detect a shark and then automatically notify beach managers. That proposal was over $200,000 to monitor just one beach.

A potentially more cost-effective option was one that we had been pursuing for the last three years: a real-time detection buoy that could become an integral part of our acoustic array and immediately notify beach managers via cellular technology of any tagged shark that swims near it. At roughly $15,000 each, the off-the-shelf version built by the tag manufacturer was cost-prohibitive, and my colleagues in California told me that the model available in 2017 was fraught with technical problems. The concept was fairly simple, but building a unit that could withstand the Cape Cod elements was a big challenge. Since 2013, John Chisholm had been working with a local company to produce such a device, and while we'd had some success, we were still without a fully functional unit in 2017. The bottom line was that we were not finding a silver-bullet technology solution to risk management and public safety.

The Atlantic White Shark Conservancy continued to play an ever-larger role when it came not just to funding our work on the water but also with public education and safety. In just five years, the conservancy had grown to become a huge asset and, quite frankly, a relief for me because I didn't have to spend as much time worrying about how I would fund the research. Instead that was largely Cynthia's job, working with her team. In addition to funding research and public outreach, the conservancy was actively engaged in the issue of public safety. During the winter of 2017, they brought in Alison Kock, the lead scientist for a respected South African shark safety and research organization called Shark Spotters, to meet with beach managers and town officials.

Shark Spotters was founded in 2004 after a series of fatal shark attacks prompted many to call for shark hunts and other lethal measures in South Africa. Shark Spotters offered an alternative based on a network of spotters with radios who could communicate directly with beach managers to clear the water in advance of an approaching shark. In Kock's meetings with Cape Cod beach managers and town officials in 2017, she emphasized the importance of addressing the presence of sharks and the risks posed to beachgoers before an attack happens. Although Shark Spotters did not eliminate the potential for a fatality (a swimmer was killed by a white shark in 2010 at a monitored beach in South Africa), the organization's work has gone a long way to creating an environment where a shark attack doesn't lead to a scene like the angry mob on the Amity dock—a scene I wanted to avoid at all costs.

Unfortunately, I knew that the sustainable coexistence of people and sharks is not the norm in many places where white sharks and beachgoers overlap. In Australia, an increased number of shark attacks beginning in 2012 and continuing through 2016 led the government to permit the use of lethal shark mitigation strategies such as shark nets and drum lines, which are prohibited under Australian law because of the white shark's protected status. The perception was that the government caved to public outcry over the danger posed by white sharks. As I saw it, it was a clear case of politics subverting science, and I was terrified it could happen in Massachusetts. Every time the issue of lethal mitigation strategies came up in discussions on Cape Cod, I stated clearly that the science does not support those strategies. Instead, lethal mitigation strategies make people feel like something tangible is being done, even when there are no data to support that they work.

Heading into the summer of 2017, Cape Cod's white sharks garnered international media attention. The British newspaper the *Guardian* warned of a "public safety issue," while *Travel + Leisure* published a "travel warning" for Cape Cod: "A scene from the shark horror movie *Jaws* may not be imminent, but popular beach spot Cape Cod is bracing for a season of increased shark sightings this summer." The media interest stemmed in part from a January article in the *Cape Cod Times* with the headline "Warning: Danger Ahead from Cape Sharks," and the dire prognosis that "[s]cientists agree it's not a matter of if—but when—there will be a fatal attack on Cape Cod." One of those scientists was George Burgess, the director of the shark research program at the University of Florida and curator of the International Shark Attack File. Burgess told the *Times* that he'd traveled the world investigating shark attacks, and he expected a call from Cape Cod in the next five years. As the *Times* reported, Burgess said: "That's when the feel-good shark story, the souvenir T-shirts, the specialty drinks at the bar, could all fade, and the public will realize exactly what's in the water just beyond the umbrellas and beach toys."

Perhaps ironically, and despite all the media hype, as the 2017 summer tourist season began in earnest, we were observing less shark activity on our twice-weekly research trips. I suspected it was related to the fact that the waters off the Cape were slower to warm that year. It had been over two years since the shark bit Kristin Orr's kayak off Plymouth and more than four years since the 2012 attack on Chris Myers. I knew the human mind could more easily process intermittent nonfatal shark attacks as an anomaly and, as a result, become somewhat complacent. The Sharktivity app helped to keep white sharks front of mind for subscribers, but the scenes at the beach suggested that shark-smart common sense was far from ubiquitous. It felt like I was always walking a fine line between telling people that there were sharks hunting seals closer to the beach than they probably thought and, at the same time, reminding them that the risk of an attack was extremely low. Finding that balance was increasingly like threading a needle, and I knew that if it felt that way for me, it had to be absolutely maddening for lifeguards, harbormasters, and other people responsible for managing risk on the beach.

Targeted, Localized Shark Hazard Mitigation Strategy

Guglielmo Marconi chose Wellfleet as the US terminus for testing transatlantic wireless communications because of the desolate landscape atop steep, forty-foot sand cliffs that provide unabated views out over the Atlantic Ocean. A few years after Marconi facilitated the first such communication, in 1903 between President Theodore Roosevelt and England's King Edward VII, Marconi Station was replaced by a newer station in

Chatham. The area atop the sand cliff became Camp Wellfleet, an artillery training facility. When Cape Cod National Seashore was created in 1961, Camp Wellfleet was absorbed by the National Park Service. The beach below the sandy cliffs was named for Marconi in 1969.

One of the things that sixty-nine-year-old Cleveland Bigelow III loves about Marconi Beach is that from the water, there's a feeling of utter isolation from the outside world. Against the backdrop of towering dunes, one sees little more than wild Cape Cod from the vantage point of a paddleboard. It was midmorning on a Wednesday in August, and Cleve was out on his paddleboard in an area locals refer to as Cleve's Cove because of how often the Chatham native paddles there. It was high tide, and the ocean was calm—a perfect morning despite the thick brown seaweed known locally as *mung*.

Cleve had ridden a wave in, paddled back out in the prone position, and was just standing back up when his board abruptly stopped dead in the water. He was about thirty yards off the beach and over a sandbar that was maybe three feet deep—plenty deep for his paddleboard. There were no rocks. Before he could process what was happening, his board veered up sharply into the air, and he lost his footing, falling against the board. Cleve quickly paddled back to the beach. He alerted the lifeguards that he just encountered a shark, and he showed them his paddleboard. Sure enough, the right rear portion of the orange and white board had the telltale tooth marks of a shark. "Didn't see it coming, didn't see it leaving," Cleve said of the shark. "It was like being on a motorcycle and getting hit by a truck. It was like boom. And then, oh." The bite mark was about a foot across, which led me to later tell the press that it was likely an eleven-foot white shark.

The incident at Marconi Beach was the first significant shark-human encounter since the incident with the kayakers off Plymouth. It was not, however, the first shark that had gotten the public's attention. Only two days before the shark bit Cleve's paddleboard, a seal was attacked very close to shore at crowded Nauset Beach. As people clamored from the bloodstained water, many thought a person had been attacked. Some of the videos of the event went viral on social media, raising concerns about just how safe it is to go swimming on the Outer Cape. As one longtime Cape Cod vacationer commented on a Facebook post about the incident:

> And all the Cape residents say is "what do you expect?" Or the ever popular "if you're afraid of sharks don't cross the bridge." Truth is, the influx of seals has brought the sharks, and who can blame them. Now what? That's what I want to hear. If conservation measures have multiplied the seal population to result in an increase in natural shark feeding cycles, what next? Close beaches? Wait for a death or dismemberment? Looking for serious marine biologist and ranger information.

She went on to say that she wanted data and a plan. It wasn't like this was a new phenomenon after all. It had been building. She didn't fault the sharks doing what sharks do, but she did worry that in absence of any mitigation, something bad was going to happen.

In response to the seal attack at Nauset, and the day before the paddleboard incident at Marconi, Barnstable County commissioner Ron Beaty called for a "targeted, localized shark hazard mitigation strategy.

"From my viewpoint," he said, "based on the sharp increase in shark-related attacks and incidents around Cape Cod in recent years, there is a clear and present danger to human life as a result of this growing problem."

His proposal, he explained, "entails use of baited drum lines being deployed near popular beaches using hooks designed to catch great white sharks. Large sharks found hooked, but still alive are shot and their bodies discarded at sea." He continued, saying, "It is only a matter of time before someone's child is killed or maimed at a Cape beach because of an encounter with one of these man-eating sharks! We need to be proactive in mitigating the problem, not reactive once it is too late."

I was glad to see an almost instantaneous backlash to Beaty's proposal, which, as I'd stated time and again, was completely unsupported by the science. The day Beaty released his shark hazard mitigation strategy, the *Boston Herald* ran a poll asking readers, "Do you back a proposal to catch and kill white sharks off Cape Cod?" to which more than 83 percent of the nearly five thousand respondents chose "No, leave nature alone." Less than 10 percent of respondents chose "I don't swim near seals," and just over 7 percent said, "Yes, it's getting scary."

The 2017 field season went from a slow start to one of the most active shark seasons I'd seen. This was the case not only in terms of the number of shark sightings but also in terms of the public's awareness of the risks posed by white sharks. People seemed more engaged in the discussion surrounding the shark situation than they'd been since the attack on Chris Myers in 2012. I suspected that in part it was because of the seal attack at Nauset, the incident at Cleve's Cove, and press generated by Ron Beaty's much-publicized shark hazard mitigation strategy.

A lot of people were perceiving the summer of 2017 as a banner year, and they assumed the number of sharks must have been increasing dramatically. But the data told a different story. What appeared to be happening, according to the data, was that the sharks were slow to arrive, but they were also moving north along the Outer Cape a couple of weeks earlier than they had in previous years. As much as I was eager to tamp down hyperbolic responses to what people perceived to be more sharks, I was also aware that there was indeed an increased risk to public safety if the sharks were leaving the relatively remote Monomoy for the more popular and crowded beaches earlier in the season.

One thing I tried to focus on when talking with the public about any increased risk was the sharks' selectivity. I had watched the videos from Nauset and interviewed people on the beach. There were surfers and swimmers in the same area where the shark attacked the seal, and the shark still selected the seal as the prey item. Clearly it wasn't because the seal was the easy target. Rather, I assumed, it was because the shark made a choice. Given the number of sharks patrolling the Outer Cape and given the number of beachgoers on Outer Cape beaches, it was clear to me that if the sharks wanted to hunt and eat humans, there would have been many, many more shark attacks.

Even still, I knew sharks make mistakes. It was what kept me up at night.

CHAPTER 12

A SLAP IN THE FACE

We have lost our liberty, innocence, and the sense of abandon one feels in diving into the Atlantic. We have entered an era that will require mindfulness and caution.

—Wellfleet Selectboard member Kathleen Bacon

We are concerned . . . about the fate and future of the white shark, and if this book has a purpose beyond exposition, explanation, and entertainment, it is to make very clear the plight of this great fish, so reviled, feared, and assaulted that it may be in danger of disappearing forever.

—Richard Ellis and John McCosker, *Great White Shark*

OCTOBER 1, 2026—41°45'08.2"N, 69°54'53.9"W (NEW ENGLAND)

The seventeen-and-a-half-foot shark swims east away from the shallows, her shadow sweeping ominously over a sandy bottom that recedes from her as she heads for open water. She's a survivor, approaching fifty years in age, and she has the scars to prove it. Each tells a story—the claw of a gray seal, a mating scar.

It's autumn, and the water is again cooling. She's heading east-northeast away from the New England coastline. She's moving with apparent purpose, with the precision of an experienced sailor who has taken a bearing and headed to sea with confidence. Of course, her ancestors had been doing this since long before the first human set out to sea. This is her realm. As the coastline disappears behind her, her finely developed senses adjust to new sounds, smells, and vibrations.

She dives effortlessly and glides over the bottom for a time before tilting her pectoral fins and slowly gaining altitude. She finds where the warm sea-surface layer meets the cold water from the depths, and she settles in there for her journey.

She's on her way.

43°26'50.3"N, 53°57'21.7"W (GRAND BANKS)

The shark is pregnant. She is not, however, matronly in anything close to the human sense. White sharks, unless reduced to a caricature, elude anthropomorphizing. Unlike so many bony fishes that produce masses of eggs and sperm, hedging their offspring's bets against a hostile environment and distributing their young over huge areas, the white shark will produce a small number of eggs and a smaller number of pups. Each pup will emerge as a miniature of its parents and will be entirely on its own from the moment it's born. Since the shark's mother will not nurse it, look out for it, or protect it, and because the shark pup will not "play" with its siblings or socialize with relatives and others of its kind, it will have to rely on its body and instincts alone. Its mission will be to reach a place of relative safety away from predators, where food is abundant. But that is still a long way off.

As the mother shark swims, her babies are developing in her womb, which is technically two wombs—or, more precisely, two uteri. In each uterus, roughly half the embryos hatch and feed on unfertilized eggs produced by the mother's single ovary. Becoming one of the ocean's apex predators is a subject Disney hasn't explored, for obvious reasons. The embryotic sharks even swallow their own teeth. It will be more than a year before the mother shark gives birth. Until then, she needs to feed often and avoid males of her own species.

On her eastward journey away from New England, she encounters the Gulf Stream and rides its arc farther east and deeper into the open Atlantic. She stays within a hundred feet of the surface some of the time, and sometimes she even ascends to the surface. But increasingly she finds clockwise-rotating eddies of warmer water that spin off the Gulf Stream, and she rides these like elevators into the depths. These depths are a place where temperature fluctuations are extreme and both salinity and the denseness of the water vacillate. It's a dynamic environment, and, despite the darkness, it's full of life—home to the largest biomass in the world's oceans, which is why she has come here along with swordfish, bigeye tuna, and beaked whales.

Most of her descents are during the day, and she descends to depths exceeding 3,000 feet, where she remains for half the day or more. These are the twilight depths extending from 650 to over 3,000 feet, known as the *mesopelagic zone*, where only a tiny percentage of sunlight penetrates. It is a vast, understudied chunk of the ocean thought to be chock-full of deepwater fishes and squid, enough to draw the attention of the white shark. Once thought to be restricted to the coastline, the white shark moves freely throughout the North Atlantic, vertically exploiting resources not available to most other species. Her unique adaptations make this possible.

She is magnificent.

39°26'38.5"N, 31°33'34.0"W (FLORES ISLAND, AZORES)

On her trip east across the Atlantic, the pregnant shark visited depths over 3,500 feet, where water temperatures fall into the midthirties and no light penetrates. She possesses a layer of mirrored crystals behind her retinas that makes her more sensitive to low-light situations than many other animals, but in these midnight depths, she must rely on her other senses to forage and navigate.

She is now, however, back near the surface, not far from Flores Island in the Azores, 1,100 miles from Portugal and more than two thousand miles from Cape Cod. She has successfully navigated more than halfway across the Atlantic using, in part, the earth's magnetic field.

The Azores are the most isolated archipelago in the North Atlantic—volcanic islands sprouting from the intersection of the North American Plate, the Eurasian Plate, and the African Plate. They straddle the Mid-Atlantic Ridge and are a migratory crossroads between the eastern and western Atlantic as well as both tropical and Arctic waters. The shark once again adapts to the ecosystem she's entered—an ecosystem she shares with more than twenty species of whales and porpoises, four out of seven species of sea turtles, and tuna, billfishes, and over sixty species of sharks and rays.

Letter to the *Cape Cod Times*, April 21, 2018:

Speaking only as one individual Barnstable County commissioner, I want to express my serious concern about an issue that is not being appropriately addressed by Cape Cod communities or the state or the feds.

On Cape Cod this summer, if a person is devoured or maimed by a great white shark at a public beach, it will cause an immediate and steady downward spiral of the tourism economy, thus killing our communities as well.

Public safety plans and contingencies must be established, or else saying "I told you so" will be the only satisfaction occurring as a result of such a tragic event, aside from satisfaction felt by a man-eating shark swimming the ocean with a belly full of human flesh.

Ron Beaty
West Barnstable

The Summer That Changed Everything

As expected, the Fourth of July 2018 holiday was accompanied by white sharks in the news. The *Boston Globe* ran the headline "Up to 7 Sharks Reported on Cape in 2 Days." I once again found myself frustrated with the media hype, and countered that "it's typical of this time of year. They get spotted by people from time to time. It's something that's been happening for years. It's nothing unusual." The team was in the fifth and final field season of the five-year population study, and I wanted to focus on collecting as much data as possible. Ironically, every time I or the Atlantic White Shark Conservancy reported on the team's success during their research trips, it provided yet another opportunity for the press to spin the story. The fact that a highly experienced team observed multiple sharks with the help of a spotter plane on designated research trips should not have been news anymore on Cape Cod. The sharks were there, and we knew how to find them. To my way of thinking, a headline like "Up to 7 Sharks Reported on Cape in 2 Days" only served to gin up shark anxiety and, of course, sell newspapers. The more responsible headline would have been about the science and what we were learning about the sharks, and while some reporters were eager to cover that story, it seemed the people who sold papers thought anxiety was better for sales.

I was pleased to observe that there were indications that, despite Ron Beaty's continued rhetoric and the media's penchant for drumming up unease, Cape Codders were making their peace with the sharks.

"Sharks have become an accepted part of the local lifestyle," declared one reporter, and it wasn't difficult to find locals at the beach who said things like, *Your drive to the beach is more dangerous than being in the water* or *There hasn't been a fatal shark attack in Massachusetts since 1936. What are the odds it will happen to me?* Some locals had changed their behavior in terms of where or when they swam or surfed. More than a few had turned to pools or the Cape's freshwater ponds. But overall, most people seemed to express a desire to coexist with the sharks instead of adopting policies to reduce their numbers.

The central question that locals continued to ask, however, was *How many sharks are there?* It was a reasonable question, for sure, and it was the goal at the center of the now much-publicized five-year population study, but I had begun to worry that any number we ultimately published would potentially be problematic. Going into the last year of the study, we had identified more than 350 individual white sharks. On the face, that seemed like a huge number compared to the five white sharks tagged in 2009. But I also knew at this point that 350 individuals was likely just part of the total population. Whatever number we published was going to get spun by those on either side of the shark debate on Cape Cod, and while the majority seemed to be in support of the sharks, I knew all too well that if . . . no . . . *when* the next attack happened, just as George Burgess had said, the feel-good shark story, the souvenir T-shirts, the specialty drinks at the bar would all likely fade, and the public would realize exactly what's in the water just beyond the umbrellas and beach toys.

A Pretty Crummy Week

It had been a pretty crummy beach week for those who scheduled their Cape Cod vacation for the last week of July 2018. Despite the less-than-ideal weather, however, Nauset Beach was packed on Friday afternoon. The temperature was in the low eighties, and the sun was trying to work its way through clouds that had hung on all morning. Several people were learning to surf on big foam boards, and a few boogie boarders sought out some space off a sandbar a little way down the beach. A dad, waist-deep in the surf, held his kids' hands as a mom farther up the beach struggled to get all the sand off her toddler's hands before handing her a peanut butter and jelly sandwich with the crusts cut off. A patchwork of colors—beach towels, umbrellas, sun shelters, and coolers—spilled out away from the beach access and lifeguard tower in both directions. It was a typical summer day at one of New England's best beaches.

The shrill sound of a lifeguard's whistle pierced the afternoon air. People turned, confused. Someone yelled, "Shark!" A few people assumed it was a joke—that's been a thing ever since the kids with the fin caused a scare in *Jaws*. But it wasn't a joke. The shark hit the seal close to shore, spreading a dark red stain over the green water. Lifeguards frantically waved in surfers near the site of the attack. Almost everyone on the beach was on their feet as the last couple of surfers made their way to shore. The lifeguards told people to get up on the berm away from the surf. Men in brightly colored trunks shielded their eyes trying to get a glimpse of the shark. Women in bikinis stood on tiptoes surveying the scene. Cell phones filmed from outstretched hands, as if their owners were at a rock concert. A set of decent-sized swells moved through the attack site, each one painted with the seal's blood. The maimed body floated into the shallows. The shark was gone, but it loomed large in everyone's memories and imaginations.

Saturday, like Friday, was mostly cloudy and hot. Jared Karlberg, aboard his boat F/V *Mildred,* caught a ten-foot white shark in a gill net a few miles off Scituate, Massachusetts. While it's illegal to target a white shark, it's not illegal to catch one accidentally while fishing legally for other species. In that situation, however, the fisherman must return the shark alive or dead to the sea, unless he has authority from someone like me to bring it to shore. As it happened, Jared previously worked for another fisherman named Jeff, who in turn had worked with John Chisholm and me back in 2009 on a History Channel show called *MonsterQuest.* I had told Jeff that if he ever landed a white shark off Massachusetts to give me a call. Jared remembered and called Jeff, who called me. I called my boss, David Pierce, and got permission to have the shark landed. Then I call Jared and told him I would love to have the dead shark for the purposes of research. Under the agency's authority,

Jared turned his boat back to Scituate. I notified the Environmental Police, and called John Chisholm and Lisa Natanson, and they met me on the town dock in Scituate.

For some reason, these kinds of events always seem to happen on a summer weekend, which puts traffic and lots of people in the way of getting work done. A large crowd surrounded us as the shark was hoisted by her tail, weighed, and then laid out on the dock. She was 657 pounds, and Lisa and I got to work dissecting her and collecting samples for colleagues. It was the first time in awhile that we had dissected a shark together, and the moment was significant for both of us. We'd worked together at the Narragansett Lab under Jack and Wes so many years ago, and we still had a relationship akin to siblings. Lisa was still working at the lab, where Nancy Kohler, who had also been at the lab the same time as me, had taken over Jack's position when he retired. Lisa asked me about Megan, whom she'd worked with long before I knew her. All those great connections and close friendships. *All because of these animals,* I thought to myself as I looked down at the shark. *All because of white sharks.*

When we opened the shark's belly, we found the remains of a seal pup and a striped bass. It reminded me of both the days working with the Apex Predators Program and then on the dock at the Monster Shark Tournaments on Martha's Vineyard. This was a comfortable space for me, and it felt good and natural to be on the dock dissecting a big fish with a crowd of enthusiastic onlookers. I knew it was something most of these people have never seen and would never see again. To me this shark was an ambassador for its species, and I enthusiastically described to the crowd, especially the children, what we were doing and why. It's a tragedy the animal died, but my focus was on making sure that the death did not go to waste. Overall, it was a great day.

I avoid social media for the most part, so I was surprised to start getting calls from reporters about the "emotional outrage" in response to the Scituate shark. Unbeknownst to me, the conservancy posted pictures on Facebook of the shark as it was hoisted off the boat, weighed, and laid out for the dissection. While many people understood that the shark was taken accidentally and that its death would at least contribute to science and a greater understanding of the animal, many others were incensed by the images and said as much in the comments.

> Barbaric! Would we display a human life like that with such disrespect . . . it's a circle of life and we need to gain respect because we are only a small portion of that circle!!!

> I can only hope she got to have a few human snacks before this day. . . . This is the kind of ignorance that is running rampant today. . . .

> I don't care if it's unintentionally or not! I just see another poor dead [shark emoji].

> So you hang it up like G@d D@mn trophy?!?! What is wrong with you?! Intentionally caught or not its criminal that it was murdered and YOU hang up like some freaking marlin!

> She was very young. The idiot people who put the gill net out need to be seriously fined and flogged!

> Nice job asshats!! Kill a female of a Vulnerable Species!! It's time for a better fishing practice.

> This . . . is terrible . . . hanging this animal in public after it was slaughtered . . . this kind of imagery just keeps feeding the lies that it is OK to kill and show off these beautiful animals. Very sick.

> AWSC, I'm a huge supporter. But I find this post very offensive. What happened to this animal is in my opinion criminal. Then to hang it like a trophy is disgusting. There are other ways for our children to learn about sharks. Wish you discouraged this and not promote it on your site. As a supporter of AWSC. I'm very disappointed. Actually a little pissed.

"I appreciate people's passion for this," I told the reporter after she read several of the Facebook posts to me over the phone. "Certainly we've come a long way in the last fifty years, where for the most part, the only good shark was a dead shark." For me this outpouring of emotion

was a double-edged sword. On the one hand, it showed how much had changed from the days of Frank Mundus and "monster fishing," but on the other hand, I worried that the pendulum could swing too far. These were apex predators more than capable of seriously injuring or killing a person. They deserved respect. Reducing them to an emoji wasn't helpful. They're not your Labrador retriever. As much as I wanted to lash out at some of these fools, I held back and once again reminded myself that everyone is entitled to an opinion and those opinions end up on social media (reminding me why I don't do social media).

Sunday morning dawned the nicest of the previous three mornings, and Roger Freeman grabbed his paddleboard and headed for Nauset Beach. The sun was shining, the offshore wind was light, the conditions were calm. It was a perfect morning. It was the fifty-four-year-old's last day on the Cape, and he felt lucky to get in one last morning session with such gorgeous weather. As he paddled out to the break, he thought back to the seal attack he'd heard happened at the same beach on Friday. He imagined that some people might think he was crazy paddling out, but after three rainy days in a cramped cottage with his in-laws, Roger decided to take his chances with the sharks. "I knew it was a measured risk," he later told a reporter at the *Cape Cod Times*, "but on this glorious Cape Cod morning, I say it's crazy not to be."

The surf session was everything he'd hoped it would be, the perfect start to his last day of vacation. He kept an eye out for sharks and seals but the morning sun in his eyes and the glare made it impossible to see what might be lurking below. After about an hour and a half of surfing, he rode one last wave shoreward. Back on the beach, a local surf instructor approached and said, "Dude, did you hear that drone? That was my buddy, and he just sent me this photo." He held up his phone. It took Roger a moment to realize the image was of him on his paddleboard with what appeared to be a large white shark passing underneath not more than a board length away.

In the days that followed, Roger thought a lot about the close encounter. Reflecting on the incident later, he said he was glad he got to go home and eat breakfast . . . rather than *be* breakfast. He fully appreciated that he was entering the shark's world when he went in the ocean. "This is their world," he told a reporter from the *Globe*, "and we have the ability to manage the interactions with them much better than they do with us and we should be doing that." He also wished those responsible for public safety would do more. "I would have certainly appreciated if there was a warning system," he told a reporter at the *Cape Cod Times*.

In 2018, drone shots of shark close encounters were a novelty, and in the days that followed, the photos went viral and traveled around the world. Roger ended up telling his story on *Good Morning America*. He received messages from friends across the country and in Europe who had seen the photo. His only regret about the incident was that the news coverage emphasized the fact he never saw the shark, and he is now known throughout the world as "the oblivious paddleboarder."

"Google it," he says with a laugh.

On Monday I was back out on the *Aleutian Dream*. We were off Wellfleet, and John King was in the boat's tower taking direction from Wayne overhead in the spotter plane.

"Comin' at you," Wayne said over the radio. "Come to port."

John turned the wheel. I was out on the end of the pulpit, wearing my standard outfit comprised of fishing boots, white shorts, and a windbreaker under my inflatable life vest. I adjusted the GoPro camera affixed to the end of the pole and began to lower it to the water's surface off the port side of the pulpit. Coming up on a white shark like this was something I'd done well over a thousand times at that point, but it was never routine. There were so many things going on at once, and everything needed to come together in an instant to successfully video-fingerprint one of these sharks. The thrill of it has never worn off.

I was leaning to port, ready to plunge the camera beneath the surface chop, but I still didn't see the shark—visibility was horrible. The water looked like coffee. I peered down, which was when the water erupted beneath me, and the shark, mouth open, breached directly under my feet. I took a couple of quick steps, a move that will later earn the name "the white shark shuffle," and then pivoted to starboard, reorienting the camera. I tried to follow the shark as it landed on its left side and slid back beneath the waves.

John, who saw the whole thing from the tower, yelled, "Holy crap! It jumped right out of the water!"

Incredulous, I turned to look up at John, "Did you see that?" I then turned to another boat off our port quarter. "Did you see that?"

At first I didn't even believe it happened. Had I imagined the whole thing? I hadn't, and there was video to prove it. For a split second I'd been looking down into the open mouth of a white shark, jaws protruded and teeth on full display. That is a view only a handful of people have ever experienced, and I know it didn't end well for most of them. It's a fine line between admiration, respect, and fear, and in that instant, I felt all those emotions at once. After much reflection, I could only conclude that the shark interpreted me, my reflection, my shadow, my image—whatever it was—through the water's surface as a potential prey item. It lunged at me. To be honest, I would like to think it wasn't trying to eat me. Nobody really wants to be eaten.

"If you're not afraid of something that could potentially kill you," I said when asked later by a reporter if I was afraid, "then you're not sane."

Driving home that evening, I thought back over the last few days—the seal attack at Nauset on Friday, the dissection of the Scituate shark on Saturday, the drone footage of the paddleboarder's close encounter on Sunday, and then my own experience with the breaching shark. It was a busy few days for white shark activity on the Cape. It was my ninth full summer season of chasing white sharks off Cape Cod, and one might think things would have normalized somewhat. The methodology, after all, is repetitive—two times a week on the water, John at the helm, Wayne overhead, Megan recording data, I'm in the pulpit, white sharks ghosting through the shallows ahead of the boat—but as repetitive as it is, no two days were ever the same. If I learned anything, it was never to be too sure about what tomorrow would bring.

On the other hand, the sharks were, to some extent, predictable, which is why I was able to pursue this research. They predictably returned to Monomoy in the late spring; some remained for days or weeks, while others used Cape Cod like a rest stop on their way north to the Gulf of Maine or Canada. They predictably came for the seals, and they'd proven themselves to be adept hunters, carefully targeting select prey even from a beach crowded with surfers and swimmers. As predictable as they were, however, they often went about their business unseen, as the drone footage proved. I had learned that they were not always hunting, not always aggressive, and sometimes not even that curious. It occurred to me that, owing to their proficient sensory array, they had assessed and accepted people in their habitat in a way that people had not accepted them.

But I got it. I got the hype and the emotional pull of the species. That breach beneath my feet was unpredictable and unnerving. It was, quite frankly, scary. We've only observed a couple of breaching displays off the Cape, where it was not nearly the effective hunting strategy that it was off Seal Island in South Africa. As much as I had come to know these animals, even specific individuals, they remained largely unknown, fearsome apex predators. Yet, as the Scituate shark proved, they were not immune to risk, and increasingly people seemed focused on their vulnerability. White sharks were no longer viewed myopically through a solitary lens of fear, animosity, and even hatred.

The fearsome monster had been replaced by a shark emoji followed by hearts and hugs emojis.

Longnook Beach

William Lytton visited the Cape every summer to conduct some of his work at the Marine Biological Laboratory in Woods Hole. For the rest of the year he lived in Scarsdale, New York, and was a professor of neurology, physiology, pharmacology, and biomedical engineering at the SUNY Downstate Health Sciences University, in Brooklyn. His research interest was computational neuroscience, which, in very simple terms, meant that he used computer models to study brain function. In even simpler terms, Bill was a very smart guy.

It was August 15, a beautiful sunny day, and Bill and his wife had decided to take their two young daughters to Longnook Beach. Longnook is a relatively remote beach in Truro backed by imposing dunes. Despite its popularity, it has a wild feel bolstered by the steep, sandy trail used to access it from the parking lot. Bill enjoyed swimming in the ocean for exercise, and because he had missed his daily workout in recent days, he decided to swim for the next half hour. As he entered the surf, all he was wearing was a red bathing suit, his watch, and his wedding ring. He was aware that sharks were in the waters off Cape Cod, but he didn't think there was much risk. Besides, the water depth dropped off quickly here, so he would remain close to shore. In the water, he alternated between breaststroke and sidestroke, gliding parallel to the shore in about eight to ten feet of water just a few yards from the beach. He passed a few people wading, but nobody else was swimming.

About twenty minutes into the swim, Bill turned south and started to breaststroke, plodding forward with his head high and legs flexing about a foot below the surface. Suddenly there was an excruciating, sharp pain piercing his left side close to his hip. He turned and immediately saw a shark coming out of the water with its mouth firmly closed around his leg. The shark's entire head was exposed, and Bill could clearly see the eyes, the gills, and the mouth, which he later described as a smile on a slate-gray face. The shark was thrashing, and Bill felt like it was trying to flip him below the surface with great torque. In that split second, a documentary about what to do when attacked by a shark flashed across Bill's mind, and he struck the gills repeatedly with heavy strokes of his closed left hand. Just as suddenly as it had grabbed him, the shark released him and disappeared.

Bill saw the expanding cloud of blood around him and knew he had to get to shore quickly. He didn't want to pass out. It took him six painstaking strokes to get back to the beach, where bystanders rushed to his aid. He was airlifted to Boston's Tufts Medical Center, where he was placed in a two-day coma and underwent six surgeries. He had nearly twelve pints of blood pumped into him. Bill later described his injuries as a wraparound bite to the left leg resulting in tears to the quadriceps as deep as reaching the bone. Tooth fragments were recovered from the wounds. According to a news article in the *Cape Cod Times*, one of the operating physicians "was amazed that Lytton was conscious and talking given that he had lacerations to most of his left leg—some as deep as an index finger, some deep enough to gouge bone—from his hip down his left leg and across his buttocks, with puncture wounds to his right leg and lacerations on his left hand." The lacerations had missed the major nerves and arteries, "but not by much."

The two doctors compared Lytton's wounds to some of the most extreme damage they had seen in victims of the Boston Marathon bombing in 2013 and to the mangled extremities seen in victims of industrial accidents and car crashes. On top of all that damage, Bill's repeated punches to the shark's gills, which likely helped to save his life, tore dorsal tendons in his left hand. It was truly a miracle he had not bled out on the beach.

Just as I'd done with Chris Myers, I once again stepped into forensics mode and eventually made the determination that the attack was indeed a white shark. The tooth fragments removed from Bill's leg possessed the telltale serrations. Quite frankly, I couldn't believe Bill had survived. He was, in my estimation, a very lucky man. There was a lot of talk about the seals and about whether Bill had used common sense. For his part, Bill said, "It's kind of terrifying thinking about it. I know it's not the best thing to say, but I didn't like sharks before, and like them even less now." His wife, who was on the beach, said she didn't think something like that could happen "so close to shore." She said she hoped the attack would be a warning for others. She hoped it would cause others to take shark safety seriously.

Labor Day was right around the corner, and beach managers and town officials were on edge in the wake of the shark attack at Longnook Beach. I reiterated at every opportunity that people must use common sense and be aware and vigilant. I always said there was no "silver bullet" to solve this issue, and the most reasonable solution, in the short term at least, was to modify our behavior, because we couldn't modify the shark's behavior. People were frustrated. The Orleans harbormaster made a public plea for beachgoers to change their behavior:

August 29, 2018

The Town of Orleans manages the protected (lifeguarded) section of Nauset Beach under the assumption that there are always sharks in the vicinity of the swim area. Lifeguards have been trained to direct visitors to stay tight to shore and are vigilantly watched during hours of operation. Staff are standing by to launch the Zodiac rescue boat from the shorebreak and EMTs are present daily throughout the summer. The challenge is trying to convey the gravity of the situation to the beachgoers. As far as I can see, local tourism and beach attendance has yet to be negatively affected by the presence of the sharks. Regardless of how much signage and information we provide, there still seems to be a concerning level of complacency. People continue to risk entering and swimming in the water even after the recent incident in Truro where the man was bitten. We are witnessing sharks feeding aggressively in shockingly shallow water on an almost daily basis over this past August. Based off of the past research data provided from Dr. Skomal and the Atlantic White Shark Conservancy, we expect this inshore behavior off Nauset Beach to continue through the Fall when the beach is not staffed. The inshore waters off of Cape Cod are truly a wild place and people should practice extreme caution while visiting. White sharks have bitten people along this coastline. Fortunately, no one has died yet. However, it can only be expected that more incidents will occur if people continue to take unnecessary risks. The beaches off the backside of the Cape are expansive and often remote. This makes it extremely challenging for Public Safety Officials and Beach Managers to provide direction to all who visit. The constant presence of white sharks in shallow water off of Nauset Beach in the months of August to October is the new norm. It is time for beachgoers to change their behavior or something terrible is going to happen. Please be safe.

Nathan Sears
Town of Orleans
Natural Resources Manager/
Harbormaster

Arthur Medici

September 15 was a Saturday, and Megan was in front of her computer working hard under an imminent deadline. Her phone burred. It was a group text from one of the local harbormasters. A surfer had been bitten by a shark at Newcomb Hollow in Wellfleet. Megan knew that before this summer, the last shark attack on a human on Cape Cod was in 2012. Now they'd had two attacks on humans in the span of a month. She knew what this could mean—how this could affect public perception of sharks and even the perception of our scientific work. She was worried about the surfer, the surfer's family, and everyone on the beach. It was a beautiful late summer weekend day. There must have been a lot of people on the beach, she thought.

Her phone chirped with a text update. The Life Flight medical transport had been called off. Megan knew what that meant, and she was completely devastated. "It was something we knew would likely happen one day," she said later, "but that doesn't make it any easier when you think about what those people went through on the beach." It was hard for Megan not to take it personally, to feel like we—like she—had failed in some way.

Arthur Medici's death was a crossroads.

Megan knew that I was aboard the *Tails* guiding the White Sharks and Whales trip. She'd led the trip the week before, and she anticipated how tough a situation it was going to be for me to navigate. Even before she started her formal studies that would lead her to become an up-and-coming shark researcher, Megan loved to tell people about sharks—sometimes whether they wanted to hear it or not. The White Sharks and Whales trip was the ultimate way to share these animals about which she was so passionate. She had given talks on white sharks while showing slides; she'd created videos and shared her expertise on social media. But to actually stand on the deck of the *Tails* and describe a shark that was swimming mere feet away from an audience of eager onlookers was like nothing else. To see the people's faces and hear their enthusiasm was to understand that she had helped them make a connection—she had helped them to see what these animals were instead of what they were not.

All the misconceptions and misrepresentations of white sharks had led Megan to feel protective of them because of just how maligned they were. The White Sharks and Whales trip allowed people to see white sharks the way we did. But how could I do that when a white shark had just killed a person? The scientist in Megan wanted to talk about fatalities as "cases of mistaken identity," but on a day like that, the rhetoric, however true it might be, didn't matter. It didn't make it any less traumatic. Nobody wanted to see someone hurt or killed by a white shark. In fact, that had become a big part of the reason we were doing the work we were doing, and it had become a big part of Megan's life, too.

Like me aboard the *Tails*, Megan knew almost immediately that Arthur's death was a game changer. It created a sense of even greater urgency. She texted me on the boat. What are you going to say? she asked.

I guess I'm going to tell them, I responded.

As the guests aboard the *Tails* gathered for the news, I cleared my throat and took up the microphone.

"I've been in touch with folks onshore who are investigating an incident that involves a young man who was boogie boarding," I said. "By all accounts so far, and there are not many, he was attacked by a shark off Newcomb Hollow Beach and he suffered injuries that were fatal. I will know a lot more when we get back in and I have a chance to investigate further." I paused. The only sound was that of the *Tails*'s diesel engines and the slap of her hull on the cold green Atlantic. The mood had changed. A plethora of emotions were emerging. I continued, "We all revere these animals, but on a rare occasion they remind us that they are powerful predators that make mistakes."

Impassioned Comments

The auditorium at the Wellfleet Elementary School is like a lot of elementary school auditoriums across America—it's a gym with a scoreboard and basketball nets, a theater with a curtained stage, and an assembly space for town meetings too large for other venues. On September 27, almost two weeks after Arthur Medici was killed in the fatal shark attack at Newcomb Hollow, early evening light spilled through large windows mounted high in the auditorium's blue-gray cinder-block walls, which echoed a hue not unlike the backs of some white sharks. Bleachers extended from one wall, fronted by rows of metal folding chairs with blue cushions lined out across the wooden plank flooring, facing a panel made up of town officials, public safety managers, and white shark and seal experts. By the meeting's start at 6 P.M., more than five hundred people had taken their seats. Television cameras lined the wall on one side, and news photographers orbited the crowd. The topic of discussion was white sharks and, specifically, what measures could realistically be taken in the wake of Arthur Medici's death to improve public safety before beaches opened in 2019.

I was not at the Wellfleet meeting, because I was giving a prescheduled talk on white sharks at the Cape Cod Museum of Natural History, in Brewster, but the director of the Division of Marine Fisheries, David Pierce, as well as Cynthia and Megan, was on the panel. Our research was the target of many of the comments that placed the blame for Arthur

Medici's death firmly in the hands of government officials, government agencies, and government policies. About half an hour into the meeting, a local mother of four, all of whom she said are ocean lifeguards at Outer Cape beaches, stepped to the microphone to address the panel:

> We have the most dangerous sharks in the ocean regularly swimming in the vicinity of anyone who swims at an Outer Cape beach. At the same time, not one single government official has taken any meaningful action to provide for our safety, for the safety of our children, our family, and our friends. Instead, certain government officials have given pet names to white sharks and prioritized the lives and safety of sharks and seals over that of those who swim in the Cape waters.

Her impassioned comments were greeted by enthusiastic applause, as were the comments of others who blamed local and federal government officials, scientists, and conservationists who advocated for what she and others characterized as "misguided and outdated government policies." Specifically, people called into question the Marine Mammal Protection Act, the 1997 prohibition on taking white sharks in federal waters, and the 2005 state ban on fishing for white sharks in Massachusetts waters. Some suggested these policies were enacted without data, and they complained that there is no mechanism whereby they can be reversed even in the face of what some claimed is a "runaway population explosion." These policies, one man said, have resulted in unintended consequences. "No sharks or seals are worth a young man's life," said another person. Another stated it more frankly: "They're eating our children."

There was a lot of vocal support in the room for culling both seals and white sharks, and the Wellfleet town administrator reminded the audience that such actions were illegal and that the removal of seals, specifically, would require the US Congress to act. The purpose of this meeting, he said, was to discuss realistic measures to mitigate the risk—things that could be accomplished before next summer. Director Pierce of the Division of Marine Fisheries echoed the town administrator, saying, "There's nothing the state can do except improve communication and other public safety issues."

Not everyone in the room wanted to address the public safety issue through lethal means. One speaker described culling as an "arrogant anthropocentric mentality." What is really needed is for people to adapt to the presence of the sharks, someone else said, which, in essence, was the approach of many beach managers who had repeatedly asked beachgoers to modify their behavior. In Orleans, for example, beach managers told the public they were going to manage the beach as if there were always a shark present. Several people talked about the ocean being the shark's habitat and not ours, and one person talked about the return of the seals and the white sharks as a conservation success story. There was applause for these positions as well, and it became clear that the room, just like the Cape, was divided.

Surveys and polls consistently showed that the majority of people were keen to find a way to coexist with the sharks, but just about everyone who spent time in the ocean on the Outer Cape was struggling on some level. Wellfleet Selectboard member Kathleen Bacon summarized attitudes well.

"We have lost our liberty, innocence, and the sense of abandon one feels in diving into the Atlantic," she said. "We have entered an era that will require mindfulness and caution."

Feasible, Actionable, and Affordable Strategies

In the wake of the meeting, the town managers from towns along the Outer Cape took a more active role in decision making about what each town would do. Since 2012, it was largely the ad hoc Regional White Shark Working Group that had been the driving force behind developing mitigation strategies, but the fatality at a town beach brought with it certain legal and liability issues that elevated the discussion to a new level. In general, the town managers agreed with my previous advice to focus on what's feasible. Many strategies sounded good until you really dug into the details. Cape Cod was a unique environment, and shark mitigation strategies that may work elsewhere may not work on the Cape. It was also important to be clear about what people meant when they said *shark mitigation*. Some of the best shark deterrence technologies are shown in studies to be effective in about 60 percent of trials. Is that enough? And what about the cost? How much is an individual or a town willing to spend to reduce the risk of a shark attack by any percentage less than 100? And what liability is associated with utilizing a town-funded, sanctioned, and operated mitigation strategy that fails to prevent a serious injury or fatality? It's a ridiculous calculus, of course. People commonly take public safety precautions like wearing a seat belt, wearing a helmet, or driving the speed limit, knowing full well that even people who take all those same precautions will still die in crashes. But there is something about being ambushed and attacked by a largely unseen predator. There's something about suddenly acknowledging that you are potential prey.

Town managers, working with their own town staff, came up with several concrete steps to take before summer. Topping the list were communication and emergency response. Because of the topography of the Cape and the remoteness of some of the beaches, cell phone communication is often spotty on the beach. In the event of a shark attack, getting the victim to advanced medical care quickly can save a limb or a life, but bystanders must be able to communicate quickly and reliably with dispatchers. In response, several towns, as well as Cape Cod National Seashore, committed to installing emergency call boxes at beaches.

Even with reliable communication, swift and effective first aid provided at the beach before help arrives is essential, especially in the case of shark attack victims whose injuries can cause them to bleed out in minutes without the proper use of a tourniquet. Before the 2019 summer season began, towns committed to making sure that beaches were stocked with trauma kits and that lifeguards had additional trauma training. In some towns, stop-the-bleed classes were also offered to the public at no charge. The conservancy played a major role in promoting and sponsoring these classes.

A bevy of other strategies were undertaken on a town-by-town basis, such as installing higher lifeguard towers, reimbursing lifeguards who purchase polarized sunglasses that make seeing sharks through the water easier, and having an ATV on-site. What the towns seemed reticent to do was to proactively engage too aggressively in the advanced technologies—things like smart buoys or drones. Town managers were even mixed on their desire to have me deploy real-time receivers at their beaches. In part, this was because there was general agreement that none of these mitigation strategies were 100 percent effective, and there was concern that having drones overhead or smart buoys in the water might give people a false sense of security. There were also the liability issues.

I found that some towns were receptive to me deploying a real-time receiver at one of their beaches at no expense to the town, but none of the towns were actively clamoring to get it done.

The other big change that beachgoers would see when they arrived in 2019 was more aggressive education and outreach. Since the formation of the Regional White Shark Working Group in 2012, education and outreach has been a central component of the response to the public safety risk posed by white sharks, but beginning in 2019, education and outreach initiatives would now amplify the message that swimming on the Outer Cape was dangerous. In this new chapter of Cape Cod's relationship with white sharks, there was less appetite for softening the message. New signs would highlight the risk both visually and in explicit messaging. To emphasize the reality of the situation, these signs showed a small figure with our raw acoustic detection data peaking in August, September, and October.

I was personally and professionally affected by Arthur's death, and I felt compelled to do more. The five-year population study was coming to a close, and the timing was right to set a research course for the next five years. Our research to date had told us a lot already about these sharks. We had identified more than five hundred individuals so far, and many of these sharks came back year after year. Our extensive tagging dataset showed they arrive as early as mid-May, but peak months were August, September, and October. As water temperatures cooled, most sharks left by mid-November, but a handful trickled through as late as mid-December. We knew their preferred areas. By wiring up most of our coastal state waters with acoustic receivers, we'd learned the Outer Cape emerged as the hotspot where white sharks spent the bulk of their time. Of course, that made a lot of sense because that was where most of the seals hung out.

We also now knew the broad-scale migration of this species in the North Atlantic. When they leave Cape Cod, almost all of them overwinter off the southeastern US, from Cape Hatteras to the Gulf of Mexico. And we now knew that some of the larger white sharks would move out into the open Atlantic as far east as the Azores, diving to depths as great as three thousand feet every day through a very broad temperature range.

This was all great information that had been used by the towns and the general public to enhance public safety, but Arthur's death was a clear message that we needed to know more. We needed to change the scale of our research by drilling down into what these sharks were doing not only every month, week, and day, but every minute and every second. I put pen to paper and drafted, with Megan's input, a five-year research plan to study fine-scale predatory behavior of the white shark off Cape Cod. Boiling it down to a simple question, we wanted to know when, where, and how white sharks hunted and preyed on seals. This required knowing exactly what they were doing in the moments before, during, and after a predatory attack. What are the exact environmental conditions at the time of an attack? What time of day? How much turbidity? What are the tide, current, and depth?

If shark attacks are essentially random events, then we wanted to be able to show that with data. Our hunch, however, was that there were patterns—very complex patterns, but patterns nonetheless. Patterns lead to statistical predictability, which leads to forecasting. If we could take all these observations and process them, not unlike the way a meteorologist processes weather pattern data, then maybe we could begin to develop forecasts of where and when white sharks were on the hunt, to provide both individuals and beach managers with another useful tool to mitigate the risk of shark attack. I ran the research plan up the agency flagpole and it was approved by my bosses, Assistant Director Mike Armstrong, Deputy Director Dan McKiernan, and Director Dave Pierce.

The new study was going to require new, sophisticated tools to observe white shark behavior directly and indirectly. Yes, we had observed white sharks eating seals over the years, but not as frequently as one would expect and usually only after the kill. Why? The answer was simple: we were likely not in the right place at the right time. The Outer Cape spans forty miles and we didn't go out at night. We needed to enhance our ability to observe behavior with drones, fine-scale acoustic monitoring, and camera tags. Of course, that required more funding, and the AWSC stepped up once again to assist. Megan and I applied for and received a grant from the Save Our Seas Foundation, and, much to my surprise, the state stepped up with additional funding. We were ready to roll.

But not everyone was pleased with the response to the death of Arthur Medici. Some expressed frustration that the mitigation strategies employed by the towns were limited to ones considered "feasible," "actionable," and "affordable." There were those who assumed a fatality would be what changed the tenor—it would be the wake-up call that finally roused environmentalists, conservationists, and government scientists from their "misguided and outdated policies." One group calling itself the Atlantic Human Conservancy advocated amending the Marine Mammal Protection Act and both federal and state protections for white sharks so that they could perform culls. Their rhetoric was couched in the idea that the government's primary responsibility should be the safety of its citizens. The founder of the group wrote in a letter published in the *Cape Cod Times*: "My opinion is that seals should be hunted for bounty money, as they were up until 1972. I also propose that the fishing for and killing of sharks be encouraged. I hate to see that 'conserving' sharks is more important than the interests of humans."

Another group, the Cape Cod Ocean Community, emerged to counter what they saw as inaction on behalf of everyone who should have been doing more. In the early days of this group, I was a frequent target of their attacks. As their cofounder put it on the group's Facebook page, they advocate for "less science, more safety." Instead of me tagging sharks and working on multiyear studies, they want to "bring proactive surveillance and detection safety measures powered by technology" to the Outer Cape to "minimize the public safety concerns and negative economic impacts associated with the increase in human encounters with white sharks on Cape Cod." They characterized my actions and the town's response as "stay out [of the ocean] to stay safe." When the cofounder heard me speak about my enthusiasm for more research funding and more tagging opportunities during a talk I was giving at Harvard University, she published on the group's public Facebook page that my tagging work was about research and not public safety and that I was not "chartered" to keep the public safe. "Are we ready to give our waters away to research and conservation?" they asked.

It was unclear how much support these groups had. Were they a vocal minority, or did they capture the frustrations and wishes of a larger, albeit less organized, swath of the population? They certainly didn't have the same level of support as the Atlantic White Shark Conservancy, which may be why they sometimes targeted the conservancy and its philanthropic success. They expressed frustration that donating to the conservancy and being pro-shark seemed to fit with a particular politic and demographic. It was "the cool thing to do."

In response to continued, vocal criticism, some of the towns, along with Cape Cod National Seashore and the Atlantic White Shark Conservancy, hired a consulting firm, the Woods Hole Group, to perform a full review and analysis of existing and potential shark mitigation options, to the tune of almost $50,000. The consultants produced a 192-page report called "Outer Cape Shark Mitigation Alternatives Analysis," which looked at technology-based alternatives like drones, barrier-based alternatives like shark nets, and biological-based alternatives like culling. They concluded, among other things, that "[i]nvesting in alternatives or strategies that may not be permittable given current regulations, may be exceedingly expensive, may have high levels of adverse environmental or human impacts, or may not be effective at mitigating shark-human interaction and may not allow the region to achieve desired outcomes." They also emphasized that "no single alternative or suite of alternatives" would "guarantee the safety of individuals who choose to enter the water."

In short, the report failed to identify a definitive solution. I'd been saying it all along, and nobody had to pay me nearly $50,000 to say it: there was no silver bullet.

"Since no mitigation alternative can provide one hundred percent safety," the report concluded, "reducing chances of unprovoked attacks on humans requires a strong commitment to education and outreach, which can result in the adoption of behaviors that may reduce the risk of a shark-human interaction."

A Scientist Studying Shark Attacks

It was a Monday night near the end of July 2020, and it was baseball night. My son Wilson played on the Gateway Seawolves, and they had a night game against the South Eastern Fire and Safety team, from Fairhaven. It had been a crazy day for me. I was out on the *Aleutian Dream* all day, and we had tagged three white sharks. Once we got back to the dock in Chatham, I had to mainline it home to get my kids fed and then get Wilson into his uniform and to the ballfield by 7:45. My wife, Kimberly, was working late. I'd been so busy, I hadn't looked at my phone until I got to the field and sat down in the bleachers. The previous game was running late, and I started scrolling through my text messages.

I'd come to learn that it was often the media that reached out to me first about a shark incident, and when I saw who had been texting me, my heart immediately sank. We'd made it through 2019, the year after Arthur Medici's death, without a significant incident concerning a white shark, and then we'd made it through the first two months of summer 2020. The COVID-19 pandemic had influenced people's summer travel that year, and businesses and the towns had bigger fish to fry than the white shark issue, so it had been a quiet summer for sharks. But clearly something had happened.

The earliest texts were from the Boston-based media that have me on speed dial, and I knew there were any number of situations that would prompt them to reach out. But then I started getting texts and calls from the national outlets. *Good Morning America* texted, asking about "the unfortunate shark attack." A reporter from WBZ, the Channel 4 CBS Boston affiliate, said it most directly. The text simply read: Can we do a Zoom tonight or FaceTime about the deadly shark attack today?

"Crap," I muttered to myself, glancing around at the other parents in the stand and wishing I could be any of them right now. The words cut deep—"deadly shark attack." I reread the texts, which is when I noticed the *Good Morning America* text didn't just say "the unfortunate shark attack"; it said "the unfortunate shark attack in Maine." I'm embarrassed to admit it, but a part of me felt a tinge of relief that this wasn't a Cape Cod fatality.

My standard operating procedure when I'm contacted by the media about a shark incident is to play dumb. Actually, I'm often not "playing," because many times the call from a reporter is the first I'm hearing of the incident. I've learned that the media are very good at getting out ahead of these types of stories. So I generally tell them that I don't know anything, and I ask them to call me back if they learn more. Reporters, in my experience, are tenacious and better at beating the bushes than I am. The problem is, they sometimes don't get all the facts before they run to the presses. As a scientist, I like to have the facts before saying anything, and that was especially the case with this incident, given that it happened in Maine. I knew I needed to let Maine's Department of Marine Resources, their equivalent of the Massachusetts Division of Marine Fisheries, take the lead on this and do their own investigation before I said anything publicly. But I already knew in my own mind that this had to be a white shark.

Just as the game was about to start, Dan McKiernan, now the director of the Massachusetts Division of Marine Fisheries, texted me. He asked if I would be willing to assist the state of Maine with the incident and speak with the commissioner of their Department of Marine Resources, Patrick Keliher, my boss Dan's counterpart in the Pine Tree State.

"Of course," I said. "I'm happy to help in any way."

A few minutes later I received a text directly from Pat. Because of COVID, all of us parents were sitting at a distance from one another in the stands, so there was no danger of someone glancing over and seeing the photograph Pat sent me. I knew immediately what it was.

It was a tooth fragment from a white shark—a tooth fragment that had been removed from the victim's body.

As the Seawolves started batting at the top of the first, I made my way down off the bleachers to a quiet spot next to the parking lot and called Pat. I told him that, under no uncertain terms, this was a white shark attack. Not surprisingly, Pat had some questions. This was, after all, not the type of event with which many people are familiar. In fact, it's a very small group of northern New England fisheries managers who have experience dealing with a shark attack—much less one that is fatal. In all the records of shark attacks going back to the eighteenth century in the New England states of Massachusetts, New Hampshire, and Maine, there have been just twelve incidents. Of these, only four were fatal, and of those four, only two occurred in the last century. The first was in 1936, which incidentally occurred in Buzzards Bay, quite close to my current house in Marion. The second was the 2018 fatality on Cape Cod, with which I am all too familiar.

Pat and I spoke for a time, and then he told me he needed to get together and discuss the incident internally with his team both from a media standpoint and from a public safety standpoint. We agreed to talk again the next day. Word spread quickly, and, by the end of the game, I was getting lots of questions from other parents. They all knew what I do for a living. Again, I largely played dumb—even though, at that point, I had all the information I needed. This had been a fatal white shark attack.

The tooth fragment was not complete, but it was obvious the dull tooth was triangular in shape, and it had the telltale serrations of a white shark. And really, what else could it have been? I knew better than most which species of sharks are off the coast of Maine in late July and which of those were likely to bite somebody. It's a really short list. That's why I was so certain it was a white shark even before I saw the tooth fragment. It was a similar situation to the Cape Cod incidents in 2012, 2017, and the two incidents in 2018, including the fatality. In the Cape Cod cases, I was willing to make the call that it was a white shark even before I saw tooth fragments or the wounds themselves because I know those waters. I know those sharks well.

My son's team won the game. As I greeted Wilson after, I couldn't help but dwell on the word "Seawolves" written across his uniform shirt. I recalled that the definition of a "sea wolf" is "a fabulous sea beast." As we drove home, I was still in dad mode, but my mind was racing ahead. I was thinking about that other sea wolf swimming 150 miles to the north. Since my kids, Wilson and Eve, were born, I had made every effort to compartmentalize my personal and my professional lives because I must, but I knew an event like this was going to be all-consuming. It wasn't just the professional and physical aspect of it. It wasn't just the media interviews and the conference calls that would no doubt ensue with colleagues in Maine. It was also the emotional aspect. The scientist in me wanted to know what had happened. I wanted facts, but the facts were typically slow to arrive. I was anxious to talk with Maine Marine Patrol and Pat to get more information. But lingering somewhere, locked away for the time being in the back of my brain, was also the simple fact that someone had just lost her life. A mother. A daughter. A wife. A person with friends who was on vacation. I was mindful of the fact that this was a human being who was just killed by a big fish, and while I'm good at compartmentalizing, that fact—the loss of life—weighed very heavily on me.

The most difficult times for me following an event like the fatality in Maine in 2020 is when I'm not on the job. It comes later, when I have time to think—when I have time to get online and learn a little about the victim. That's when it really hits me. I search Google Earth in the general location of the attack and zoom in. I look at the local houses, closest town, wharves, marinas, and fishing boats. I read news reports, look up the victim, and find out more about her—family, friends, job. The pictures of the victim in life hit me particularly hard—attending a social event, traveling with his or her family, laughing on a beach, never anticipating what was to come. Although it may look like it to others sometimes, I'm not a droid just gathering data and spitting out facts. Like anyone, I'm obviously horrified by an event like this. It changes you. It truly does. While it may not seem like it to the media outlets with whom I speak—or even to my boss—I'm emotional about it. But I've found there is often this lag between responding professionally and then responding emotionally. In part, it's because the process takes time—time between when the incident occurs and when I see photographs of the body and read the reports. There's a whole mental bureaucracy associated with trauma and death, which, in many ways, staves off the emotional response. I need permission to meet with the medical examiner and review the autopsy photographs and report. There is a lot of waiting, but then when I'm sitting there by myself looking at the photographs—looking at a dead human being that was killed by an animal I've literally dedicated my life to studying—that's when it hits me hard.

I've noticed that the way it impacts me has changed over time. I've only directly dealt with two white shark fatalities in my career, but I've spent most of my life thinking about these animals and what they are capable of. What they will and won't do. I think back to watching *Jaws* for the first time all those years ago. When I saw the film, I knew even then that much of what the shark was doing was fake. Part of me wanted to be the scientist, to be Matt Hooper, because it looked like a fun, cool job. I really wasn't afraid of the shark. I knew that the probability of a shark biting a person was incredibly low—so low that it didn't really enter my mind. Wanting to study white sharks wasn't about fear or some adrenaline-fueled aspiration. The more I learned about white sharks, the more I wanted to know what made these animals tick. I wanted to be Matt Hooper and study these fish. I wanted to figure out how they live, how they go about their business, and ultimately, I wanted to use that information to help conserve them.

As a young shark scientist, I was a bright-eyed, energetic guy who downplayed shark attacks. Any lecture I gave on sharks, and I've been giving them for forty years now, always downplayed the shark attack side of things. But that's changed. Even when I started working on white sharks off Cape Cod, I still thought the risk of shark attack was minimal. It wasn't something to be really concerned about. I wanted to figure out the shark's broad-scale movements. I wanted to figure out its biology. I wanted to figure out how to conserve them. It wasn't about shark attacks.

But then the movie and my professional career started to converge and, for the first time, I started looking back to what I'd seen in the movie not so much as the scientist studying sharks, but instead the scientist studying shark attacks. I hadn't anticipated that. I also hadn't anticipated that fear would become part of the equation—but I now must admit that anxiety starts creeping into my world. Into my dreams.

I'm floating in the open ocean. The sky is clear and bright, the seas are calm, I can see for miles. I look down. The water is deep and very dark, quiet, still. But I can sense the presence of something big, alive, powerful, and dangerous. I don't see it. I never see it, but it's there. I know it's there. Am I safe? I think so, but doubt slowly enters my mind and cracks the door for fear to creep in. I'm alone, completely alone, except for the presence. Getting closer. Terror is knocking. I need to get out, but I can't. I can't.

I wake up and don't go back to sleep. I can't go back to sleep. I don't want to.

Seeing these shark attacks—seeing tooth fragments and autopsy photographs, and hearing firsthand accounts—has started to tap into my own innate fear of being bitten. It hasn't really changed my fundamental view of these animals, but it certainly has broadened it.

My perspective has changed, as I've seen what white sharks can do to people. It's very rare, but it happens.

I'm no longer the scientist that says it doesn't happen.

After an Attack

In the wake of a shark attack, I feel pressure—pressure to produce meaningful information that can affect public safety. Throughout my career, there have been many incidents that I think of as being slapped-in-the-face moments—times when something dramatic occurs, catching me off guard, and then I think back and realize that I should have seen it coming. The moment, the slap in the face, gets my attention, wakes me up and gets me back on track. Gretel in the salt pond was one such moment. The five sharks we tagged on Labor Day weekend 2009 was another. The

2020 Maine fatality was not only a slap in my face but also in the collective face of the region.

Directly after the incident, there was a lot of talk that made it seem like people in Maine were unaware of the risk. In much the same way Ida Parker and Kristin Orr thought of the white shark issue as a Cape Cod issue and not a Plymouth one before the attack on Kristin's kayak, people in Maine largely thought that their state's coastline was somehow immune. The presence of white sharks in the Gulf of Maine was not a new phenomenon in 2020. We know that for as long as humans have inhabited the coastlines of Massachusetts, New Hampshire, Maine, Nova Scotia, and even Newfoundland, white sharks have been present. Archaeological evidence has revealed that the very first humans living in the area observed and respected white sharks, as evidenced by the existence of white shark teeth in funeral rituals and sacred ceremonies. What had changed, perhaps, was the size of the population or how that population responded to changing ecological factors, such as the extirpation of seals and then their return. If the sharks were responding to changes in the ecosystem, and if those changes were bringing them closer to shore, then the level of risk may also have been changing. Our research is studying exactly those possibilities, but the reality is that the sharks have always been there—the risk has always existed.

The 2020 fatality in Maine shocked the state into taking a more proactive approach to sharks. While the efforts to deploy acoustic receivers and initiate a tagging program similar to Massachusetts were motivated by public safety, the greater effect is that white shark research in the northwest Atlantic is expanding and becoming more collaborative. The New England White Shark Research Consortium was formed soon after the fatal attack to provide a coordinated regional approach to studying white sharks, and includes the Rhode Island Department of Environmental Management, the Massachusetts Division of Marine Fisheries, the Atlantic White Shark Conservancy, the Center for Coastal Studies, the University of Massachusetts Amherst, the University of Massachusetts Dartmouth's School for Marine Science and Technology, the New England Aquarium, the New Hampshire Department of Natural and Cultural Resources, the Maine Department of Marine Resources, Arizona State University, the Atlantic Shark Institute, the NOAA Fisheries' Apex Predators Program, and Fisheries and Oceans Canada. Eventually the consortium will maintain an impressive array of acoustic receivers stretching from Rhode Island to Canada, allowing scientists to gather even more data about where these sharks go and what they are doing.

Increased tagging effort with newer tagging technologies will provide ever-widening glimpses into the largely secret lives of these apex predators—a thing I never could have imagined when I graduated from college and went to work at the Apex Predators Program with Jack Casey in 1983. It will be one of the most robust white shark research efforts anywhere in the world, and while it will certainly enhance public safety throughout New England and into Canada, it will also provide invaluable information about a species that is still listed as "vulnerable," according to the International Union for Conservation of Nature (IUCN) Red List of Threatened Species. I have to remind myself that 1983 was only forty years ago. In many ways it feels like it was much longer, and makes me wonder what young scientists like Megan will learn in the next forty years.

Restoring an Apex Predator

When Richard Ellis and John McCosker published their 1991 treatise titled *Great White Shark*, they wrote, "We are concerned . . . about the fate and future of the white shark, and if this book has a purpose beyond exposition, explanation, and entertainment, it is to make very clear the plight of this great fish, so reviled, feared, and assaulted that it may be in danger of disappearing forever." For those living on Cape Cod today, Ellis and McCosker's caution may seem unbelievable, but at the time of the writing, there were no shark management measures at all on the Atlantic coast of the United States. It wouldn't be until 1997 that the white shark received federal protection and even later until the species received protection in various states along the eastern seaboard. Cape Cod's white sharks are some of the first tangible evidence that white shark numbers in the northwest Atlantic may be rebounding from low levels. We've come a long way, but we still have a challenging path ahead.

When people advocate for an imperiled species, the situation may seem so desperate that it's hard to imagine what success might look like. It's even harder to imagine how that success could trigger a return to the very same mentality that caused the species to be imperiled in the first place. But that may be what we are seeing in the northwest Atlantic with white sharks—a shift in perception. Restoring an apex predator to an ecosystem is a monumental task, and there are few successful examples. In part this is because we, as humans, are loath to give up territory we've claimed. We returned the gray wolf to Yellowstone, but we shoot it when it leaves the park boundaries and enters "our territory."

Since the second decade of the twentieth century, we've claimed the ocean as our escape, our respite, and our playground. We slayed the monsters that inhabited it, and we tamed it with lifeguards and a bevy of amenities ranging from toilets to food huts. In our collective view, the ocean became "safer," and over a century, we slowly altered our behavior toward it. We developed more seaside resorts and beaches for recreation. We spent more time in the water. We introduced our children to it. We learned to surf and boogie board and paddleboard. We designed wet suits, allowing us to extend our range northward and longer into the colder months. Ultimately we gave ourselves the impression that we had conquered it. We made it both suit and serve us.

Now the white shark has returned to one of America's most iconic summertime destinations, and it's challenging our perception of what the ocean is to us. For the first time in a long time, we have a hazy sense of what it means not to be the top predator. For the first time in a long time, we've had to consider what it means to be prey, even if we're only mistaken as such. What do we do with those emotions? Do we celebrate our ecological success—the restoration of an apex predator to an ecosystem—or do we defend our hard-won territory?

Is there a middle ground?

We have some hard decisions to make, and how we make those decisions will say a lot about us and how we perceive our place on this planet. Ellis and McCosker struggled to find optimism at the end of their book. "Extinction and endangerment are our legacy," they write, "and included on the long list of the world's depleted species is the great white shark." There are many more reasons to be optimistic about the white shark today, but that optimism will only prevail if scientists like Jack Casey, me, and Megan Winton continue to carry the torch from one generation to the next, motivated by our admiration, respect, and even a bit of fear for these remarkable animals.

⚓

JUNE 3, 2027—38°41'11.1"N, 45°41'27.3"W (MIDDLE OF THE NORTH ATLANTIC)

As summer approaches, the large female white shark will leave the Azores and head back west across the Mid-Atlantic Ridge toward the east coast of the United States. It's unknown where she will have her pups. Nobody knows for sure why she dives as deep as she does, or why she stays down so long. Is it related to feeding or reproduction? These mysteries and others contribute to the mythology that surrounds her.

Wherever she goes, she is only viewed fleetingly, but the stories that emerge from those brief observations are embellished, as they have been since the first man saw the first white shark. A fleeting glimpse of her bulk drifting under a boat becomes an aggression. Unseen teeth are imagined. Her eyes become "impenetrable and empty as the eye of God." Humans fill the void with imagination originating in an instinctive place of fear. Stories aggrandize the myth. Her unyielding, crescent-shaped caudal fin propels her deeper into the human psyche, causing us to overlook the genius of her migratory journeys.

We should marvel. By the time the Vikings learned to navigate the North Atlantic between continents, the white shark had been making just as long a journey for millennia. At the time Champlain relied on an astrolabe, cross staff, and de Medina's tables to reach New England, the white shark was making similar passages with enviable accuracy and precision. Silently. Purposefully. Mostly invisible to sailors on the water. Yet when she's seen, she's labeled "monster" and "man-eater."

She is "like a locomotive with a mouthful of butcher knives," says Matt Hooper.

To which Benchley chastises, "She only obeys her own instincts."

ACKNOWLEDGMENTS

My life with sharks would not have been possible without the love and support of my parents, Bernie and Irene, and my siblings, Vickie, Bernie, Lenore, Burt, Maggie, and Paul. I've worked with so many people along this incredible journey and would like to particularly thank: my mentors Jack Casey, Wes Pratt, Chuck Stillwell, and Frank Carey; my colleagues Nancy Kohler, Lisa Natanson, Brad Chase, George Benz, and John Chisholm; fishing buddies Chuck Walker, Billy Chaprales, and Luke Gurney; my assistants Kevin Johnson, Twice Tougas, Bobby Fuller, Robbie Goodwin, and Erin Rechisky; tournament organizers Clint and Rick Allen, Bobby Jackson, and Steve James; filmmakers Nick Caloyianis, Clarita Berger, Nick Stringer, and Sarah Cunliffe; and my agency leadership, including Randy Fairbanks, Phil Coates, Paul Diodati, David Pierce, Dan McKiernan, and Mike Armstrong.

This research would not be possible without the support of the Atlantic White Shark Conservancy and, specifically, Cynthia and Ben Wigren. Our tagging team—Captain John King, Brian Hanson, Wayne Davis, Ken Johnson, and fellow scientist Megan Winton—is responsible for the most successful white shark research program in the Atlantic.

Above all, I am indebted to my wife, Kimberly, and children, Wilson and Eve, who taught me how important it is to leave my job on the front porch when I get home.

—Greg Skomal

As someone who has the privilege of making my living telling stories where, as nature writer Tom Dee puts it, "science and poetry meet," I am forever indebted to Sue Standing and James Lawrence. As a companion and collaborator with whom I've explored so many wild places, Karen Talbot has my deepest respect and appreciation. Finally, Greg and I are beyond grateful to our agent, Susan Canavan, for inspiring this project and for her thoughtful and diligent support in bringing it to life.

—Ret Talbot

ADDENDUM

POPULATION STUDY RESULTS

On July 27, 2023, two weeks after the hardcover publication of this book, a population study (p. 268) was published in the journal *Marine Ecology Progress Series.* The paper, titled "An Open Spatial Capture–Recapture Framework for Estimating the Abundance and Seasonal Dynamics of White Sharks at Aggregation Sites," is authored by Megan Winton, with Greg as a coauthor. In the book, we tell the story of Cape Cod becoming the newest white shark hotspot worldwide (p. 215), and this research now supports that narrative.

The paper produced the first estimate of white shark abundance in any portion of the North Atlantic. As you read in this book, the data were collected during weekly surveys by "video fingerprinting" (p. 268) individual white sharks. Across 130 research trips, 393 individual white sharks were identified. Using Megan's model, this number was extrapolated to a population size of 800 individual white sharks that visited Cape Cod from 2015 to 2018. These data place Cape Cod among the largest white shark hotspots worldwide. In comparison, population estimates from South Africa, central California, South Australia, and Guadalupe Island, Mexico, range from 78 to 1,279 individuals over similar time frames.

As we state in *Chasing Shadows*, we view the resurgence in the white shark population along the coast of New England as a conservation success story. We also acknowledge, however, that having such a large shark population adjacent to popular beach destinations presents increased risk that public safety officials, beach managers, politicians, and others must address. As a scientist and a science writer, we believe data are essential to informing management measures, and the baseline data presented in this paper will be essential to the assessment of both risk management and conservation measures moving forward. As nature reestablishes itself in areas where it was extirpated, we will need to come to terms with inherent risks present in an ecologically healthier world.

—*Greg & Ret*

INDEX

ABC, 21, 29, 69, 72, 93, 163–64, 171, 209, 232
Adventure II, 97–98
Adventures of Huckleberry Finn (Twain), 44
Agassiz, Louis, 108, 120
aggregation sites, 135, 217
Air Jaws (documentary), 217–19
Aleutian Dream, 269–71, 286–91, 295, 316–18, 330–31
Ali, Muhammad, 11
Allen, Clint, 143, 144–46
Allen, Rick, 143, 145–46, 163–66
Almeida, Frank, 184
American Elasmobranch Society, 156–57, 274
American Fisheries Society, 99
American Institute of Biological Sciences, 59–60
American Museum of Natural History, 58
American Sportsman (TV show), 69, 72, 93, 163, 232
Andrea Doria: The Final Chapter (TV show), 167
Annear, Steve, 276
annuli, 117–18
apex predator, 16, 90, 136, 161, 219, 277
  restoring, 300, 337–38
Apex Predators Program, xiii, 31–37, 38–39, 60, 72–73, 75, 80–82, 89, 103–4, 106–7, 124–26, 142, 152, 288, 336–37
  on Martha's Vineyard, 141–48, 152
Apex Shark Expeditions, 220–21
Aquatic Resources Trust Fund, 127–29
Aquinnah, 136–37, 140
Arctic Circle, 160
Armstrong, Mike, 328
Asmutis-Silva, Regina, 13, 16–17, 21–22
Atlantic Human Conservancy, 329
Atlantic White Shark Conservancy (AWSC), 5, 258–59, 263, 265–69, 287, 294, 301–2, 310, 313, 314, 321, 326, 328, 329–30, 336
attacks. *See* shark attacks
Australia, 15, 84, 174, 213, 217
  shark attacks, 211, 302
  shark cage diving in, 156–58, 260
Azores, 309–10, 327, 339

Backlinie, Susan, 247
Backus, Richard, 59
Bacon, Kathleen, 307, 325
Ball, Sheldon W., 255
Ballard, Robert, 166
Ballston Beach, shark attack, 255–57
Baltimore Orioles, 83, 180
Barkalow, Clifton T., 64
basking sharks, 159, 167, 170–73, 176–77
Bay Shore Mako Shark Tournament, 66, 73–77, 83, 148
Bay Shore Tuna Club, 66
Bean, Bruce, 195–200
Beaty, Ron, 3, 305, 306, 310, 311
Beaufort Laboratory, 177
Bellows, Clark, 48
Benchley, Peter, 78, 93, 247
  Montauk Monster in Noank, 93–94, 98, 100–101
Benchley, Wendy, 247
BenDavid, Arthur, 143
Benway, Bob, 119
Benz, George, 108, 126, 296
  Arctic Circle sharks, 168–70
  background of, 99–100
  Montauk Monster in Noank, 98–101, 102–3, 146–47
  parasite and copepod research, 99–100, 122, 146–47, 170, 231, 259

shark airplane spotting, 203–5, 207, 208, 209–10, 227–31, 249
shark research cruise on *Wieczno*, 99, 118–19, 121, 122–23
white shark sighting, 203–5, 207, 208, 209–10
Bernal, Diego, 274
Bigelow, Cleveland, III, 304–5
Bigelow, Henry Bryant, 124
Big Wave Productions, 170, 248, 296
"bite and spit," 277
bite radius, 87
Block Island, 93–97, 100, 134, 135, 194
bluefin tuna, 50, 171–72, 203, 206–7, 229, 274, 295
blue sharks, 37, 38, 68, 83, 93–94, 107, 114, 146, 153, 154, 166, 169, 234, 235
age and growth studies, 117–20, 124–25, 126–27
Blue Water, White Death (documentary), 84, 104, 111–12, 115, 157, 167, 232–36
Boats, John, 4–5
Body Snatchers (documentary), 170, 296
Booth, Joe, 3
Boston Big Game Fishing Club, 143, 145
Boston Globe, 37, 131, 156, 200, 206, 215, 222, 226–27, 230, 276, 294, 298, 310, 316
Boston Herald, 251, 305
Boston Marathon bombing of 2013, 320
Boston Red Sox, 180
Boston Sea Rovers, 166–70
Boston University Marine Program (BUMP), 176–77
Bowles, Ian, 222, 226–27
Braddick, Donnie, 144
breaching, 161, 217–21, 317, 318
Breaux, John, 127–28
Breen, George, 227–30
Brodeur, John, 50–54, 61–63, 64
bronze whalers, 158
Brooklawn Country Club, 29–30, 43–45, 73
Brunswick Surf Club, 62–63
Bryant, Nelson, 76, 84, 148
buccal pumping, 194
Buck, John, 98–101
Bull Island, 178–79
Bunker Hill Community College, 10
Bureau of Commercial Fisheries, 40, 59
Bureau of Sport Fisheries and Wildlife, 40, 50, 65, 73, 128
Burgess, George, 160, 175, 298, 303, 311
Burgess, Russ, 4–5, 7, 9–10, 12–13, 16–17, 19, 21–22
Burke, John C., 180–83
Buzzards Bay, 178, 332
bycatch, 81, 149–50, 272

Cabby Shack (Plymouth), 5–6
Caezza, Cynthia, 7–9, 19, 23
cage diving. *See* shark cage diving
Cahoon Hollow Beach, 17
Cailliet, Gregor, 117, 122, 125, 273–74
Caloyianis, Nick, 116, 167, 169–73, 176, 210, 231, 232
Camp, Raymond, 43
Campbell, Billy, 164–65
Camp Wellfleet, 304
Cape Cod
geography of, 11
geology of, 135–36
history of, 10–11
Cape Cod Bay, 9, 205, 278
Cape Cod Canal, 10, 11, 205
Cape Codder, 250
Cape Cod Hospital, 256
Cape Cod Museum of Natural History, 323–24
Cape Cod National Seashore, 9–10, 223, 263, 304, 326
Cape Cod Ocean Community, 329
Cape Cod Times, 21, 22, 33, 175, 208, 302–3, 310, 315, 316, 319
Cape Hatteras, 56, 92, 328
Cape Haze Marine Laboratory, 58–60
Cape Mallebarre, 197
Cape May, R/V, xi
Capone, Lisa, 193, 207
Capsized: Blood in the Water (movie), 60
Carcharodon carcharias, 28, 30, 79. *See also* white sharks
Carey, Frank, xii, 20, 69–72, 105, 171, 211, 229–31, 254

Casey (shark), 299
Casey, John G. "Jack," xi–xiv, 33–37, 133
  background of, 40–41
  George Benz and, 99, 102–3, 126
  Lisa Natanson and, 125–26, 313
  Montauk Monster in Noank, 102–3, 106
  retirement of, 288, 313
  shark fishing, 50, 55, 64–67, 73–77, 78, 106, 122, 152
  shark management and protection, 84–85, 148–49
  shark research and shark tagging, xi–xii, 35–37, 38–39, 57–58, 64–77, 80–81, 87, 91–92, 102–3, 106, 107, 110, 118, 125–26, 142, 144, 163, 211, 219, 254, 260, 289, 299, 337
  volleyball games, 33–35
Castro, José, 133, 149
Cataldo, Paul, 295
Catalina Island, 41, 46–47
Cat Cove Marine Laboratory, 129, 158–59
Caterpillar, 262–63, 291
caudal fins, 55–56, 79, *82*, 163, 207, 216, 245, 289, 339
Celotto, Ernie, 78, 93–96, 98–101, 102–7, 135
ceratotrichia, 68
Chadwick, Beryl, 59
Champlain, Samuel de, 10, 197
Chaprales, Billy, 184, 248
  bluefin tuna tagging, 171–72, 206–7
  shark tagging, 171–72, 177–78, 184, 205–8, 230–32, 267, 269–70
  white shark sightings, 203–10, 213, 228
Chase, Brad, 127–29, 158–59, 171–72
Chatham Chamber of Commerce, 243, 246
Chatham Clothing Bar, 247
Chatham Fish Pier, 195
Chatham Harbor, 195–96, 237
Chatham Light, 196, 198
Chatham Parks and Recreation Department, 206, 264
Chatham Shark Center, 294, 297
Chilmark, 140
China trade, 60
Chisholm, John, 301, 312, 313
  AWSC and, 266, 268
  Naushon Island white shark, 184–85, 186–87
  shark attacks, 250, 285
  shark cage diving, 233
  shark research and shark tagging, 205, 270–71, 275, 287, 294, 298
Clark, Eugenie ("Shark Lady"), xii, 58–59, 60, 99, 166, 167, 262
Clausen, Bob, 165
Cleve's Cove, 304, 306
cloaca, 244
Coast Guard, US, 61, 63, 97–98, 223–24
Coast Guard Beach, 15
Colella, Drew, 178, 180, 181
Collins, Michelle, 4–6, 12–13, 16–17, 19, 21–22, 23
Connecticut Department of the Environment's Fisheries Bureau, 99–100
Convention on International Trade in Endangered Species of Wild Fauna and Flora (CITES), 221
Cooperative Tagging Program, 66–77, 80–81, 83, 91, 118, 151, 289
Coop's Bait and Tackle (Edgartown), 138–40, 145
Copeia (journal), 71
Coppleson, Victor, 62
Cornish, Dick, 48
Cortés, Enric, 273
Cousteau, Jacques, 25, 29–30, 43, 44, 80, 84, 102, 111–12, 166
COVID-19 pandemic, 331
Cricket II, 43, 45–46
culling, 324–24, 329, 330
Curly (shark), 233, 234–36, 268
Cuttyhunk Invitational Tournament, 153–54, 155

Darenberg, Carl, Sr., 69
da Vinci, Leonardo, 4
Davis, Wayne, 7, 12, 13, 17, 18–19, 233, 269, 287, 316–17
Dayton Daily News, 55, 64
Deadliest Catch (TV show), 225

Delaware II, 137
dermal denticles, 109–10
Dingell-Johnson Act of 1950, 127
dinosaurs, 4, 134–35, 140–41, 272
Diodati, Paul, 182, 184, 202–5, 251–52
Discovery Channel, 167, 176, 225, 248, 267
  Shark Week, 6, 60, 152, 169–70, 217–18, 231, 247–48, 267
dogfish, 56, 159, 176
Domeier, Michael, 224–26
Domeier, Mike, 238
Donilon, Charles E., III, 90–92, 94, 113–16
dorsal fins, 3, 23, 26, 67, 68, *82*, 245, 249, 250
Drew, Shawn, 35
Dreyfuss, Richard, 20, 30–31
Drumm, Russell, 46
Dubrule, Greg, 94–96, 98–101, 102–7
Dunaway, Vic, 49
Dust Bowl, 82
Duxbury Beach, 278–79, 281

Earth (documentary), 201–2, 219
Ebert, David, 273
ectotherm, 70, 133–34
Edward VII of United Kingdom, 303–4
elasmobranch, 59
Eldredge, Ernie, 187–89, 191
elephant seals, 71, 245–46
Elizabeth Islands, 121, 140, 153, 173, 178–79
Ellis, Richard, 55, 307
Emory University, 272
endotherm, 70–71, 133–34
energy efficiency of sharks, 15, 244
Entenmann, Charles, 73
Eocene Epoch, 136
ESPN, 163–66, 167, 238
Evans, Linda, 280
Everett High School, 9
evolution, 134–36
  of sharks, 4, 116, 134–35, 140–41, 160–61
extinctions, 4, 136, 278, 338
Ezyduzit, 172, 177–78, 205–10, 230–31, 232, 266, 267, 269

Fairbanks, Randy, 128–29, 142, 165
Fallows, Chris, 218–22, 250
Fallows, Monique, 219–22
Falmouth Airpark, 204, 205
Falmouth Fire Department, 191
Falmouth Harbor, 180, 183–84
false albacore, 137, 146, 147, 178–79
False Bay, 15, 219–23
Farallon Islands, 15, 37, 75, 83–84, 125, 217, 231, 239, 245
"fast-twitch" muscles, 134
Fay, Gavin, 275
fear of sharks, 1, 8, 92, 100–101, 111–12, 185, 201, 208–9, 219–20, 229, 257
Federal Aid in Sport Fish Restoration Act of 1950, 127–28
Federal Aid in Wildlife Restoration Act of 1937, 127
Federal Emergency Management Agency (FEMA), 11
"feeding frenzy," 111–12
Filoramo, Jean, 50–54
fin whales, 69, 80, 93, 135
Fire Island Inlet, 75
Fischer, Chris, 238–40, 250–54, 259–63, 291–94
  shark expeditions, 259–63
  Shark Men, 224–25, 238, 239–40
  Shark Wranglers, 250–51, 259
  violation of scientific trust, 297–99
Fish, Charles, 32
Fish, Marie Poland, 32
Fish and Wildlife Act of 1956, 40, 41
Fish and Wildlife Service, 40, 41
Fisher, Carl, 42
Fishermen Helping Scientists, 187–88
Fishery Management Plan for Sharks of the Atlantic Ocean, 151, 159
fishing. *See* shark fishing; sportfishing
Fishing with Charlie Moore (TV show), 167
Fish N Boat, 153–55
Flores Island, 309
Florida Program for Shark Research, 160, 175, 298
Fothergill, Alastair, 202
Franz, Lisa, 243, 246
Freeman, Roger, 315–16

Fuller, Bobby, 173
Fuller, Jesse, 173
fur seals, 28, 218

Galilee Docks, 90–91, 113
gangions, 120
Gateway Seawolves, 330, 332, 333
Gay Head, 136–37
Gay Head Cliffs, 136
Gay Head Light, 121
Geiger, Adam, 165
Gelsleichter, Jim, 259
gender and racial inequality in shark research, 288
Gilbert, Perry, 59, 63
Gilkes, Cooper, 138–40
gill slits, *82*, 193, 239
Gimbel, Peter, 104, 111–12, 115, 157, 167, 232–34, 235
Gioseffi, Glenn, 32–34
Girl of the Sea of Cortez, The (Benchley), 93
Glacial Lake Cape Cod, 4
Goats Neck, 179
Goldman, Ken, 157
Good Morning America (TV show), 21, 209, 257, 316, 331
Gordon, Ian "Shark," 156–58
Gosnold, Bartholomew, 10
Gottlieb, Carl, 247
Gray, William, 62
gray seals, 197–200, 211, 222–23, 278
Great Depression, 42
Great Point Lighthouse, 196
great white sharks. *See* white sharks
Greenland Fisheries Investigations, 67
Greenland sharks, 67, 168–70
Gretel (shark), 178–92, 296, 335
Grey, Zane, 46, 84
Griffin, Mary, 204
Guadalupe Island, 15, 217–18, 226, 267
Gulf of Maine, 134, 173, 193, 194, 197, 277, 317
Gulf of Mexico, 160, 327
Gulf Stream, 119, 133–34, 243, 308–9
Gurnet Light, 278–79
Gurney, Luke, 295–96

Hadley Harbor, 178–79
Hansen, Paul, 67
Hare Harbor, 161, 162
harpooning sharks, 47, 69, 72, 86, 91, 95, 101, 106, 109–10, 162–63
Haskell, Brad, 38
Hawkins, Thomas, 178
Hayashi, Tuck, 178–83
Hemingway, Ernest, 46
Henderson, A. C., 117
Higgins Beach, 11
History Channel, 239, 240, 250, 312
Holder, Charles Frederick, 41, 46–47
Holland, Kim, 157
Homo sapiens, 116
Horna, Jim, 222
Hudson, Henry, 10
Hudson Anglers, 149
Hudson Canyon, 41, 70
Hudson River, 41
Hueter, Bob, 240, 259, 293, 297–98
Hunter, Chad, 284–85
Hurricane Florence, 11

ichthyologist, 125
Indianapolis, USS, 112
Individual Fish Records, 103
International Shark Attack File, 303
International Union for Conservation of Nature's (IUCN) Red List, 221, 337
In the Slick of the Cricket (Drumm), 46

Jackson, Michael, 83
Jameson (shark), 296–97
Jaws (Benchley), 93, 100
Jaws (movie), 8, 20, 30–31, 33, 34, 43, 60, 75, 78, 81, 83–87, 93, 95, 97, 102, 110, 111–12, 121, 152, 175, 180, 200–201, 219–20, 227, 247, 280, 334–35. *See also* Matt Hooper; Quint
"*Jaws* effect," 93–96, 100–101, 156, 201–2
Jaws Comes Home (documentary), 231–36, 267
Jeremy Point, 279
Johnson, Kevin, 142, 165
Johnston, Jack, 12
Jurassic Period, 134–35, 140–41
Jurassic Shark (documentary), 169–70

Karlberg, Jared, 312–13
Kelly, Gene, 69–70
Kenney, Jerry, 33
Kerry, John, 178
Kerwin, John, 163–66
Kerwin Communications, 163–66, 167
Kidd, William, 178
King, John, 269, 290, 295, 316–18
Kneebone, Jeff, 205, 209–10, 271, 274–75, 286, 288
Knight, Richard, 49
Kock, Alison, 301–2
Kohler, Nancy, xii, 74, 113, 114–15, 124–25, 173, 184, 288, 313
Krueger, Bill, 89–90, 127
Kulmatiski, Anna, 279–80, 284–85

Labrador Current, 133
Lackeys Bay, 179, 186, 189–91
Lady Pat, 75–76
Lake, Peter, 233–34
Lakehurst Naval Air Station, 61–62
Lake Worth US Open Shark Tournament, 85
Lamniformes, 141
L'Anse aux Meadows, 160–61
lateral lines, 216–17
Lazarus, Adam, 181–82
Leonardo, Joe, 165
liability issues, 257, 264–65, 325–27
Lighthouse Beach, 196
Little Green Island, 194
Lobel, Phil, 176
Lobsterville Beach, 136–37, 139, 140–41
Log from the Sea of Cortez, The (Steinbeck), 93
Long Island Rail Road, 42
Long Island Sound, 29, 45, 102
longline surveys, 26, 65, 120–24
Longnook Beach, 17, 20, 318–21
  shark attack, 318–21
Lorenzini, 161
Lowe, Chris, 157
Luke (shark), 296
Lupovitz, Ronna, 32–33
Lutcavage, Molly, 171
Lyman, Ed, 176
Lytton, William, 17, 318–21
McBride, Brett, 260–61
McCarthy, Kevin, 113, 114–16
McCarthy, Morgan, 248–49
McCosker, John, 37, 277, 307, 337, 338
McCully, Katie, 229
Macdonald, Catherine, 288
MacDonald, Stephen, 97–98
McGrory, Brian, 215, 226–27
McKiernan, Dan, 184, 187, 192, 328, 332
MacKinnon, Rod, 195–200
Maine Department of Marine Resources, 331–32, 336
Maine shark attack of 2020, 330–34, 335–37
mako sharks, 46, 47, 50, 66, 68, 89, 93–94, 107, 136, 141, 153–55, 166
  Bay Shore Mako Shark Tournament, 66, 73–77, 83, 148
management, xii, 36, 148–49, 149–51, 159
Manning, Joanne, 178, 180
Manomet Beach, 278–79, 283
Manomet Point, 278–79
Maranatha Christian Church, 10
Marconi, Guglielmo, 303–4
Marconi Beach, 304–5
Marconi Station, 303–4
Marine Fisheries Advisory Commission, 202–3
Marine Mammal Protection Act of 1972, 20, 84, 324, 329
"mark-recapture" methodology, 68
Martha's Vineyard, 135–48
  author's regional biologist job, 137–48, 175, 176
  geography of, 136–37
  geology of, 135–36
  JawsFest, 247
  mini Apex Predators Program, 141–48
  Monster Shark Tournament, 143–46, 154, 156, 163–66, 176, 185, 313
  white shark in Naushon Island salt pond (Gretel), 178–92, 296
Martha's Vineyard Striped Bass and Bluefish Derby, 44, 146–48
Martha's Vineyard White Marlin Tournament, 143
Martinson, Chuck, 184

Mary Lee (shark), 262–63, 292
Massachusetts Department of Fish and Game, 184, 204, 206, 278
Massachusetts Division of Marine Fisheries, 7, 45, 127, 136–40, 165, 184, 202–3, 206, 213, 245–46, 251, 323–24
Massachusetts Energy and Environmental Affairs, 206, 222, 226
Massachusetts Environmental Police, 184, 185–86, 188–89
Massachusetts Environmental Trust (MET), 176, 207
Massachusetts Saltwater Fishing Guide, 184
Massachusetts Shark Research Program, xiii, 20, 152–53, 190, 213, 246
Massachusetts Sportfishing Tournament Monitoring Program, 145
Mather, Frank, xii, 66, 67–68
Matt Hooper, 20, 30–31, 33, 34, 40, 57, 81, 86, 87, 110, 152, 334
Mayhew, Gregory, 173–74
Medici, Arthur, 9–14, 17–18, 21–23, 321–25, 327–29, 331
megalodons, 135, 140–41
Menemsha Harbor, 173
Menemsha Hills, 136
Menemsha Pond, 139, 140
menhaden, 55–56, 75, 114, 143, 216, 277
mesopelagic zone, 309
metabolism, 70–71, 133–34, 277
Miami Seaquarium, 62
Mickley, Gerald, 85
Mid-Atlantic Bight, 56, 173, 299
Mid-Atlantic Fishery Management Council, 148–49
Mid-Atlantic Ridge, 263, 309, 339
Mid-Cape Highway, 10
Mi'kmaw First Nation, 161–63
Mildred, F/V, 312–13
misogyny in shark research, 288
Miss Costa (shark), 298–99
Monmouth Beach, 55–56
Monohansett Island, 179–80
Monomoy National Wildlife Refuge, 196–200
Monomoy Point, 196, 204–5, 207, 222–23
"monster fishing," 46–47, 48, 315
MonsterQuest (TV show), 312
Monster Shark Tournament, 143–46, 154, 156, 163–66, 176, 185, 313
Montauk, 41
  history of sportfishing, 41, 42–43, 48
Montauk Boatmen and Captains Association, 94–95, 151
"Montauk Monster" (shark), 71, 93–96, 98–101, 102–7
Monterey Bay Aquarium, 273
"Morph Cards" ("morphometrics"), 103
Moss Landing Marine Laboratories, 125, 273–74
Mote Marine Laboratory, 58, 240
Mueller (shark), 23
Mundus, Frank, 25, 43, 45–46, 48–49, 96, 144, 149, 181, 260, 315
Mutiny II, 48
Myers, Chris, 255–57, 264, 303, 306, 320
Myers, J.J., 255–57
Mystic Aquarium, 93–94, 96, 98–99, 102–3

Nakano, Hideki, 117
Nantasket Beach, 8
Nantucket Sound, 196
Naples Daily News, 85
Narragansett Laboratory, 31–37, 72–73, 80, 83, 113, 119, 124–26, 142, 273–74, 313
Natanson, Lisa, 125–26, 184, 199, 273–74, 289, 313, 314
National Data Buoy Center, 11
National Geographic, 167, 169, 170, 176, 224–25, 250, 289
National Hurricane Center, 11
National Marine Fisheries Service (NMFS), 32, 36, 40, 90, 148, 149–51, 159
  Narragansett Lab. *See* Narragansett Laboratory
  Shark Population Assessment Group, 272–73
  Shark Tagging Program, xii, 69–70, 73

National Oceanic and Atmospheric Administration (NOAA), 40, 73, 149, 213, 239–40, 251, 252, 268, 336
Atlantic Highly Migratory Species, 240, 298
Cetacean and Sea Turtle Team, 177
Nauset Beach, 15, 17, 224, 311–12, 315, 320–21
seal attack, 304, 305, 306, 317
shark attack of Walter Szulc, 248–50
white shark sighting, 227–28, 229
Naushon Island, 121
white shark in salt pond on, 178–92
Naylor, Gavin, 259
neoselachians, 140–41
Neptune Islands, 15, 157, 217
Newcomb Hollow Beach, 9–10, 13–14, 23
shark arrival, 14–16
shark attack of Arthur, 17–18, 21–23, 321–25
New England Institute for Medical Research, 59
New England White Shark Research Consortium, 336–37
New Smyrna Beach, 271–72
New York Bight, 40–41, 42, 65, 78–79, 194, 299
New York Daily News, 33, 48
New York Sportfishing Federation, 149
New York Times, 43, 46, 47, 76, 83, 84, 93, 144, 148
New York University, 58
New York Yankees, 61
Ninigret National Wildlife Refuge, 102, 105
Nonamesset Island, 121, 178, 179
Nord, Kevin, 278
Normally Fun, 165–66
North Beach Island, 15
Northcross, Wendy, 226
North Truro Air Force Station, 16–17

Oak Bluffs Harbor, 144–45
obligate ram ventilation, 193–94
Ocean, M/V (R/V *Ocearch*), 259–62
Ocean City, 63–64
OCEARCH, 224, 226, 238–40, 248, 250–54, 266–67, 291–94
shark expeditions, 258–63
violation of scientific trust, 297–99
O'Connell, Tom, 143–44
Office of Naval Research, 59
Offshore Adventures (TV show), 225, 238
O'Keefe, Cate, 184
Old Man and the Sea, The (Hemingway), 46, 181
Oleander, M/V, 119
olfactory receptors, 79–80, 216
oophagy, 25–26
Orleans Town Hall, 263–64
Orr, Bobby, 145
Orr, Kristin, 279–86, 303, 336
otoliths, 117–18
oxytetracycline (OTC), 118, 126–27

Pacific Shark Research Center, 273
Parker, Ida, 278–86, 336
Pearce, Richard, 97–98
Penobscot Bay, 194
Penobscot River, 66–67
permitting, 240, 251, 260, 291, 292–93, 297–98
Peters, Dave, 184
Petersen disc tags, 67–68
Pierce, David, 292–93, 312–13, 323–25, 328
Pilgrim Monument, 16
Pilgrims, 10, 278
Plankton Ecology Program, 32–33, 37–38
Pleasant Bay, 15
Pleistocene, 136
Plymouth Beach, 9
Plymouth Harbor, 9–10
Plymouth Town Wharf, 4–6, 23–24
Point Judith, 90–91, 113
Pollock Rip, 196, 200
porbeagle sharks, 141, 150, 173
Port au Choix, 162–63
Porter, Norman, 61
Pound, Thomas, 178
Pratt, Harold "Wes," 33, 121, 122, 164, 167
author's job interview with, 32
Montauk Monster in Noank, 99, 104–5, 106–7

Pratt, Harold "Wes" (*cont.*)
shark cage diving, 112–17, 164, 169, 234
shark research and shark tagging, xii, 35–36, 38, 39, 69, 71–75, 78, 80, 87, 89, 104–5, 106–7, 124, 125, 144, 173, 188, 211, 254, 289, 299, 313
President's Award in Academic Achievement, 90
Project Echo, 51
Provincetown, 10, 16, 205, 278
PSAT tags, 171, 172, 180–81, 189, 207–8, 262
Puijila darwini, 135

Quint, 43, 95, 112, 201

Race Point Lighthouse, 12, 278
racial inequality in shark research, 288
"ram and bite," 16, 216
"rare-event species," 81, 83
Reelin, 93–96
Regional White Shark Working Group, 325–25, 327
regulations, 137, 149–51, 156, 159, 182, 202–3, 291, 292–93, 297–98, 324
Reliable Fish Company, 6
Renegade, 143–44
rete mirabile, 134
Rettenmeyer, Carl, 98–99, 100, 104–5, 106
Return of Jaws (documentary), 267
Rhode Island State College, 32
right whales, 176–78, 244–45, 276–77
Ristori, Al, 65
Rocha, Isaac, 9–10, 12, 13–14, 17–18, 21–23
"rogue shark" theory, 62, 63
Roosevelt, Theodore, 303–4
Rueby, F/V, 171–72
Rutzen, Mike, 201

salmon, 66–67
Saltonstall-Kennedy Grants, 213
sandbars, 14, 304
sandbar sharks, 35–38, 58, 59–60, 107, 152, 166, 173, 174–75
Sandy Hook Marine Laboratory, 40–41, 50, 64–65, 72–73, 106
Saquatucket Harbor, 206, 207
Save Our Seas Foundation, 328
Scheider, Roy, 86–87
Schroeder, William C., 124
Scratchy (shark), 295
Scripps Institution of Oceanography, 58
Scylla (whale), 12–13, 16, 23
Sea Girt, New Jersey, 50–56, 61–62
Seal Island, 28, 217, 218–22
seals, 135, 257
shark feeding, 15–16, 20, 28, 79, 194–95, 211, 222–23, 245–46, 278, 317–18
Sears, Nathan, 276, 294, 320–21
Seaweed Too, 94–95
September 11 attacks (2001), 170–71
Serling, Rod, 29–30
Shanty Rose (Plymouth), 6
sharks. *See also* white sharks
age and growth studies, 117–20, 124–25, 126–27
anatomy of, 55–56, 82, *82*
classification of, 58
difficulty of studying, 27–28
evolution of, 4, 116, 134–35, 140–41
fear of. *See* fear of sharks
identification of, 103–4
populations, 37, 148–51, 159
public engagement with, 151–55
senses of, 79–80, 109, 216
Tails of the Sea cruise, 4–9, 12–13, 16–19, 21–22, 322–23
shark attacks, xiii, 3, 37, 211, 333–37
1916 incident, 50–51
1936 incident, 8, 175, 229, 263, 311, 332
1989 incident, 152–53
aftermaths of, 60–64, 335–37
Arthur Medici, 17–18, 21–23, 321–25, 327–29, 331
"bite and spit," 277
California of 2008, 211
Chris Myers, 255–57, 264
fear of. *See* fear of sharks
Ida Parker and Kristin Orr, 278–86
John Brodeur, 50–54, 61–63, 64

Maine death of 2020, 330–34, 335–37
maps of, *xi–xiii*
mitigation strategies, 190–91, 265, 300–306, 325–30
public safety issues, 20, 65, 92, 174–75, 184, 190–91, 210–11, 222–24, 226–29, 231, 237–38, 245–46, 263–65, 290–91, 294, 300–301, 320–21, 323–25
"rogue shark" theory, 62, 63
of sailors, 60, 111–12
Walter Szulc, 248–50, 264
William Lytton, 17, 318–21
Shark Attacks (Coppleson), 62
shark cage diving, 111–16, 164–66, 232–36
in Australia, 156–58
in South Africa, 218–22
SharkCam, 248, 250, 254–55, 263, 267, 268
shark conservation, xii, 36, 84–86, 117, 146–49, 151, 156, 221, 325, 337. *See also* Atlantic White Shark Conservancy
shark dissections, 20, 66, 74, 80, 81–82, 86, 105–6, 110, 111, 158, 272–73, 312–15
Sharkers, The, 48–49
shark fins, 60, 150, 156
shark fishing, xi, 40–41, 45–49, 60, 64–67, 84–85, 88–89, 93–96, 151–56, 158, 272
regulations, 148–51, 150, 202–3, 291, 292–93
shark fishing tournaments, 47–49, 66, 85–89
Bay Shore Mako Shark Tournament, 66, 73–77, 83, 148
Monster Shark Tournament, 143–46, 154, 156, 163–66, 176, 185, 313
Snug Harbor Shark Tournament, 96–98
Shark Handbook, The (Skomal), 8
Shark Industries Inc., 58, 59
shark meat, 88–89, 151, 285. *See also* shark fins
Shark Men (documentary), 224–26, 238, 239–40
"shark menace," 45–46, 48, 64, 84, 88–89, 247, 284
Sharks Around Us, The (Skocik), 85
Shark! Shark! (Young), 47
Shark Shield, 186–87
"shark smart," 300–301
Sharks of the Deep Blue (documentary), 167
Shark Spotters, 302
shark tagging (shark tagging programs), xi–xiv, 23, 36–37, 64–71, 83–84, 153–54, 171–73, 205–8, 210, 211–13, 217, 230–36, 248, 254–55, 261–62, 266, 295–99, 327
Charles Donilon's encounter, 90–92, 94, 113
Cooperative Tagging Program, 66–77, 80–81, 83, 91, 118, 151, 289
Megan Winton's first summer, 286–91
shark teeth, 15, 16, 26, 55–56, 79, 98, 140–41, 161–63, 195, 276, 336
Sharktivity, 294, 297, 303
Shark Tracker (newsletter), 144, 145
Shark Trackers (TV show), 167
Shark Week, 6, 60, 152, 169–70, 217–18, 247–48, 267
Shark Working Group, 299–301
Shark Wranglers (TV show), 250–51, 259
Shaw, Robert, 201, 247
Sherman, Ken, 159
Shrewsbury Rocks, 56, 62
Silent World, The (documentary), 111
Silva, H. M., 117
Simonitsch, Mark, 187–89
Skerry, Brian, 116, 164–65
Skocik, Regina, 85
Skomal, Bernard (father), 29, 39, 44
Skomal, Bernie, 43–44, 45
Skomal, Burt, 29, 43–44, 45
Skomal, Eve, 333
Skomal, Irene (mother), 29, 39, 44–45
Skomal, Kimberly, 330–31
Skomal, Wilson, 232, 330, 332, 333
Smith, Leslie, 147
Smith, Stuart, 223–24
Smith Research Corporation, xi
snakes, xiii

Snappa, 90–92, 113–16, 169
Snug Harbor Shark Tournament, 96–98
soupfin sharks, 58
South Africa, 28, 156, 213, 217–23
Chris Fischer and OCEARCH, 238, 250–51, 252, 254, 291
shark cage diving in, 218–22
Shark Spotters, 301–2
Spargo, Abbey, 173
Spaulding Rehabilitation Hospital, 17
Species Recovery Grants, 213
spiders, xiii
Spielberg, Steven, 30, 86–87, 121, 201
sportfishing, 40–45, 45–46, 136–43. *See also* shark fishing
Sport Fish Restoration Program, 127
Springer, Stewart, 58, 59–60
Squibnocket Pond, 140
Stage Point, 279–86
Starr, Jay, 233
Starrfish, 232–33
Steinbeck, John, 93
Steinhart Aquarium, 37, 125
Stellwagen Bank, 7, 9
Stevens, John, 117
Stewart, Lance, 164
Stillwell, Chuck, xii, 33, 35–36, 38, 74, 106, 121, 122, 124–25
Stockton Hotel (Sea Girt), 50–51, 63
Stretch, D. Allen, 63
Stringer, Nick, 170, 231, 232–36, 296
striped bass, 42, 44, 50, 137, 141, 142–48, 180, 313
Summers, Erin, 177–78, 184, 187, 288
sunglasses, 19, 22, 281, 326
Szulc, Walter, 248–50, 264

tagging. *See* shark tagging
Tails of the Sea, 4–9, 12–13, 16–17, 18–19, 21–22, 322–23
Tanaka, Sho, 117
"taste buds," 109
Taylor, Valerie, 84
Tern Island, 195–96
Tester, Albert, 59
thermocline, 70
Thompson, Jerome, 91
Thoreau, Henry David, 11
Thorrold, Simon, 259
tiger sharks, 38, 60, 65, 94, 95, 143
Tobin, Dan, 206, 264
Truro Police Department, 256
Tufts Medical Center, 319
Tuna Club of Avalon, 41–42
tuna fishing, 41–42
Twitter, 21, 292
Tyson, Mike, 11

Undersea World of Jacques Cousteau, The (documentary), 25, 29–30, 43, 44, 80
Unicorn, F/V, 173–74
University of Connecticut's Marine Sciences Institute, 98–99
University of Maryland, 99, 167
University of Massachusetts, 40, 271, 275
University of Rhode Island (URI), xii, 31–32, 37–39, 57, 72–73, 89–90, 110, 125, 153
US Atlantic Shark Tournament, 48–49
US Atlantic Tuna Tournament, 47, 48

Vanderbilt, Anne and William H., 58
Veckatimest Island, 179
Vikings, 10
Vineyard Gazette, 173–74
volleyball games, 33–35
Voorheis, Tim, 171–72

Walker, Ben, 154
Walker, Chuck, 153–55, 170–71, 295–96
Wallop, Malcolm, 127–28
Wallop-Breaux Amendments, 127–29
Wampanoag people, 10, 136, 140, 141
Waterman, Stan, 111, 232
Weeks, Ann, 1
Wellfleet Elementary School, 323
Wellfleet Fire Department, 17–18
Wellfleet Motel, 9–10, 12
Western Hemisphere Shorebird Reserve Network, 197
West Gutter, 179–80, 183, 184, 188, 189
Whale and Dolphin Conservation (WDC), 5, 6–9
whale blubber, 56, 71, 79, 108–9, 235, 245
whale watching, onboard *Tails of the Sea*, 4–9, 12–13, 16–17
White Pointer II, 219–22
white sharks, xiii–xiv

anatomy and physiology of, 55–56, 70–71, 79, 82, *82*, 109–10, 117–18, 133–34, 193–95, 216–17, 244, 276–77, 278
attacks. *See* shark attacks
biology and life history of, 84–85
cage diving. *See* shark cage diving
Charles Donilon's shark tagging encounter, 90–92, 94, 113
ecological research, 69–72
evolution of, 4, 16, 134–35, 140–41, 160–61
feeding and diet of, 15–16, 20, 25–27, 55–56, 78–80, 108–9, 161, 174, 194–95, 215–17, 229–30, 234–35, 244, 276–78
hotspots, 15–16, 75, 174, 217–18, 245, 291
life span of, 133
mainstreaming of, 87–89
mating and reproduction of, 26, 308–9
migration, 56, 79, 133–35, 160–61, 193–95, 211, 212, 243–45, 277–78, 307–9, 327–28, 339
misconceptions and misrepresentations of, 59, 60, 201, 308, 322
mitigation strategies, 190–91, 265, 300–306, 325–30
Montauk Monster, 93–96, 98–101, 102–7
mythologies of, 3–4
naming of, 28, 79
in Naushon Island salt pond, 178–92
populations, 37, 83–84, 299–300, 324, 337
population study, five-year, 268–71, 275, 278, 289, 290, 292–93, 297–98, 300, 310–11, 327
public safety issues, 20, 65, 92, 174–75, 184, 190–91, 210–11, 222–24, 226–29, 231, 237–38, 245–46, 263–65, 290–91, 294, 300–301, 320–21, 323–25
rare and valuable, 82–85
restoring an apex predator, 337–38
senses of, 161, 216–17
sightings. *See* white shark sightings
tagging. *See* shark tagging
White Sharks and Whales trip, on *Tails of the Sea*, 4–9, 12–13, 16–19, 21–22
white shark sightings, 173–74, 202–14, 227–30, 237, 245–46, 278–79, 281, 302–3, 310–11
seal attack observed by Bruce and Rod, 199–201, 202
White Shark Symposium at California State University, 125
Whitney, Nick, 259
Wieczno, F/V, 99, 119–23, 126–27, 137, 155, 286
Wigren, Ben, 219, 224, 251
Wigren, Cynthia
background of, 218, 219–20, 224
shark cage dive in South Africa, 218–22
shark conservation and AWSC, 258–59, 265–68, 294, 301–2
shark conservation and OCEARCH, 224–26, 250–54, 258–59, 263, 266–67
Wilderness Act, 196
Winton, Megan, 271–75
background of, 271–73
shark attack of Arthur Medici, 321–25, 328–29
shark research, 243, 272–75, 300, 328
shark research cruise on *Aleutian Dream*, 286–91, 295, 317
wolves of Yellowstone, 27–28, 197, 237, 294, 338
Woods Hole, 40, 121, 123, 179, 181, 184
Woods Hole Group, 330
Woods Hole Marine Biological Laboratory, 69, 123, 318
Woods Hole Oceanographic Institution, 20, 40, 58, 66, 69–71, 123, 178, 249
Wood's Seafood (Plymouth), 23–24
World War II, 42, 61, 112

Yellow Jacket, 97–98
Yellowstone wolves, 27–28, 197, 237, 294, 338
Young, William, 46–47, 48

Zeeman, Stephan, 177

ABOUT THE AUTHORS

DR. GREG SKOMAL—in addition to being an accomplished marine biologist, underwater explorer, photographer, and author—is a leading white shark expert in the Atlantic. He is a senior fisheries biologist with the Massachusetts Division of Marine Fisheries and currently directs the Massachusetts Shark Research Program. He is an adjunct professor at the University of Massachusetts Intercampus Marine Science graduate program; an adjunct scientist at the Woods Hole Oceanographic Institution in Woods Hole, Massachusetts; and a member of the Explorers Club and the Boston Sea Rovers. Greg has authored dozens of scientific research papers and has appeared in a number of film and television documentaries, including programs for National Geographic, Discovery Channel, PBS, and numerous television networks. He is a regular on *Shark Week* and *Shark Fest* and is the author of *The Shark Handbook*. He holds a master's degree from the University of Rhode Island and a PhD from Boston University. He lives with his family in Marion, Massachusetts.

RET TALBOT is an award-winning freelance journalist who covers ocean issues at the intersection of science and sustainability. His work can be found in publications such as *National Geographic, Discover, Mongabay, Yale Environment 360*, and other venues. As a science writer, he has embedded with marine scientists around the world, including places like Papua New Guinea, Sulawesi, and Belize, and he frequently works closely with scientists like Greg to bring compelling stories about science to a general audience. He lives on the coast of Maine with his wife, Karen Talbot, who provided the illustrations for this book.